T0215317

Experimental Methods in Heat Transfer and Fluid Mechanics

Experimental Methods in Heat Transfer and Fluid Mechanics

Je-Chin Han and Lesley M. Wright
Mechanical Engineering Department
Texas A&M University

CRC Press
Taylor & Francis Group
Boca Raton London New York

CRC Press is an imprint of the
Taylor & Francis Group, an **informa** business

First edition published 2020
by CRC Press
6000 Broken Sound Parkway NW, Suite 300, Boca Raton, FL 33487-2742

and by CRC Press
2 Park Square, Milton Park, Abingdon, Oxon, OX14 4RN

© 2020 Taylor & Francis Group, LLC

CRC Press is an imprint of Taylor & Francis Group, LLC

ISBN: 978-0-367-89792-5 (hbk)
ISBN: 978-0-367-49780-4 (pbk)
ISBN: 978-1-003-02117-9 (ebk)

Typeset in Times
by Lumina Datamatics Limited

eResource material is available for this title at https://www.crcpress.com/9780367897925.

Contents

Preface

Undergraduate Fluid Mechanics and Heat Transfer Laboratories are required courses for all mechanical, chemical, nuclear, and aerospace engineering undergraduate students. These senior-level undergraduate courses typically cover basic measurements of pressure, velocity, flow rate, conduction, convection, and radiation heat transfer. Advanced Fluid Mechanics and Heat Transfer Laboratories are also required for most engineering graduate students. These graduate-level experimental courses are typically taught as individual courses named Turbulence Measurements or Convection Heat Transfer Measurements. Many universities also offer a single experimental course of heat transfer and fluid mechanics for engineering graduate students. However, in these cases, there are not many textbooks available that cover both heat transfer and fluid mechanics experimental techniques together in one book at the graduate level.

Experimental Methods in Heat Transfer and Fluid Mechanics focuses on how to analyze and solve the classic heat transfer and fluid mechanics measurement problems in one book. This book places emphasis on fundamental principles, detailed measurement techniques, data presentation, and uncertainty analysis. This book provides many well-known advanced experimental methods and their engineering applications such as steady state and transient measurement techniques of Thermographic Liquid Crystal (TLC), Infrared Thermography (IR), Temperature-Sensitive Paint (TSP), Pressure-Sensitive Paint (PSP), Naphthalene Sublimation Mass Transfer Analogy, Hot-Wire and Cold-Wire Anemometry, and Particle Image Velocimetry (PIV) for engineering graduate students and researchers. It is unique in that it provides (1) detailed step-by-step measurement principles, (2) detailed step-by-step measurement procedures, and (3) many detailed step-by-step measurement examples. This experimental knowledge will equip graduate students and researchers with the much-needed capability to read and understand the heat transfer and fluid mechanics-related research papers in open literature and give them a strong experimental background with which to tackle and solve the complex engineering heat transfer and fluid flow problems they will encounter in their professional careers.

JC Han began teaching the Experimental Methods in Heat Transfer and Fluid Mechanics course in the Department of Mechanical Engineering at Texas A&M University in 2000, and Lesley M. Wright began teaching the course in 2009. This book has evolved from a series of lecture notes for this graduate-level course over the past 20 years. Many MS and PhD students specializing in the thermal and fluid sciences have taken this course for their graduate-level experimental heat transfer-fluid flow course as well as for their measurement technique preparation to conduct their heat transfer and fluid mechanics experimental research projects. This book bridges the gap between undergraduate level and graduate level heat transfer-fluid flow experimental methods as well as serves the need of graduate students and researchers looking for advanced measurement techniques for their thermal, flow, and heat transfer engineering applications.

This book covers most updated experimental methods in heat transfer and fluid mechanics. The book is divided into 12 chapters:

Chapter 1: Introduction
Chapter 2: Velocity and Flow Rate Measurements
Chapter 3: Temperature and Heat Flux Measurements
Chapter 4: Experimental Planning and Analysis of Results
Chapter 5: Steady-State Heat Transfer Measurement Techniques
Chapter 6: Time Dependent Heat Transfer Measurement Techniques
Chapter 7: Liquid Crystal Thermography Techniques
Chapter 8: Optical Thermography Techniques
Chapter 9: Pressure-Sensitive Paint (PSP) and Temperature-Sensitive Paint (TSP) Techniques
Chapter 10: Mass Transfer Analogy Measurement Techniques
Chapter 11: Flow and Thermal Field Measurement Techniques
Chapter 12: Flow Field Measurements by Particle Image Velocimetry (PIV) Techniques

There are many excellent experimental fluid mechanics and heat transfer books available. Although we do not claim any new ideas in this book, we do attempt to present the subject in a systematic and logical manner. We hope this book is a unique compilation and is useful for graduate students and researchers.

Finally, we would like to sincerely express special thanks to former Texas A&M University students. Dr. Shantanu Mhetras (PhD, 2006) spent a lot of time and effort to input most of the manuscript from the original Experimental Methods in Heat Transfer and Fluid Mechanics Class Notes in 2000–2006. Without his diligent and persistent contributions, this book would be impossible. Recently, Dr. Chao-Cheng Shiau (PhD, 2017) worked to create many of the figures seen throughout the text. By converting scanned figures and hand sketches into a digital format, he has made a significant contribution to this work. Finally, we would like to extend appreciation to other former students, Drs. Diganta Narzary (PhD, 2009), Nafiz Chowdhury (PhD, 2017), and Andrew Chen (PhD, 2018) for their help in completing the book and drawings.

Je-Chin Han
Lesley M. Wright

Authors

Je-Chin Han is currently a university distinguished professor and Marcus Easterling Endowed chair professor at Texas A&M University. He earned a BS degree at National Taiwan University in 1970, an MS degree at Lehigh University in 1973, and a ScD at MIT in 1976, all in mechanical engineering. He has been working on turbine blade cooling, film cooling, and rotating coolant-passage heat transfer research for the past 40 years. He is the co-author of 250 journal papers, lead author of the book *Gas Turbine Heat Transfer and Cooling Technology*, and author of the book *Analytical Heat Transfer.* He has served as editor, associate editor, and honorary board member for eight heat transfer-related journals. He received the 2002 ASME Heat Transfer Memorial Award, the 2004 International Rotating Machinery Award, the 2004 AIAA Thermophysics Award, the 2013 ASME Heat Transfer Division 75th Anniversary Medal, the 2016 ASME IGTI Aircraft Engine Technology Award, and the 2016 ASME and AICHE Max Jakob Memorial Award. He is a fellow of ASME and AIAA.

Lesley M. Wright is associate professor and Jana and Quentin A. Baker'78 faculty fellow at Texas A&M University. Prior to joining Texas A&M, she was a member of the mechanical engineering faculty at Baylor University for ten years. She earned a BS in engineering in 2001 at Arkansas State University and an MS and a PhD in mechanical engineering at Texas A&M University in 2003 and 2006, respectively. Currently she is investigating enhanced convective cooling technology, including heat transfer enhancement for gas turbine cooling applications. This experimental research has led to the development of innovative cooling technology for both turbine blade film cooling and internal heat transfer enhancement. In addition, Dr. Wright continues to investigate the effect of rotation on the thermal performance of rotor blade cooling passages. Her research interests have also led to the development of novel experimental methods for the acquisition of detailed surface and flow measurements in highly turbulent flows.

1 Introduction

1.1 CONDUCTION, CONVECTION, AND RADIATION HEAT TRANSFER

The transfer of energy, in the form of heat, can be seen in countless applications while affecting the lives of people around the world. From heat engines used for transportation and power generation, to photovoltaic solar cells; from heating and air conditioning systems, to radiative space heaters; from thermal management of the avionics in an airplane cockpit to cooking on an open fire, heat is being transferred all around us. Therefore, it is necessary to understand the physical phenomena causing energy to be transported from one entity to another. By measuring the amount of heat transfer in various systems, engineers and researchers can optimize these processes, leading to a more efficient transfer of thermal energy. Heat transfer is classified into three basic modes: conduction, convection, and radiation. Depending on the system, these modes may be isolated from one another or occur simultaneously. In the following sections, each mode is briefly discussed, so the foundation is laid to discuss the design and execution of heat transfer experiments in later chapters.

1.1.1 CONDUCTION

When a temperature difference exists within a solid, heat is transferred from high temperature to low temperature. For example, Figure 1.1 shows heat is conducted from the high temperature side to the low temperature side through a building or container wall. This is a one-dimensional, steady-state, heat conduction problem if T_1 and T_2 are uniform. According to Fourier's conduction law, the temperature profile is linear through the plane wall.

Fourier's Conduction Law:

$$q'' = -k \frac{dT}{dx} = k \frac{T_1 - T_2}{L} \tag{1.1}$$

and

$$q'' \equiv \frac{q}{A_c} \text{ or } q = q'' A_c$$

where:
   q'' is the heat flux, W/m^2
   q is the heat rate, W or J/s
   k is the thermal conductivity of solid material, W/m × K
   A_c is the cross-sectional area for conduction, perpendicular to heat flow, m^2
   L is the conduction length, m

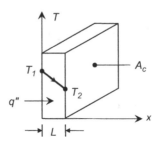

FIGURE 1.1   One-dimensional heat conduction through a building or container wall.

One can predict the heat rate or heat loss, q, through the plane wall by knowing T_1, T_2, k, L, and A_c. This is the simplest, one-dimensional, steady-state problem. However, in real-life applications, there are many complicated conduction problems such as two-dimensional or three-dimensional, unsteady heat conduction.

For conduction problems, it is important to know the thermal conductivity of the solid material, k. The thermal conductivity of a solid material can be determined from Equation (1.1) by measuring T_1, T_2, q, for a given L, and A_c, from a simple conduction experiment as shown in Figure 1.2. The temperatures, T_1 and T_2 can be measured by thermocouples and the heat rate can be measured by the power source (q (Watts) $= V \times I$).

The thermal conductivity depends on the material and slightly depends on the operating temperature. In general, the thermal conductivity of metallic materials is greater than that of non-metallic materials, for example, k (copper ~ 400) > k (aluminum ~ 200) > k (steel ~ 50) > k (stainless steel ~ 15) > k (insulation material = 0.01 ~ 1). The thermal conductivity of solids is greater than liquids and subsequently higher than gases, for example, k (solid) > k (water ~ 0.6) > k (air ~ 0.03). The detailed thermal conductivity of many engineering materials can be found from any heat transfer textbook [1].

Given: A_c, L
Measured: T_1, T_2
$$q = V \cdot I$$

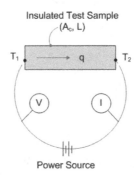

FIGURE 1.2   One-dimensional, steady-state heat conduction experiment.

To calculate k:

$$k \equiv \frac{\dfrac{q}{A_c}}{\dfrac{T_1 - T_2}{L}}$$

- k depends on material and temperature
- $k_{Solid} > k_{Water}$ (~0.6) $> k_{air}$(~0.03)
- $k_{Metal} > k_{Non\text{-}metal}$
- k_{Copper} (~400) $> k_{Aluminum}$ (~200) $> k_{Steel}$ (~50) $>$
- $k_{S.Steel}$ (~15) $> k_{Insulation\ Material}$ (~0.01~0.1~1)

Figure 1.3 shows how to design a test setup for thermal conductivity measurements and the measured temperature profiles for the conductivity calculations for copper and carbon steel rods. It shows that copper has a smaller temperature gradient than carbon steel. This indicates copper has a higher thermal conductivity than carbon

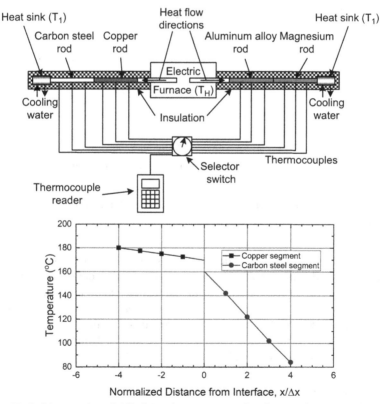

Typical temperature distributions along copper and carbon steel segments in thermal conductivity experiment.

FIGURE 1.3 One-dimensional, steady-state test section for thermal conductivity measurement.

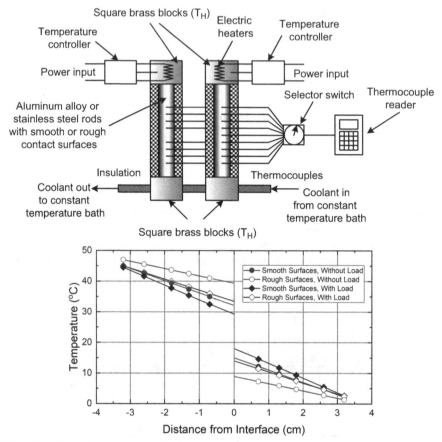

Effect of applied load on temperature distributions along rod segments with smooth and rough surfaces in thermal contact conductance experiment.

FIGURE 1.4 One-dimensional, steady-state test section for thermal contact resistance measurement.

steel [2]. Figure 1.4 shows how to design a test setup for thermal contact resistance measurement and the measured temperature profiles for the contact resistance calculations for smooth and rough surfaces with and without external loads. It shows that the contact between the rough surfaces without an external load yields a greater temperature drop which means a higher contact resistance, while the contact between the smooth surfaces with an external load has the smaller temperature drop which means a lower contact resistance [2].

1.1.2 CONVECTION

Convection is caused by fluid motion over a surface. For example, Figure 1.5 shows heat is removed from a heated solid surface to a cooling fluid. This is a two-dimensional boundary layer flow and heat transfer problem. According to Newton,

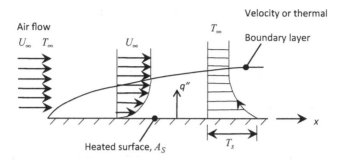

FIGURE 1.5 Velocity and thermal boundary layer.

the heat removal rate from the heated surface is proportional to the temperature difference between the heated wall and the cooling fluid. The proportionality constant is called the heat transfer coefficient; the same heat rate from the heated surface can be determined by applying Fourier's law of heat conduction to the cooling fluid due to the no slip condition at the wall.

Newton's Cooling Law:

$$q'' = -k_f \left.\frac{dT}{dy}\right|_{y=0} = h(T_s - T_\infty) \tag{1.2}$$

Also,

$$h = \frac{q''}{T_s - T_\infty} = \frac{-k_f \left.\frac{dT}{dy}\right|_{y=0}}{T_s - T_\infty} \tag{1.3}$$

and $q'' \equiv \dfrac{q}{A_s}$ or $q = q'' A_s$

where:
T_s is the surface temperature, °C or K
T_∞ is the fluid temperature, °C or K
h is the heat transfer coefficient, W/m^2 × K
k_f is the thermal conductivity of fluid, W/m × K
A_s is the surface area for convection, exposure to flow, m^2

One can predict the heat rate or heat loss, q, from a flat surface by knowing T_s, T_∞, h, and A_s. This is the simplest convection problem for flow over a flat surface. However, in actual applications, there are many complicated convection problems such as flow over a circular cylinder or turbine blade, as well as flow through a circular tube or rectangular channel.

For the convection problem, it is important to know the heat transfer coefficient, h. It is noted that the heat transfer coefficient depends on the fluid properties (such as air or water as the fluid), flow conditions (i.e., laminar or turbulent flows), and surface configurations (such as flat surface or circular tube). The heat transfer coefficient can be determined experimentally or analytically. From Equation (1.3), the heat transfer coefficient can be analytically determined by knowing the temperature profile in the fluid during convection and then taking the cooling fluid temperature gradient near the wall. However, this requires solving two-dimensional boundary-layer equations.

Similarly, from Equation (1.3), the heat transfer coefficient can be experimentally determined by measuring T_s, T_∞, and q, for a given surface area A_s, from a simple convection experiment as shown in Figure 1.6. The temperatures, T_s and T_∞, can be measured by thermocouples and the heat rate can be measured by the power source (q (Watts) $= V \times I$). Heat transfer coefficients are normally shown as Nusselt numbers (related to fluid conductivity and geometry), and Nusselt numbers are typically correlated with Reynolds Numbers (related to cooling flow velocity and geometry) and Prandtl Numbers (related to coolant fluid properties) as shown in Figure 1.7.

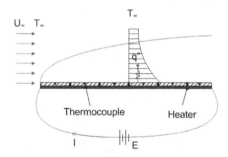

FIGURE 1.6   Experiment for measuring the local and average convection heat transfer coefficient.

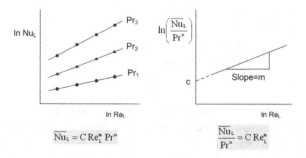

FIGURE 1.7   Dimensionless representation of convection heat transfer measurements.

$$\text{Nu}_X = f_4(x^*, \text{Re}_X, \text{Pr})$$

$$\overline{\text{Nu}_X} = f_5(\text{Re}_L, \text{Pr})$$

Heat Input $q = I \cdot E$ measured

$$h_L = \frac{\sum q}{A_S(T_S - T_\infty)}$$

$$\overline{\text{Nu}_L} = \frac{\overline{h_L} \cdot L}{k}$$

Fluid properties are based on the film temperature, $T_f = \dfrac{1}{2}(T_S + T_\infty)$

Table 1.1 provides typical values for heat transfer coefficients in many convection problems [1]. As can be seen, in general, forced convection has more heat transfer than natural convection; water as the coolant removes much more heat than air; and boiling or condensation, involving phase change, have much higher heat transfer coefficients than single phase convection.

Figure 1.8 shows a typical steady-state wind tunnel for an external flow and heat transfer experiment [2]. The sketch shows the test section for air cross flow over a long, heated, cylindrical rod. The copper rod surface heat transfer coefficients (Nusselt numbers based on rod diameter and air thermal conductivity) versus flow velocity (Reynolds numbers based on rod diameter and measured air velocity by Pitot-static tube) can be obtained from the measured rod temperatures, using four thermocouples attached to the rod, the power input to the rod from the electric resistance heater, and the thermocouple placed in the mainstream to measure the air

TABLE 1.1

Typical Values of Heat Transfer Coefficient

Type of Convection	h, W / m^2 × K
Natural convection	
Caused by ΔT : air	5
Caused by ΔT : water	25
Forced convection	
Caused by fan, blower: air	25–250
Caused by pump: water	50–20,000
Boiling or condensation	
Caused by phase change	10,000–100,000
Water → Steam	2,500–50,000
Freon → Vapor	

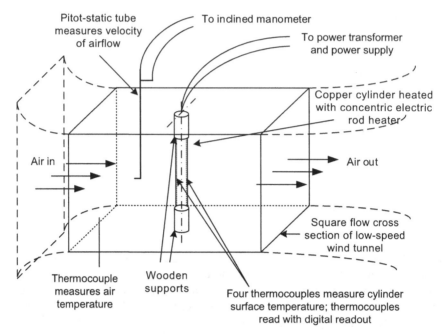

Schematic of test apparatus for cross flow over cylinder experiment.

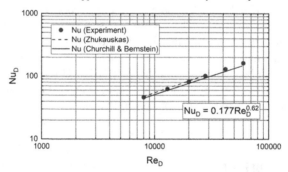

Convective heat transfer for cross flow over a long cylinder.

FIGURE 1.8   Steady-state wind tunnel for external flow heat transfer experiment.

temperature. It shows that the rod surface heat transfer coefficients (Nusselt numbers) increase with increasing flow velocity (Reynolds numbers).

A transient (time-dependent) experiment to measure heat transfer coefficients from an impinging jet is shown in Figure 1.9. The sketch shows the test section for pressurized air flow impinging on a thin, heated, stainless film, supported by a thick, low conductivity acrylic sheet, with thermocouples attached to the heater film [2]. The heated thin film surface temperature versus time can be recorded during the transient experiment. The surface heat transfer coefficients (Nusselt numbers based on jet diameter and air thermal conductivity) versus impinging jet velocity

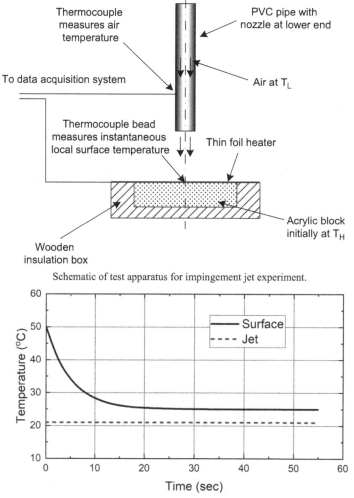

Schematic of test apparatus for impingement jet experiment.

Variations of jet and surface temperatures with time in impingement jet experiment.

FIGURE 1.9   Test loop for jet impingement cooling using the transient technique.

(Reynolds numbers based on jet diameter and measured air velocity) can be obtained by the measured thin film temperatures versus time using the surface temperature history solutions of the one-dimensional, semi-infinite, solid transient conduction equation with convection boundary conditions [1]. As expected, the surface heat transfer coefficients increase with increasing impinging jet velocity.

Heat transfer coefficients can be measured internally, as represented in Figure 1.10. This figure depicts a traditional, steady-state experiment to measure heat transfer coefficients within a square channel [2]. The sketch shows the test section for air flow moving inside a heated, copper duct. The channel heat transfer coefficients (Nusselt numbers based on duct hydraulic diameter and air thermal conductivity) versus air flow velocity (Reynolds numbers based on duct hydraulic diameter and measured

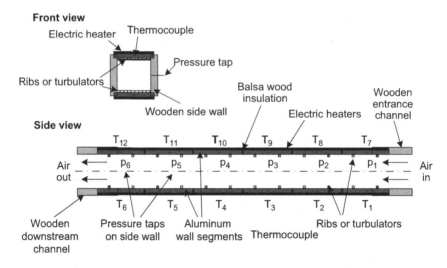

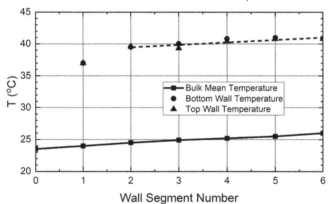

Schematic of test section for internal flow experiment.

Streamwise variation of temperature along test section in internal flow heat transfer experiment

FIGURE 1.10 Typical test loop for internal flow heat transfer experiment.

air velocity or flow rate) can be obtained by the measured copper wall temperature distributions from the thermocouples, the power input from electric heaters, and the measured inlet and outlet bulk mean air flow temperatures by thermocouples. It is expected that the copper wall heat transfer coefficients (Nusselt numbers) increase with increasing flow velocity (Reynolds numbers).

1.1.3 RADIATION

Radiation is caused by electromagnetic waves from solids, liquids, or gases. For example, Figure 1.11 shows heat is radiated from a solid surface at a temperature greater than absolute zero. According to Stefan-Boltzmann, the radiation heat rate

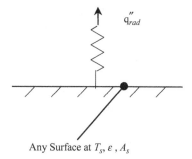

Any Surface at T_s, ε, A_s

FIGURE 1.11   Radiation from a solid.

is proportional to the surface's absolute temperature raised to the fourth power, the Stefan-Boltzmann constant, and the surface emissivity.

Stefan-Boltzmann Law:
For a real surface,

$$q'' = \varepsilon \sigma T_s^{\,4} \tag{1.4}$$

For an ideal (black) surface, $\varepsilon = 1$

$$q'' = \sigma T_s^4$$

and

$$q'' \equiv \frac{q}{A_s} \text{ or } q = q'' A_s$$

where:

ε is the emissivity of real surface, $\varepsilon = 0-1$ ($\varepsilon_{\text{metal}} < \varepsilon_{\text{non-metal}}$)

T_s is the absolute temperature of the surface, K $\left(K = {}^\circ C + 273.15\right)$

σ is the Stefan-Boltzmann constant, $\sigma = 5.67 \times 10^{-8} \, \text{W/m}^2 \times \text{K}^4$

A_s is the surface area for radiation, m^2

One can predict the heat rate or heat loss, q, from a solid material by knowing T_s, ε, and A_s. This is the simplest radiation problem from a flat surface. However, in actual scenarios, there are many complicated radiation problems such as radiation exchange between many surfaces as well as radiation with a participating medium (gas radiation).

For surface radiation problems, it is important to know the emissivity of the solid material, ε. The emissivity of a surface can be determined from the above Equation (1.4) by measuring T_s and q, for given surface area, A_s, from a simple radiation experiment. The temperature, T_s, can be measured by the thermocouples and the heat rate can be collected by a radiation pyrometer placed at the top of an enclosure.

The surface total emissivity primarily depends on the material, wavelength, and temperature. It is between 0 and 1. In general, the emissivity of metals (around 0.1) is much less than that of non-metals (around 0.9). Note that radiation from a surface can go through air as well as a vacuum environment. The detailed hemispherical emissivity (normal emissivity) of many engineering materials (gray and diffuse surfaces) can be found from any heat transfer textbook [1]. For many engineering materials, surface emissivity equals absorptivity. However, for non-gray surfaces, they are not the same. From this measured surface emissivity (and absorptivity), one can solve typical radiation problems in a variety of furnaces.

Emissivity: ε is a surface radiation property

$$\varepsilon\left(\theta,\varphi,\lambda,T\right) - \text{Monochromatic Emissivity} \quad \varepsilon\left(T\right) - \text{Total Emissivity}$$

For gray and diffuse surface - Real surface

The total hemispherical emissivity is given by,

$$\varepsilon\left(T\right) \equiv \frac{E(T)}{E_b(T)} = \frac{E(T)}{\sigma\,T^4} = \frac{\text{Measured}}{\text{Theory}} \left(\begin{array}{c} \text{Experimental data based on} \\ \text{normal emissivity, } \varepsilon\left(T\right) \cong \underbrace{\varepsilon\left(T\right)}_{\text{normal}} \end{array} \right)$$

1.2   EXPERIMENTAL METHODS IN HEAT TRANSFER AND FLUID MECHANICS

From the introduction of heat transfer, it was shown the following parameters are needed for experiments in heat transfer and fluid mechanics. For experimental conduction heat transfer, one needs to measure material temperature distributions and heat flux (heat input) in order to determine thermal conductivity and to determine thermal contact resistance if conduction occurs between two different materials. In convection heat transfer experiments, one needs to measure flow velocity or flow rate, fluid and surface temperature distributions, and heat flux (heat input) in order to calculate surface heat transfer coefficients for a given fluid. In experimental radiation heat transfer, one needs to measure surface temperature and radiation flux in order to determine surface emissivity.

Surface temperatures can be measured by thermocouples, thermochromic liquid crystals, infrared thermography, and temperature sensitive paint, among others. Surfaces can be heated using a variety of resistance heaters. Static pressure distributions can be measured using static taps, pressure probes, and pressure sensitive paint. Thermocouples, thermistors, RTDs, and hot / cold wires can be used to measure fluid temperatures. In addition, fluid velocity and turbulence can be measured with a variety of instruments, including multi-hole pressure probes, hot wire anemometers, or particle image velocimetry (PIV). Mass or volumetric flow rates can be measured by obstruction flow meters or rotameters. The experiments can be conducted under steady-state or transient conditions. There are many experimental methods in thermal and fluid flow applications [3–6]. This book covers basic (standard) experimental methods to advanced (detailed) measurement techniques.

The basic experimental methods include pressure and flow measurements by standard pressure probes and flow meters, temperature and heat flux measurements by standard thermocouples and electric heaters. The advanced experimental methods include pressure and flow measurements by hotwire anemometers and PIV, temperature and heat flux measurements by liquid crystal paints, infrared thermal imaging cameras, temperature sensitive paint, and thin-film heaters. This book covers several state-of-the-art experimental methods in measuring heat transfer and fluid mechanics engineering problems. The book is divided into 12 chapters, Chapters 1–6 cover the basic (standard) experimental methods, and Chapters 7–12 cover the advanced (detailed) measurement techniques.

"Chapter 1: Introduction." This chapter provides an introduction to conduction, convection, and radiation heat transfer problems. From simplified experimental concepts, the readers are led into the experimental methods in heat transfer and fluid flow measurements.

"Chapter 2: Velocity and Flow Rate Measurements." This chapter provides available measurement tools for fluid mechanics experiments. It begins with the fundamentals of pressure probes, and discusses their implementation through fabrication of static pressure taps and pitot-static probes and calibration of transducers. Various obstruction flow meters, including orifice and Venturi meters, are presented for mass flow rate measurements. The chapter concludes with a discussion of error estimates for pressure probes and flow meters.

"Chapter 3: Temperature and Heat Flux Measurements." Basic temperature measurements are introduced in this section. Thermocouple laws, fabrication, calibration, and selection are explained. Heat flux gauges are also presented in this chapter. Finally, error estimates for thermocouples and heat flux gauges are discussed.

"Chapter 4: Experimental Planning and Analysis of Results." This chapter takes the experimentalist through the planning, setup, and commissioning of heat transfer experiments. Planning begins with a thorough literature survey and identification of the required equipment. The test section is then designed, fabricated, and instrumented. Finally, testing commences with required calibrations, so the setup can be validated against data available in open literature. This chapter also includes the presentation of results, experimental uncertainty, and conclusions.

"Chapter 5: Steady-State Heat Transfer Measurement Techniques." This chapter presents well-established, steady-state heat transfer methods. Using copper plates, thermocouples, and resistance heaters, surface heat transfer coefficients can be measured. This approach is applied to acquisition of overall, channel averaged heat transfer coefficients, regionally averaged coefficients, or local heat transfer coefficients. The fundamental principles for each of these methods is discussed along with detailed, step-by-step examples.

"Chapter 6: Time Dependent Heat Transfer Measurement Techniques." This chapter begins with implementation of a transient, lumped capacitance technique to measure surface averaged, heat transfer coefficients. The measurement of local heat transfer coefficients is also discussed using both thin film heat flux gauges and thin foil heaters with thermocouples. Detailed examples are provided for each of these techniques. The fundamental principles to obtain detailed heat transfer coefficient distributions with a one-dimensional, semi-infinite solid model are also presented, with additional detailed examples to follow in later chapters.

"Chapter 7: Liquid Crystal Thermography Techniques." This chapter discusses the fundamental principle of the steady-state, local heat transfer measurement technique known as yellow-band tracking (including a detailed step-by-step experiment example). The hue-saturation-intensity (HSI) method is presented as an alternative, steady-state technique. Transient methods using liquid crystal thermography are also discussed: single color capturing method and the transient HSI method. Detailed examples are included to demonstrate the differences in each of these methods.

"Chapter 8: Optical Thermography Techniques." The use of infrared (IR) cameras for heat transfer coefficient measurements is discussed in this chapter. Both steady-state and transient wall temperature distributions can be obtained with an IR camera, both methods are presented in detail. In addition, step-by-step examples are included for each of these experiments.

"Chapter 9: Pressure Sensitive Paint (PSP) and Temperature Sensitive Paint (TSP) Techniques." Pressure sensitive paint (PSP) is used to obtain detailed pressure and film cooling effectiveness distributions on a surface. Temperature sensitive paint (TSP) can be used to measure detailed surface temperature distributions. Similar to IR and liquid crystal techniques, TSP can be used to measure detailed heat transfer coefficient distributions in both steady-state and transient tests. To demonstrate the use of these photo-luminescent paints, detailed descriptions of the experimental setups, data acquisition, and data analysis are provided.

"Chapter 10: Mass Transfer Analogy Measurement Techniques." This chapter discusses the fundamental principle of the naphthalene sublimation technique for the measurement of detailed surface mass/heat transfer coefficient distributions, and the process is captured in a step-by-step example. This is followed by the presentation of the foreign gas concentration sampling technique to measure the local film cooling effectiveness. It also includes the discussion of the ammonia-diazo technique for the detailed film cooling effectiveness distributions.

"Chapter 11: Flow and Thermal Field Measurement Techniques." This chapter includes the use of multi-hole probes for velocity and flow direction (swirl) measurements. Hotwire anemometry is also discussed for measurement of local, instantaneous velocities and turbulence quantities. Local fluid temperatures are also considered using miniature thermocouple probes and cold wire anemometers. Examples for each of these probes are provided for the reader.

"Chapter 12: Flow Field Measurements by Particle Image Velocimetry (PIV) Techniques." This chapter discusses the fundamental principles of particle image velocimetry (PIV) for detailed, instantaneous velocity measurements. Basic system components for traditional PIV experiments are discussed. In addition, the general methodology for image processing is included alongside detailed examples. Recent advancements in PIV are also mentioned.

PROBLEMS

1. Run a simple experiment, as sketched in Figure 1.3, to determine the thermal conductivity of copper, aluminum, stainless steel, and carbon steel, respectively. Compare the measured results with the available values in a textbook.
2. Run a simple experiment, as sketched in Figure 1.6, to determine the convection heat transfer coefficients of air flow over a flat plate with various velocities. Compare the measured results with an available correlation in a textbook.
3. Run a simple experiment, as sketched in Figure 1.8, to determine the convection heat transfer coefficients of air flow across a cylinder with various velocities. Compare the measured results with an available correlation in a textbook.
4. Run a simple experiment, as sketched in Figure 1.10, to determine the convection heat transfer coefficients of air flow through a smooth-surface circular tube with various velocities. Compare the measured results with available correlations in a textbook.
5. Run a simple experiment to determine the total emissivity of copper, aluminum, stainless steel, and carbon steel. Compare the measured results with the available values in a textbook.

REFERENCES

1. Incropera, F. and Dewitt, D., *Fundamentals of Heat and Mass Transfer*, 5th ed., John Wiley & Sons, New York, 2002.
2. Lau, S.C., *MEEN 464-Heat Transfer Laboratory Instruction Manual*, Mechanical Engineering Department, Texas A&M University, College Station, TX, September, 2002.
3. Eckert, E.R.G. and Goldstein, R.J., *Measurements in Heat Transfer*, 2nd ed., Hemisphere Publishing Corporation, Washington, DC, 1976.
4. Goldstein, R.J., *Fluid Mechanics Measurements*, Hemisphere Publishing Corporation, New York, 1983.
5. Han, J.C., Dutta, S., and Ekkad, S.V., *Gas Turbine Heat Transfer and Cooling Technology: Chapter 6 Experimental Methods*, Taylor & Francis Group, New York, December, 2000.
6. Lee, T.-W., *Thermal and Flow Measurements*, CRC Press, Taylor & Francis Group, New York, 2008.

2 Velocity and Flow Rate Measurements

2.1 PRESSURE MEASUREMENTS

There are several standard velocity measurement devices such as static pressure probes and total pressure probes. The proper velocity measurement device can be chosen depending upon the specific application. Static pressure probes include wall-mounted taps and the Prandtl tube; total pressure probes can be referred to as a Pitot tube, Pitot-static tube, Kiel probe, or multiple-hole probes (such as Cobra, 5-hole, and 7-hole). These probes provide time-averaged velocity [1,2].

There are several advanced velocity measurement methods such as hot wire anemometry, laser doppler anemometry (LDA), and particle image velocimetry (PIV). These advanced velocity measurement techniques, which provide time-averaged as well as time-dependent velocity measurements, will be discussed in later chapters.

2.1.1 PRESSURE UNITS AND QUANTITIES

Flow pressure is defined as force per unit area. Several common pressure units have been used, such as Pascal (Pa, Newton per square meter), bar, atmosphere (atm), Torr, pound force per square inch (psi), inches of water (in H_2O), inches of mercury (inHg), etc. It is useful to know their relationships and conversion quantities. It is also important to know the difference between the gauge pressure (flow pressure measured by a gauge) and absolute pressure, i.e., gauge pressure (psig) equals absolute pressure minus atmosphere pressure. Flow pressure (gauge pressure or absolute pressure) can be referred as total pressure, static pressure, or dynamic pressure, i.e., total pressure = static pressure + dynamic pressure.

- 1 Pa (1 N/m^2), 1 psi (1 lbf/in^2), 1 bar = 100 kPa, 1 atm = 101.325 kPa = 1.01 bar;
- 1 atm (absolute) = 760 Torr = 760 mmHg = 29.92 inHg = 10.3508 mH$_2$O = 407.513 inH$_2$O;
- 1 Pa = 1.4504 × 10^{-4} psi;
- 1 in H$_2$O = 0.0361 psi = 5.20 psf;
- 1 in Hg = 0.491 psi = 70.73 psf = 13.62 in H$_2$O;
- Gauge pressure = absolute pressure − atmosphere pressure; $P_g = P_{abs} - P_{atm}$
- Total pressure = static pressure + dynamic pressure; $P_t = P_s + \frac{1}{2}\rho V^2$

2.1.2 STATIC PRESSURE PROBES

Wall-Mounted Taps: It is often convenient to measure the flow static pressure by using wall-mounted taps as shown in Figure 2.1. A 1/32″~1/16″ round hole can be drilled through the wind tunnel wall for an external flow study or through the channel wall for an internal flow test. The flow static pressure at the tap region can be measured using a transducer or manometer. The ratio of the estimated boundary layer thickness (δ) to the tap diameter (D) is very important. If $D < \delta$, the flow should not be affected; however, if $D > \delta$, the recirculation zone supports a pressure gradient, and the mainstream flow dips down into the tap altering the flow and yielding an incorrect measurement of the static pressure [1].

Prandtl Tube: The static pressure can also be measured using a Prandtl tube, as shown in Figure 2.2. The thin-wall, hollow probe diameter is around 3/8″~1/4″ with several taps (approximately 1/32″ diameter) drilled downstream from the leading nose. The distance between the nose and taps is generally 5–10 probe diameters. The flow static pressure at the tap can be measured using a transducer or manometer. The probe is generally bent with a 90° angle for easy insertion through the wind tunnel or channel wall [2].

Flush-mounted Strain Gauge: An accurate way for fluctuating pressure measurements is with a flush-mounted, pressure strain gauge. There is a miniature silicon diaphragm in which a fully active 4-arm Wheatstone bridge has

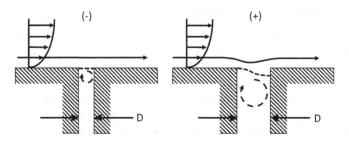

FIGURE 2.1   Schematic of wall-mounted taps.

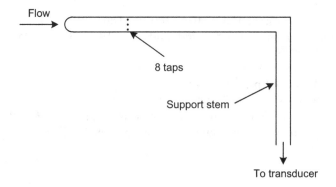

FIGURE 2.2   Schematic of a Prandtl tube.

been atomically bonded using solid state diffusion. The silicon diaphragm deformation increases with flow static pressure acting from the flush-mounted surface, and its resistance (or its output voltage from the bridge circuit) increases with deformation. The strain gauge can be as small as 0.030″ diameter; this small pressure sensitive area provides a high frequency response (>40 kHz). The DC voltage output can be averaged over time.

Flush-mounted Piezoelectric Gauge: This gauge is based on the idea that squeezing a quartz crystal produces an electrical charge, and its output voltage increases with the static pressure acting from the flush-mounted surface. The piezoelectric gauge is sensitive for transient phenomena. It actually measures pressure versus time, dP/dt, and must be integrated with time in order to obtain the time-averaged static pressure [1].

2.1.3 TOTAL AND STATIC PRESSURE PROBES

Pitot Tube: The flow total pressure is often measured with a Pitot tube, as shown in Figure 2.3. The thin-wall hollow probe diameter is around 1/8″~1/4″ with a small opening (around 1/32″ diameter) at the nose. The flow total pressure, including static and dynamic at the leading edge region, can be measured using a transducer or manometer.

Pitot-Static Tube: By combining the above-mentioned Pitot tube and Prandtl tube, a Pitot-static tube can be used to simultaneously measure the flow total and static pressures, as shown in Figure 2.4. There are coaxial tubes, a center hollow tube with small opening for the total pressure, and an annulus tube with several taps downstream from the nose for the static pressure. Both the total pressure at the leading edge and static pressure (about 5 ~ 10 diameter downstream) can be measured with a transducer or manometer. From the total and static pressure measurements, the time-averaged velocity can be calculated by using Bernoulli's equation. The total and static pressures can be measured separately using individual pressure transducers, or the measurements can be coupled in a differential transducer to directly measure the pressure difference [2].

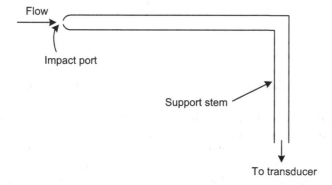

FIGURE 2.3 Schematic of the Pitot tube.

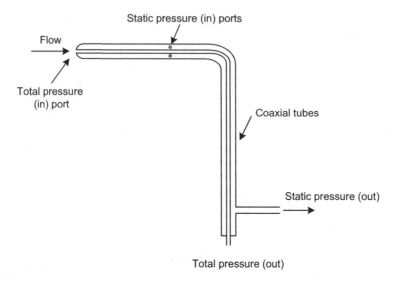

FIGURE 2.4   Schematic of the Pitot-static tube.

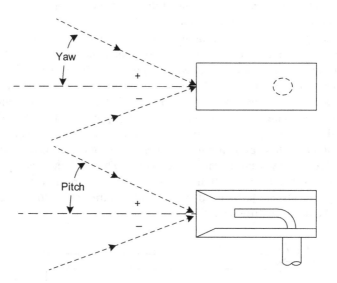

FIGURE 2.5   Kiel probe.

Kiel Probe: To eliminate flow angle effects due to probe misalignment, the Kiel probe can be used, as show in Figure 2.5. An external hood surrounds a traditional Pitot tube. In swirling flows, the probe can be used without knowledge of the local flow direction. The hood acts to align the flow perpendicular to the Pitot tube, providing and accurate measurement of the total pressure. The accuracy of the Kiel probe is unaffected for large yaw angles of the flow, ±40°. With the addition of the external housing on the probe, the flow field may be altered due to the increased size of the probe [1].

Multiple-Hole Probes: The Pitot-static probe is convenient to measure the difference between the total and static pressure at a location in the flow. With this pressure difference, the fluid velocity is easily calculated. However, it is often necessary to know both the flow direction and velocity of a fluid (i.e., swirling flows or areas of recirculation). To simultaneously measure flow direction and velocity, more measurement points are required.

A three-hole probe is used to determine the flow direction. These pressure probes are often called "Cobra" probes as the three probes give the appearance of a flattened cobra's head with its hood extended. The probe consists of a center probe and two probes, angled symmetrically from the center probe. The probe is inserted into the flow and rotated until the pressures measured on the outer two probes are equal. When these pressures are balanced, the probe is pointed directly into the flow. The center probe will provide the total pressure. Therefore, the three-hole probe provides researchers with the flow direction and total pressure. The outer pressure probes do not yield the static pressure, as they are angled to the flow; therefore, the three-hole probe cannot be used to measure the flow velocity.

Increasing the number of pressure ports increases the number of measurements; Figure 2.6 shows both five- and seven-hole pressure probes. Five-hole probes have a central port surrounded by four other ports. When first introduced, these probes were used to measure the total pressure, yaw angle, and pitch angle ($\pm 35°$). In recent years, using an extensive calibration, these probes have also been used for static pressure measurements. With the static and total pressures, the velocity can

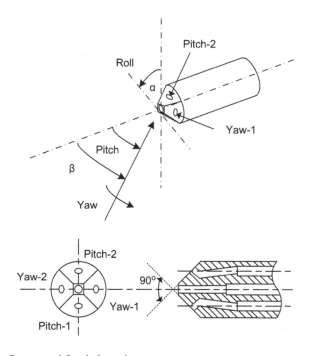

FIGURE 2.6   Seven- and five-hole probes.

be determined. Surrounding the center port by six additional ports, creates a seven-hole probe. These probes can capture a wider range of yaw angles: $\pm 60°$. With a rigorous calibration, these probes can be used to determine local flow direction, total pressure, static pressure, and fluid velocity. The calibration and application are discussed in Chapter 11.

2.2   VELOCITY MEASUREMENTS

The fluid velocity can be calculated from the measured static and total pressures in the flow field. Pressure transducers or manometers can be used with the above-mentioned probes for static and total pressure measurements. With the difference between the total and static pressures, the fluid velocity can be calculated using the principle of Bernoulli's equation.

There are several advanced velocity measurement methods such as hot wire ane-mometry, laser Doppler anemometry (LDA), and particle image velocimetry (PIV). These advanced velocity measurement techniques, which provide time-averaged and transient velocities, will be discussed in the later chapters.

2.2.1   MANOMETERS

Manometers are widely used due to their ease of implementation and relatively low cost. Simply, a manometer consists of a column of a known fluid. The height of the fluid column results from the pressure difference imposed on each side of the column. One side of the fluid column can be exposed to a total pressure port and the opposite side connected to a static pressure port. This would be a direct, differential pressure measurement. The manometer can also be used to measure a pressure relative to atmospheric pressure, by leaving one side of the fluid column directly exposed to the atmosphere.

> **U-Tube Manometer:** The most basic manometer is the U-tube manometer. Figure 2.7 shows a u-shape transparent tube with 3/8″~1/2″ diameter. Each leg of the manometer is routed to a different pressure source. For the scenario shown in Figure 2.7, the pressure difference, $P_1 - P_2$, can be measured.

$$P_1 - P_2 = (\rho_{\text{water}} - \rho_{\text{air}}) \times g \times h$$

> In this case, the process fluid is air, and the manometer fluid is water. The density of air is $\rho_{\text{air}} \sim 1.225$ kg/m^3, and the density of water is approximately $\rho_{\text{water}} \sim 999$ kg/m^3. With this density difference, the density of air can be neglected, and the pressure difference is approximated:

$$P_1 - P_2 = (\rho_{\text{water}}) \times g \times h$$

> In SI units (matching the water density given above), the height of the water, h, is measured in meters, and g is the local acceleration of gravity, $g = 9.81$ m/s^2.

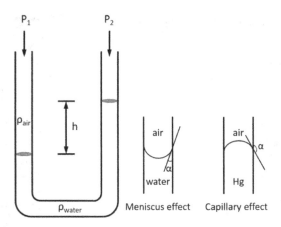

FIGURE 2.7   U-tube manometer.

Oil, water, and mercury are common manometer fluids. Be careful of capillary and meniscus effects when reading the tube scale. Read the instruction
manuals from the manufacturers to ensure what fluid should be used in the
U-tube manometer. For example, to improve the accuracy of the manometer, manufacturers will require the use of oil as the manometer fluid, but
they supply a linear scale calibrated to inches of water. If the fluid is filled
with a different fluid, and the density varies from that specified by the manufacturer, the scale is no longer valid, and the pressure differential will be
measured incorrectly.

Well Type Manometer: Figure 2.8 shows a well type manometer. One leg of
the U-tube manometer is enlarged so the change in height (h_2) is small compared to h_1 in the small tube. Only one scale is installed in the small tube
which is read in inches or centimeters.

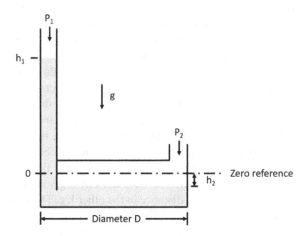

FIGURE 2.8   Well type manometer.

FIGURE 2.9   (a) Inclined manometer (upper), (b) micromanometer (lower).

Inclined Manometer: Figure 2.9a shows a typical inclined manometer. The small diameter leg of a well type manometer is inclined from vertical. This causes the fluid to travel further within the tube for a given pressure making the reading more accurate. The length scale is installed on the small diameter, inclined tube which is read in inches or centimeters. In this figure, red gauge oil is used per the manufacturer's requirements for the calibrated scale.

Micromanometer: Large diameter wells are used in micromanometers to minimize meniscus effects. Figure 2.9b shows a typical micromanometer. For small pressure differences, these manometers utilize a built-in micrometer to accurately measure very small height changes of the manometer fluid. The accuracy of the micrometer is generally around 0.001 inch. The device shown in Figure 2.9b utilizes an electronic detector. With no pressure difference across the manometer, the micrometer is set to its zero position. The micrometer is raised away from the manometer fluid, and the pressure difference is applied across the manometer. The micrometer is lowered until it touches the surface of the manometer fluid. As the metal micrometer contacts the fluid, and electrical signal is detected, and the micrometer length can be recorded. With this type of manometer, the user must use caution. The micrometer reads the distance one leg of the manometer has moved. This length must be doubled to account for the fluid displacement on both sides of the manometer. The fluid rises by a given distance in one leg and goes down by the same distance in the other leg. Therefore, the total manometer height, h, is double the distance measured in one leg. With the manometer shown in the figure, water is used as the manometer fluid, and green dye is used to minimize the meniscus while improving visualization.

2.2.2 Pressure Transducers

Manometers are a convenient and cost-effective way to measure pressure differentials. While they are typically used to measure relatively low-pressure differences, with a proper gauge fluid, manometers can measure larger pressure differences. A variety of pressure transducers are available for measuring large pressures. Unlike manometers, many of these transducers can also be used to measure pressure fluctuations at high frequencies.

Pressure Gauge: The Bourdon tube type pressure gauge, as shown in Figure 2.10, is the most common dial gauge for pressure measurements [1]. The basis of operation is the oval cross section tube begins to straighten when pressure is applied to it. This motion is then detected by the linkage attached to its end which drives a small rack and pinion resulting in the rotation of the gauge indicator needle. The gauge pressure range has been pre-calibrated by the manufacturers. There are different gauges for measuring different ranges of pressure, for example, from 0 ~ 5 psig, 0 ~ 20 psig, 0 ~ 50 psig, 0 ~ 100 psig, 0 ~ 500 psig, 0 ~ 1000 psig, etc. The pressure gauge can be directly inserted

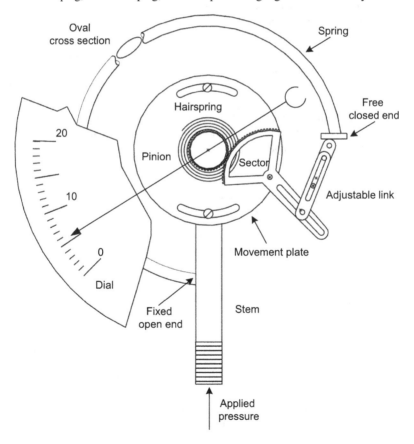

FIGURE 2.10 Bourdon type pressure gauge.

or attached to a pressure port at any desired location along a pipe. Pressure gauges provide stable measurements and are therefore suitable for time-averaged measurements. Researchers should note the pressure provided by a standard pressure gauge is the "gauge" pressure, P_g ($P_{abs} = P_g + P_{atm}$).

Pressure Transducers: Traditional pressure gauges described above rely on mechanical changes to provide researchers with pressure measurements. While pressure transducers also rely on a mechanical change within the device, these changes are usually converted into a voltage or current output for the user. This electrical output can be calibrated (either by the manufacturer or within the laboratory) to associate the measured voltage/current to pressure. The operating principle for two different pressure transducers is shown in Figure 2.11 [1,2].

The transducer in Figure 2.11a uses a small diaphragm in contact with the pressurized fluid. The opposite side of the transducer can be exposed to atmosphere or it could be coupled with a second pressure to provide a differential pressure measurement. With the pressure difference across the diaphragm, the diaphragm expands or contracts. A strain gauge is attached to the diaphragm to capture the deformation. The strain gauge output (voltage) can be calibrated and will work as a device for measuring pressure.

Figure 2.11b shows a piezoelectric pressure sensor. Housed inside the transducer is a piezoelectric crystal. When a force (pressure) is applied to the crystal, an electrical charge is generated. The magnitude of this electrical signal is proportional to the force applied to the faces of the crystal. With the voltage measured across two faces of the crystal, the transducer can provide a differential pressure measurement or the pressure relative to atmospheric pressure. The voltage produced by the piezoelectric crystal must be calibrated over the range of desired pressures.

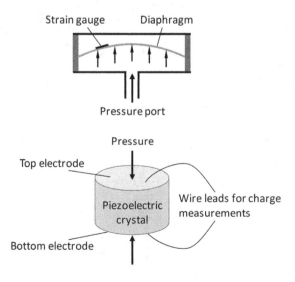

FIGURE 2.11   Concept of pressure transducer (upper: strain gauge, lower: piezoelectric gauge).

Scanivalve®: In 1955, the Scanivalve Corporation was founded out of a need for measuring many pressure points from a single pressure transducer. Originally servicing laboratories with large scale, wind tunnel testing, the Scanivalve has evolved and spread into a variety of industries. Early Scanivalve products utilized a single transducer multiplexed to dozens of individual pressure ports. As shown in Figures 2.12 and 2.13, the 48 J Scanivalve utilizes one integral pressure transducer and its zeroing circuit

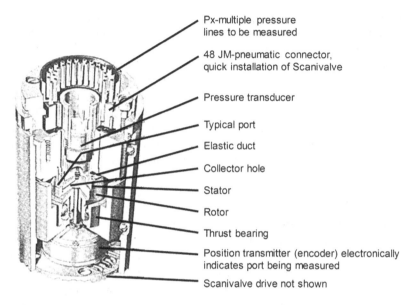

Px-multiple pressure lines to be measured

48 JM-pneumatic connector, quick installation of Scanivalve

Pressure transducer

Typical port

Elastic duct

Collector hole

Stator

Rotor

Thrust bearing

Position transmitter (encoder) electronically indicates port being measured

Scanivalve drive not shown

FIGURE 2.12 Cut away drawing of a Scanivalve.

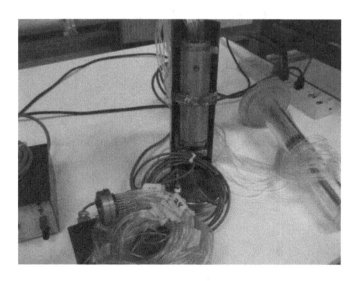

FIGURE 2.13 Scanivalve.

to do the work of 48 transducers and 48 zero circuits [3]. The transducer is sequentially connected to the various pressure ports via a radial hole in the rotor which terminates at the collector hole. As the rotor rotates, this collector hole passes under the separate pressure ports in the stator. Because the rotor must rotate from port-to-port, the pressure is measured at one port at a time. This device cannot be used for simultaneous measurement of the 48 pressure points. As only one transducer is used in the device, only one calibration is required for the 48 pressure ports.

The reach of the Scanivalve Corporation has extended over the years, and their current pressure transducers are capable of simultaneous pressure measurements. The self-contained DSA line can measure up to 16 pressure simultaneously. Miniature pressure scanners are also available with 64 pressure channels. For a high volume of measurements, the Scanivalve ERAD system is modular and can be expanded for as many of 512 pressure channels. These new systems use individual transducers for the simultaneous measurements. Also, they are capable of recording at 2.5 kHz and include temperature measurements for improved accuracy. The electronic pressure scanning devices are versatile and costly as compared to traditional single point transducers.

2.2.3 Velocity Calculations

Pitot-Static Pressure Probe: A Pitot-static probe, as shown in Figure 2.4, can be inserted into a flow channel for velocity measurement. The static pressure, at location 1, and the total (stagnation) pressure at location 2, can be measured by connecting the ports to a pressure transducer or manometer. By applying Bernoulli's equation to locations 1 and 2, assuming no viscous loss, one obtains the following:

$$\frac{P_1}{\rho} + \frac{1}{2}V_1^2 + gz_1 = \frac{P_2}{\rho} + \frac{1}{2}V_2^2 + gz_2$$

Since $V_2 = 0$ at a stagnation point, one obtains:

$$\frac{P_1}{\rho} + \frac{1}{2}V_1^2 = \frac{P_2}{\rho} + g(z_2 - z_1)$$

Also, there is no elevation change along the Pitot-static probe, so $z_1 = z_2$. Bernoulli's equation is further simplified

$$\frac{P_1}{\rho} + \frac{1}{2}V_1^2 = \frac{P_2}{\rho}$$

or

$$P_2 = P_1 + \frac{1}{2}\rho V_1^2$$

This equation should be familiar, as P_2 is equivalent to the total pressure of the fluid with P_1 being the static pressure.

$$P_t = P_s + \frac{1}{2}\rho V^2$$

Rearranging the pressure equation, it is possible to explicitly solve for the velocity of the fluid. The following equation clearly shows the need to accurately measure the difference between the static and stagnation pressure to have an accurate measurement of the fluid velocity.

$$V_1 = \sqrt{\frac{2(P_2 - P_1)}{\rho}}$$

By attaching the outlets of the Pitot-static pressure probe to a manometer or transducer, one can have an accurate, inexpensive ($100–$1000) setup which provides the velocity of the fluid at a selected point in the flow field. Note that the probe should not be yawed or pitched in the flow (i.e., parallel to flow velocity) or the measured velocity will be incorrect.

For compressible flow, where the Mach number is greater than 0.3, Bernoulli's equation should not be used. In this case, the velocity can be calculated based on the isentropic flow equations.

$$\text{Ma} = \frac{V}{C}$$

$$C = \sqrt{\gamma RT} \ \left(\text{Speed of sound}\right)$$

$$\gamma = \frac{c_p}{c_v} \ \left(1.4 \text{ for air}\right)$$

$$V = \sqrt{2\frac{\gamma}{\gamma-1}\left(\frac{P_t}{P}\right)\left[\left(\frac{P_t}{P}\right)^{\frac{\gamma-1}{\gamma}} - 1\right]}$$

Pressure Probe Traversing: A Pitot-static probe provides a point measurement. To measure a velocity distribution, one needs to design a 1-D, 2-D, or 3-D traversing system. Due to the probe size, the Pitot-static probe is normally traversed for mainstream flow velocity measurements. For boundary layer velocity profile measurement, one can use a boundary layer probe. The boundary layer probe measures total pressure, similar to the Pitot tube shown in Figure 2.3 but with a smaller probe diameter. The local velocity can be calculated from the above-mentioned

Bernoulli's equation by using the measured local total pressure from the boundary layer probe and the measured static pressure from a wall-mounted pressure tap at the traversing location. From boundary layer theory, the static pressure inside the boundary layer is uniform in the y-direction at the given x-location. In other words, the measured static pressure with the surface, wall-mounted pressure tap is the same static pressure across the boundary layer thickness. Therefore, the local velocity profile inside the boundary layer can be calculated from the measured local total pressure profile (P_t) and the static pressure (P_s) at the selected x-location by using Bernoulli's equation.

Combined Pressure and Temperature Probes: By attaching a thermocouple (see Chapter 3) to the Pitot-static pressure probe, both the time averaged pressure (velocity) and temperature at a given point or over a given traversing profile can be measured. There are several reasons why the temperature may be measured along with the pressure within the fluid. For high speed flows, it is necessary to know the local recovery temperature. In a scenario where the flow might be heated or cooled, it is necessary to measure the temperature near the pressure probe, so the proper density of the fluid can be included in the velocity calculation.

2.3 FLOW RATE MEASUREMENTS

It is often required to measure the flow rate for a given experiment. A flow meter can be included in the flow loop for flow rate measurements. Various flow meters can be chosen depending upon the specific application. There are differential-pressure meters such as an orifice plate, Venturi meter, and flow nozzle; force meters such as rotameters; and momentum meters such as turbine flow meters.

2.3.1 DIFFERENTIAL PRESSURE METERS

Obstruction Flow Meters: Commonly used obstruction flow meters include the orifice plate, Venturi meter, and flow nozzle. They measure the pressure drop due to an intentional reduction of flow area, as shown in Figure 2.14. They can be placed in the pipe for direct flow rate measurements. An example is included demonstrating how to calculate the flow rate from the measured pressure drop $(P_1 - P_2)$ across an orifice flow meter.

An orifice flow meter (the top and bottom drawings of Figure 2.14) is used to measure a fluid flow rate. Different sizes of the orifice bore diameter can be used to measure various ranges of the flow rate. The mass flow rate of the fluid is directly proportional to the pressure drop across the orifice: increasing the flow rate increases the pressure drop. The size of the orifice plate must be selected in conjunction with the manometer or transducer used for this differential pressure measurement. If the flowrate is too large, the differential pressure could exceed the range of the transducer. If the flowrate is too small, measurement of the small pressure difference

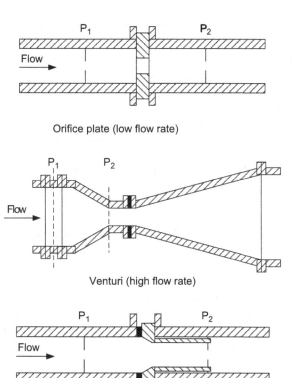

Orifice plate (low flow rate)

Venturi (high flow rate)

Flow nozzle (high flow rate)

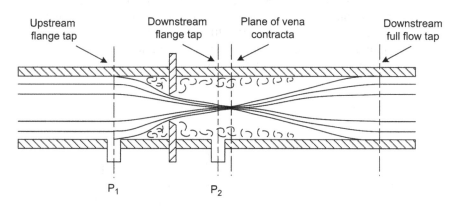

FIGURE 2.14 Obstruction-type flow meters.

may be inaccurate. Therefore, preliminary calculations should be completed to estimate the differential pressure range for your experiments with various size orifice plates. In addition to the manometer or differential transducer, the orifice flow meter is also instrumented with a "pressure gauge" to measure upstream pressure before flow enters orifice plate. The upstream pressure must be measured separately from the pressure drop; this static pressure is needed to calculate the density of the fluid (which is needed to determine the mass flow rate).

Because orifice meters are widely used for flow rate measurements in many industries, a number of standards are available for their construction, implementation, and usage. For example, the location of the upstream and downstream pressure taps is not arbitrary. As the fluid travels through the orifice, an area of separation is formed downstream of the plate. The pressure difference across the orifice will vary depending on where the downstream pressure measurement is taken. Also, the orifice plate should be constructed so the loss through the plate is predictable. With the calculation of the mass flow rate, multiple constants are used based on the construction of the meter, and if the plate is not constructed according to industry standards, the constants may not be applicable, and therefore, the flow rate will be measured incorrectly.

A well-known company supplying flow meters across the oil and gas industry is Emerson. While this company deals with many automation tasks, their Daniel flow meters are well-regarded for their performance and ease of operation. The Daniel orifice plates are fabricated according to industry standards with tight manufacturing tolerances. To support these meters, they have made available the Daniel Orifice Flow Calculator software. This calculator can be used with a wide variety of fluids, pipe sizes, and orifice sizes. While this software package is convenient to use, researchers should be aware of what is required to accurately calculate the mass flow rate from this type of meter.

2.3.2 Force Meters

Rotameters: Rotameters, or floater meters, as shown in Figure 2.15, are based on a force balance: drag force + buoyancy force = floater weight. They can be purchased commercially with various ranges of volume flow rates, for example, 0 ~ 100 scfs, 0 ~ 200 scfs, 0 ~ 500 scfs, etc. They are easily plumbed into a flow loop for direct measurement of the flow rate. They are convenient for small volumetric flow measurements. Because

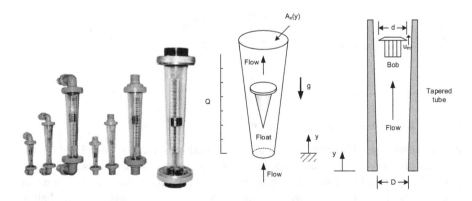

FIGURE 2.15 Rotameters.

the meters are generally calibrated for standard atmospheric conditions (scfs = standard, cubic feet per second), they must be corrected if the fluid pressure and/or temperature vary from standard temperature and pressure (STP).

Dwyer Instruments manufactures a variety of flow devices. The Dwyer Series RM Rate-Master Flowmeters (rotameter) can be used to measure volumetric flow rates. The installation manual supplied by Dwyer provides a correction when using their meters at non-standard conditions, $Q_{corrected}$.

$$Q_{corrected} = Q_{reading} \sqrt{\frac{P_{actual}}{P_{standard}} \cdot \frac{T_{standard}}{T_{actual}} \cdot \frac{1}{SG}}$$

where:

$Q_{corrected}$ is the true volume flow rate with corrected pressure, temperature, density

$Q_{reading}$ is the volume flow rate reading directly from rotameter at standard conditions

P_{actual} is the pressure reading from pressure gauge near the exit of the rotameter

T_{actual} is the temperature reading from thermocouples near the exit of the rotameter (absolute)

SG is the molecular weight of foreign gas divided by molecular weight of air.

Turbine Flow Meters: Turbine flow meters, as shown in Figure 2.16, are also used for flow rate measurements. The meter operates using a turbine rotor. The rotational speed of the shaft is proportional to the mass flow rate through the meter. Therefore, these meters require a sensor to accurately measure the rotational speed of the shaft. The optical or magnetic sensor is calibrated, so the output from the sensor is used to determine the fluid mass flow rate. With proper calibration, high accuracy can be reached for wide range of flow rates (±0.5%). Similar to the other flow meters, turbine meters can also be placed directly into the flow loop.

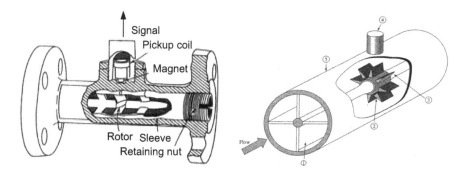

FIGURE 2.16 Turbine flow meters.

2.3.3 FLOW RATE CALCULATIONS

For Obstruction Flow Meters: From the top and bottom drawings of Figure 2.14, the following mass conservation concept can serve as the basis to calculate the volume flow rate and mass flow rate.

$$P_1 + \frac{1}{2}\rho v_1{}^2 = P_2 + \frac{1}{2}\rho v_2{}^2 + \Delta P_{\text{friction loss}}$$

$$Q = \text{Volumetric flow}$$

$$= v_1 A_1 = v_2 A_2$$

$$= \frac{A_2}{\sqrt{1 - \frac{A_2^2}{A_1^2}}} \sqrt{\frac{2\left[P_1 - P_2 - \Delta P_{\text{friction loss}}\right]}{\rho}}$$

$$= C_D E Y A_2 \sqrt{\frac{2\left(P_1 - P_2\right)}{\rho_1}}$$

Flow coefficient: $K = C_D E = \dfrac{C_D}{\sqrt{1 - \left(\dfrac{D_2}{D_1}\right)^2}}$; $P_1 = \rho_1 R T_1$

C_D = Discharge coefficient = $Q_{\text{actual}}/Q_{\text{ideal}} \approx 0.65$ to 0.62

E = Area ratio factor = $\dfrac{1}{\sqrt{1 - \left(\dfrac{A_2}{A_1}\right)^2}}$

Y = Expansion coefficient due to density change ~ 0.98

For Compressible Flow:

Let $P_2 = P_0$

$$\frac{P_1}{P_0} = \left(1 + \frac{\gamma - 1}{2}M_0{}^2\right)^{\frac{\gamma}{\gamma - 1}} = \left(\frac{\gamma + 1}{2}\right)^{\frac{\gamma}{\gamma - 1}} = 1.89$$

If $M_0 = 1$ at throat chocked, $Y = C_p / C_r = 1.4$; C = sonic velocity = $\sqrt{\gamma R T}$.

$$\frac{T_1}{T_0} = 1 + \frac{\gamma - 1}{2}M_0{}^2$$

$$\dot{m} = \rho_0 A_0 V_0$$

$$= \frac{P_0 A_0}{RT_0} M_0 C_0 \quad C_0 : \text{sonic velocity at throat}$$

$$= \frac{P_0 A_0}{RT_0} M_0 \sqrt{\gamma RT_0}$$

$$= \frac{P_1 A_0}{\sqrt{RT_1}} \frac{\sqrt{\gamma} M_0}{\left[1 + \frac{\gamma - 1}{2} M_0^2\right]^{\frac{\gamma+1}{2(\gamma-1)}}}$$

$$= \frac{P_1 A_0}{\sqrt{RT_1}} \sqrt{\gamma} \left(\frac{\gamma + 1}{2}\right)^{\frac{2(\gamma-1)}{\gamma+1}}$$

If $M_0 = 1$ at throat, chocked flow.

For Rotameters: From Figure 2.15, the following force balance concepts can serve as the basis to calculate the volume flow rate (and mass flow rate).

$$v = \text{Average velocity at vertical position } v(y)$$

"Float" force balance:

$$F_D + F_B = W_f,$$

Drag force + Buoyancy force = Weight of the float

where

A_f is the cross-sectional area of float
V_f is the volume of float
A_{gap} is the annular area between float and wall

$$F_D = \frac{1}{2} \rho_g v^2 A_f C_D$$

$$= W_f - F_B$$

$$= \rho_f V_f g - \rho_g V_f g$$

$$= (\rho_f - \rho_g) V_f g$$

$$Q = v A_{\text{gap}} = C A_{\text{gap}} \sqrt{\frac{2g(\rho_f - \rho_g)V_f}{A_f \rho_g}}$$

$$= \text{Volume flow}$$

$$\left(C \sim \sqrt{C_D}, \; C_D : \text{drag coefficient of float}\right)$$

2.3.4   Experiment Examples

Flow Rate Calculations: The flow rate of air through the test channel, depending on the Reynolds number, was monitored by the pressure drop across the orifice meter. The following discussion refers to the pressure drop calculations corresponding to the desired flow rate (Reynolds number). These calculations are applicable only to the square-edged orifice plate with flange taps as shown in the top and bottom drawings of Figure 2.14.

Orifice Equation: The equation for the flow rate of a gas through the ASME square-edged orifice meter is given by [4]:

$$\dot{m} = SD_2^{\,2}KY\sqrt{(p_1/T_1)Gy\Delta p}\qquad(2.1)$$

where:

$\dot{m}$ is the mass flow rate, lbm/sec

D_2 is the orifice diameter, inches

K is the flow coefficient, dimensionless

p_1 is the static pressure before orifice, inches Hg absolute

T_1 is the temperature before orifice, °F absolute

G is the specific gravity of gas (for air, = 1.00)

y is the supercompressibility factor, dimensionless (for air = 1.00)

Δp is the pressure drop across orifice, inches H_2O

S is a constant, = 0.1145 (if Δp is in inches of H_2O)

    = 0.4219 (if Δp is in inches of Hg)

Flow Coefficient, K: The quantity K is given by:

$$K = C/\sqrt{1-\beta^4}\qquad(2.2)$$

where:

C is the discharge coefficient = (actual flow rate/theoretical flow rate)

β is the diameter ratio, D_2/D_1, and D_1 is the pipe inside diameter

Experimental values of K, corresponding to various pipe diameters, values of β, and Reynolds numbers can be referred to in the ASME Research Report on fluid meters [4]. For the present study (β =0.5 and Re = 10,000–60,000), the K value was in the range of 0.62–0.65.

Expansion Factor, Y: This factor takes into account there is an uncontrolled expansion (both longitudinal and lateral) of gas after the orifice due to the reduced pressure in that region. The value of Y, for any gas, may be computed from the following empirical equation:

$$Y = 1 - \left[0.41 + 0.35\left(D_2/D_1\right)^4\right]\left[\left(\Delta p/p_1\right)\left(1/k\right)\right]\qquad(2.3)$$

Here, $k = C_p/C_v$; ratio of specific heats.

If k and D_2/D_1 are held constant in Equation (2.3), Y is a linear function of $\Delta p/p_1$, hence the graph of Y versus $\Delta p/p_1$ is a straight line. A graphical solution of Equation (2.3) for any gas can be found in the ASME Research Report on fluid meters. For the present study, the value of Y was approximately 0.98 [4].

Pressure, p_1, and Pressure Drop, Δp: The static pressure, designated by p_1 in Equation (2.1), is measured at the upstream pressure tap. The pressure drops across the orifice, Δp, is the difference between the pressure measured at this tap and the pressure measured at a similar tap located on the downstream side of the orifice plate.

Super Compressibility Factor, y: If Equation (2.1) was restricted to be used with perfect gases only, the term y would be unnecessary. That is, y is a factor which corrects for the fact that all real gases deviate from the perfect gas equation.

Example: Measuring Air Flow:

The use of Equation (2.1) involves a trial and error procedure. For instance, suppose that flow measurements are being made with a 0.515-inch diameter orifice in a 2-inch pipe and that the measured quantities are:

p_1 is the 29.9 inch Hg. Abs.
T_1 is the 60°F (520°R)
Δp is the 9.78 inch Hg

The corresponding value of Y for $\Delta p / p_1 = 9.78/29.9 = 0.327$ is 0.904. The specific gravity, G and the super compressibility factor, y are both unity for air. Hence, from Equation (2.1):

$$\dot{m} = 0.4219(0.515)^2 K(0.904)\sqrt{(29.9/520)(9.78)}$$

The mass flow rate, $\dot{m}$, cannot be found until K is known and K cannot be found until the Reynolds number, a function of $\dot{m}$, is known. Trial and error method is required to solve the above expression.

For a 2-inch pipe ($D_1 = 2.067$ inches) and $\beta = 0.250$, the variation in K with Reynolds number is small. By selecting a mean value of K, which is 0.6048, the approximate value of $\dot{m}$, may be used to compute an approximate Reynolds number, from which a new value of K will be obtained. This new value of K may be used in Equation (2.1) to get a better value of $\dot{m}$. The procedure may be repeated until there is no appreciable change in the values of $\dot{m}$ and K.

Example: Reynolds Number and Mass Flow Rate Determination

The mass flow rate of air through the test section is set by the desired Reynolds number. The flow rate is monitored by measuring the pressure drop across a square edged orifice meter.

The required mass flow rate (based on the Reynolds Number) can be determined using the following equation:

$$\dot{m}_{required} = \frac{Re}{D_h} \mu A_c$$

where:

Re is the Reynolds number

D_h is the hydraulic diameter

μ is the viscosity of air $\left(1.033 \times 10^{-6} \frac{lbm}{s \times in} \text{ at } 78°F\right)$

A_c is the cross-sectional area of the test section

The above-mentioned mass flow rate Equation (2.1) can be used for actual mass flow rate calculations, $\dot{m}_{actual}$. The following assumptions have been made. These values have been used over time and are generally accepted.

$D_{orf} = 1$ inch   (orifice diameter)

$K = 0.62$ (flow coefficient)

$G = 1$ (specific gravity of air)

$y = 1.0$ (supercompressibility factor of air)

$S = 0.1145$ (a constant, if p is in inches of H_2O)

$Y = 0.98$ (expansion factor)

The pressures measured at the orifice meter should be adjusted until the actual mass flow rate matches the required mass flow rate.

Once the mass flow rates are approximately equal, the actual Reynolds Number can be calculated using the following equation:

$$Re_{actual} = \frac{\dot{m}_{actual} D_h}{m A_c}$$

The actual Reynolds Number should be used when reducing the experimental data. The following table shows the values of Δp for each test. These values were calculated for a 4:1 channel aspect ratio ($2'' \times 0.5''$), $p_1 = 29.9$ inch Hg. Abs., and $T_1 = 60°F$ (520 °R).

Reynolds Number (required)	Δp (inches of H_2O)	
	Stationary Test	Rotational Test
5000	0.15	0.155
10,000	0.6	0.62
20,000	2.35	2.4
40,000	8.4	8.5

PROBLEMS

1. Write down the typical pressure ranges for the following manometers/barometer in both inH_2O and psig:
   a. A U-tube manometer using water as the working fluid.
   b. A U-tube manometer using mercury as the working fluid.
   c. An inclined manometer using a fluid (oil) with a specific gravity of 0.826.
   d. A micromanometer using water as the working fluid.
   e. A Bourdon-type gauge.

2. Find the pressure of a tank using a U-tube manometer, as shown in the Figure 2.7. If the water level for both legs are at $"0"$ inch before it is connected to the tank, and the water column height, $h = 4$ inches after it is connected to the tank.

3. Describe how you will measure the pressure drop (P_1-P_2) along a circular pipe with a steady, fully developed air flow from point 1 to point 2 using a U-tube manometer, with sketches. Indicate how to implement pressure taps.

4. Describe how to measure the total pressure (P_t) and static pressure (P_s) at the center of a circular pipe using a Pitot-static probe, with simple sketches.

5. Derive the equation for flow velocity (V) inside a circular tube by applying the Bernoulli's equation, after you have obtained the total pressure (P_t) and static pressure (P_s). Assume the density of air is ρ.

6. Determine the velocity (m/s), mass flow rate (kg/s) and Reynolds number of a steady, uniform-velocity air flow in a rectangular duct. The height and width of the duct are 0.5 m and 1 m, respectively. The flow temperature is 50°C, the static pressure (P_s) is 30 in H_2O (gauge), and the differential pressure (P_t-P_s) is 20 inH_2O. The gas constant R $= 87$ J/(Kg×K), and the dynamic viscosity (μ) of air at 50°C is 1.98×10^{-5} kg/(m×s).

7. We know that the reading of a rotameter is going to change if the pressure and temperature at the exit of the meter is not close to the standard temperature and pressure (conventionally 70°F, 1 atm). Given the required air volume flow rate of 2 SCFM, please obtain the reading (observed) on the rotameter when the exit pressure and temperature are 2 psig and 50°C, in SCFM.

8. The reading of a rotameter will also change if the specific gravity of the working fluid is not 1 (i.e., not air). Given the required volume flow rate is 2 SCFM at standard temperature and pressure, please obtain the reading on the rotameter when CO_2 is used as the working fluid.

9. There is an orifice flow meter. Given the following parameters: diameter of pipe $D_h = 2.067$ in, diameter of orifice bore $D_{orf} = 1.5$ in, typical discharge coefficient for an orifice plate $C = 0.61$, specific gravity $G = 1$ for air, super compressibility factor $y = 1$, $S = 0.03103$ (a constant for pressure unit in inH_2O), expansion factor $Y = 0.98$, and fluid temperature $T = 120°F$.

Calculate the actual mass flow rate $\dot{m}_{actual}$ (in lbm/s) when the static pressure before orifice plate $P_1 = 40$ inH$_2$O (gauge pressure) and the pressure drop across orifice plate $\Delta P = 20$ inH$_2$O.

10. Continue from question 9, if you need an actual mass flow rate $\dot{m}_{actual}$ of 0.15 lbm/s, what will be the pressure drop, ΔP, in inH$_2$O when the fluid temperature is 120°F, the static pressure before orifice plate $P_1 = 450$ inH$_2$O (absolute) and the diameter of orifice bore $D_{orf} = 1.0$ in?

REFERENCES

1. Morrison, G.L., *MEEN 637-Turbulence Measurement and Analysis*, Mechanical Engineering Department, Texas A&M University, College Station, TX, August, 1994.
2. Lee, T.W., *Thermal and Flow Measurements*, CRC Press, Taylor & Francis Group, New York, 2008.
3. Russo, G.P., *Aerodynamic Measurements*, 1st ed., Woodhead Publishing, Cambridge, UK, 2011.
4. Leary, W.A., and Tsai, D.H., *Metering of Gases by Means of the ASME Square-Edged Orifice with Flange Taps*, Sloan Laboratory for Automotive and Aircraft Engines, MIT, July, 1951.

3 Temperature and Heat Flux Measurements

3.1 TEMPERATURE MEASUREMENTS

There are several standard temperature measurement devices such as thermal expansion temperature sensors, electrical resistance temperature sensors, and thermoelectric temperature measurement sensors. The proper temperature measurement device can be chosen depending upon the specific applications. Thermal expansion temperature sensors include liquids in glass casings and bi-metallic measuring devices; electrical resistance temperature sensors can be thermistors or resistance temperature detectors (RTD); thermoelectric temperature measurement sensors are commonly named thermocouples [1–4].

There are several optical temperature measurement methods such as liquid crystals, temperature sensitive paints, and infrared thermography. These advanced temperature measurement techniques, which provide detailed surface temperature distributions, will be discussed in the later chapters.

3.1.1 THERMAL EXPANSION TEMPERATURE SENSORS

Liquid-in-Glass Thermometers: Across the general public, thermometers are the most common type of temperature measurement device. The fundamental principle of liquid-in-glass thermometry is based on the thermal expansion of a liquid with a high volumetric expansion coefficient ($\beta = \Delta V/(V_o \Delta T)$). The large bulb holds the majority of liquid; when heated, the liquid expands and rises into the capillary tube. The tube is calibrated and has an etched scale providing the bulb temperature.

The two most commonly used liquids in thermometers are mercury and alcohol. The volumetric expansion of mercury is $\beta = 18 \times 10^{-5}$ (1/°C). Figure 3.1 shows a typical mercury thermometer. Mercury thermometers are commonly used at relatively high temperatures, as the boiling point of mercury is 315°C (600°F); however, mercury thermometers must be used above their freezing point (−38°C = −39°F). On the other hand, alcohol has a low freezing point (−117°C = −179°F). With a relatively low boiling point (70°C = 158°F), alcohol thermometers are better suited for low temperature applications. Generally, red dye is added to the alcohol to improve the visibility of the fluid.

The expansion registered by a thermometer is the difference between the expansion of the liquid and the expansion of the glass stem. This difference is dependent on the heat conducted from the stem to the bulb (or vice versa).

FIGURE 3.1   Mercury thermometer.

Therefore, when high accuracy is required, the researcher is cautioned to note the "immersion level" shown on the thermometer. The scale etched on the stem has been calibrated based on the immersion of the device. The scale can be calibrated with only the bulb immersed in the fluid or for the entire thermometer to be immersed within the fluid.

Bimetallic Thermometers: Figure 3.2 illustrates the deformation that occurs when two dissimilar metals are joined together. When two metals with varying thermal expansion coefficients are joined together, the assembly will deform when the temperature deviates from the temperature at which they were joined together. When the bonded strip is subjected to a temperature higher than the bonding temperature, it will bend in one direction, and when it is exposed to a lower temperature, the strip will bend in the opposite direction.

The most effective bimetallic thermometers utilize one material with a low thermal expansion coefficient and one material with a high coefficient.

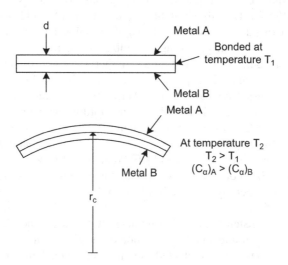

FIGURE 3.2   Concept of bimetallic thermometry.

Steel has a relatively high thermal expansion coefficient ($\beta = 2 \times 10^{-5}$– 20×10^{-5} 1/°C), and it is cost efficient. Nickel is easily added to steel to create an alloy with a relatively low expansion coefficient ($\beta = 1.7 \times 10^{-8}$ 1/°C).
Due to the simplicity of these devices, they have a long life with negligible maintenance expenses. They exhibit stable operation over long periods of time (± 1°C). These types of devices have been used for decades in traditional "on/off" devices (thermostats). As the metal deforms, after a certain amount of growth, the strip will touch a contact point, completing an electric circuit, and activating a device. The growth, or curvature, of the bimetallic thermometer can be approximated from the metal thickness, d, the temperature change from the bonding temperature, ΔT, and the expansion coefficient difference between the two metals: $r \, \alpha \, d/[(\beta_1 - \beta_2) \times \Delta T)]$.

3.1.2   ELECTRICAL RESISTANCE TEMPERATURE SENSORS

Electrical methods of temperature measurement are very convenient because they furnish a signal that is easily detected, amplified, or used for control purposes. When properly calibrated, these devices are very accurate in a wide variety of applications. The devices are constructed of a material that experiences a change in resistance proportional to a temperature change. Depending on how the resistance of the device changes with temperature allows the sensor to be classified as either an RTD or thermistor [2].

Resistance Temperature Detectors (RTDs): Resistance temperature detectors utilize a metallic resistance element. Figure 3.3 shows the fundamental principle of RTD thermometry. As the temperature of the metal increases, the resistance of the wire also increases. By measuring the resistance change of the metal, its temperature change (relative to a reference temperature) can be determined. RTDs are often referred to as electrical resistance thermometers.
To improve the accuracy of an RTD, using a metal that has a linear response to temperature change is ideal. Platinum is often viewed as the standard for RTDs as its resistance is linear over a very wide range of temperatures. Because platinum exhibits a predictable and reproducible change in

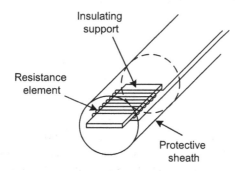

FIGURE 3.3   Fundamental principle of RTD thermometry.

electrical resistance with temperature, the temperature can be calibrated with a high degree of accuracy ($\pm0.005°C$). The linear relationship between measured resistance and temperature for an RTD is shown below. If other metals are used, the relationship between resistance and temperature could exhibit polynomial characteristics.

$$R = R_0\left[1+\alpha\left(T-T_0\right)\right]$$

where

R is the metal conductor resistance at temperature, T

R_0 is the resistance at reference temperature, T_0

α is the temperature coefficient of resistivity

The construction of an RTD can lead to a large and bulky sensor head. The metal wire is wrapped around an insulating support to eliminate mechanical strains. With thermal cycling of the wire, it must be free to expand and contract without inducing strain; therefore, the support must also expand with the wire. The wire and support are encased within a protective sheath to protect the wire from harsh environments and potential corrosion. In addition to being relatively large, yet fragile, researchers are cautioned about the calibration of an RTD. As the device is based on resistance changes of a short, small diameter wire, the lead wire connecting the RTD to the data acquisition system can introduce a significant amount of resistance into the circuit. Ideally, the RTD should be calibrated within the actual data acquisition setup.

Thermally Sensitive Resistors (Thermistors): Based on the name, thermistors operate according to a similar principle as RTDs. However, thermistors are constructed from materials that generally see a decrease in electrical resistance with increasing temperature. The relationship between resistance and temperature for thermistors is shown below. These devices are very sensitive as the resistance changes rapidly (exponentially) with temperature. A small change in temperature creates a large change in resistance, which can be measured with high accuracy; for a platinum RTD, a small change in resistance occurs with a large temperature change. This non-linear behavior is often viewed as a disadvantage for thermistors, as they must be calibrated over the entire range of expected temperatures.

$$R = R_0\exp\left[c\left(\frac{1}{T}-\frac{1}{T_0}\right)\right]$$

where

R is the thermistor resistance at temperature, T

R_0 is the resistance at reference temperature, T_0

c is the experimentally determined constant (350–460 K)

Thermistors are constructed from semiconductor materials. Figure 3.4 shows sample thermistors. The material has an intermediate conductivity falling between metals and insulators. This ceramic-like material has a negative

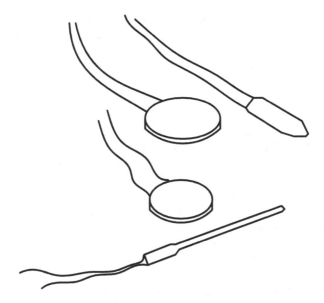

FIGURE 3.4   Fundamental principle of thermistor thermometry.

temperature coefficient of resistance (opposite of RTDs). The elements are also usually encapsulated, so they can be used in a wide variety of environments. Generally, the construction of the thermistors is viewed as very rugged. With a high resistance element, the lead wires used in the circuit add minimal resistance to the data acquisition circuit. In addition, due to the higher resistance of the sensor, low currents are passed through the sensor for measurements; this minimizes self-heating of the resistor and improves accuracy. The disadvantages of thermistors are the non-linear resistance-to-temperature relationship and the restriction to relatively low temperature (<300°C). The performance of thermistors is generally stated to be within ±0.01°C.

3.2   THERMOCOUPLE THERMOMETRY

Thermocouples are the most common method for temperature measurement. Thermocouple thermometry is based on two thermoelectric effects: (i) a voltage potential is created when two dissimilar metals are joined together and (ii) the voltage potential is primarily a function of junction temperature. The thermoelectric phenomena result from simultaneous flows of heat and electricity (i.e., coupled flows of entropy and electricity). Basically, a thermocouple junction is created by joining two dissimilar conductors. A circuit is formed with the two metals and an output voltage is generated within the circuit. A definite relationship exists between the voltage and the temperature of the thermocouple junction. Therefore, the output voltage can be calibrated over a range of temperatures to yield a compact, cheap device for measuring temperatures.

Three basic principles occur in thermocouple circuits: (i) Seebeck effect, (ii) Peltier effect, and (iii) Thomson effect. In addition, three basic laws apply to thermocouple construction: (i) Law of Homogeneous Materials, (ii) Law of Intermediate Materials, and (iii) Law of Successive or Intermediate Temperatures. Each of these effects and laws will be discussed in detail [1–4].

3.2.1 Fundamental Principles for Thermocouples

Seebeck Effect (Thomas Johann Seebeck, 1770–1821): A voltage (or emf) is generated in an open thermocouple circuit due to a temperature difference between junctions in the circuit (Figure 3.5). When two wires, dissimilar metals, are joined at both ends and one end of the circuit is heated, there is a continuous current which flows in the thermoelectric circuit. If the circuit is opened, the net open circuit voltage (Seebeck voltage) is a function of the junction temperature and the composition of the two metals. This voltage is proportional to the temperature difference between two junction points and the coefficient of two dissimilar materials A and B that are used to connect these two junction points ($\alpha_{A,B}$). For small temperature changes, voltage potential is linearly related to the temperature change of the circuit. The constant of proportionality, $\alpha_{A,B}$, is known as the Seebeck coefficient.

$$dE_s = \alpha_{A,B}dT$$

$$E_s = \int_{T_1}^{T_2} \alpha_{A,B}dT$$

Peltier Effect (Jean Charles Athanase Peltier, 1785–1845): Peltier heat is the quantity of heat (in addition to I^2R) that must be removed from the junction to maintain the junction at a constant temperature (Figure 3.6). Additional heat is created in the circuit due to the externally supplied power. This generation of heat is proportional to the temperature increment at the junction and the coefficient of two dissimilar materials A and B that are used to make the junction point ($\pi_{A,B}$).

$$dQ_P = \pi_{A,B}dT$$

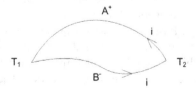

FIGURE 3.5 Seebeck effect.

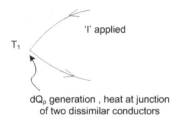

T_1

'I' applied

dQ$_p$ generation , heat at junction
of two dissimilar conductors

FIGURE 3.6   Concept of Peltier effect.

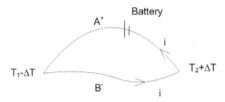

Battery

A$^+$

i

$T_1-\Delta T$

B$^-$

i

$T_2+\Delta T$

FIGURE 3.7   Concept of Thomson effect.

Thomson Effect (William Thomson, 1824–1907, Lord Kelvin from 1892):
To maintain a constant temperature of a conductor (through which a temperature gradient exists), energy must be removed from the conductor (Figure 3.7). With an external voltage source, a longitudinal temperature gradient exists through a conductor. To maintain the conductor at a constant temperature, energy must be removed (I^2R). This energy is proportional to the temperature difference between two junction points and the properties of two dissimilar materials A and B that are used to connect these two junction points $\left(\sigma_A - \sigma_B\right)$.

$$\pi_{A,B} = \left(\sigma_A - \sigma_B\right)$$

$$E_{T_1,T_2} = \int_{T_1}^{T_2}(\sigma_A - \sigma_B)dT$$

Law of Homogeneous Materials: A thermoelectric current cannot be sustained in a circuit of a single homogeneous material by the application of heat alone, regardless of how it might vary in cross section. This implies there must be two dissimilar homogeneous materials, A and B, in order to produce the voltage (emf) between two junction points. The third soldering or welding material cannot generate a voltage (emf) at the junction point by itself (Figure 3.8). Therefore, the soldering or welding material would not affect the junction point temperature.

Law of Intermediate Materials: The algebraic sum of the thermoelectric forces in a circuit composed of any number of dissimilar materials is zero

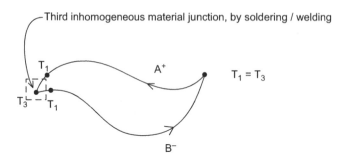

FIGURE 3.8   Concept of law of homogeneous materials.

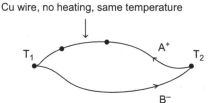

FIGURE 3.9   Concept of law of intermediate materials.

if all of the circuit is at a uniform temperature. A copper wire can be added to material A or material B and it does not affect the temperature difference between T_1 and T_2 (Figure 3.9). The third material must be isothermal. This is important, because the distance between T_1 and T_2 can be increased by adding copper wires which are relatively inexpensive compared to the thermocouple wires. This also implies a third material, such as solder, can be used to create a junction between the two dissimilar metals. With a small volume, the junction will be at a uniform temperature, and not create an additional voltage drop in the circuit.

Law of Successive or Intermediate Temperatures: If two dissimilar homogeneous materials produce thermal emf_1 when the junctions are at T_1 and T_2 and produces thermal emf_2 when the junctions are at T_2 and T_3, the emf generated when the junctions are at T_1 and T_3 will be $\text{emf}_1 + \text{emf}_2$ (Figure 3.10).

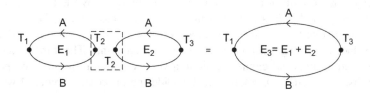

FIGURE 3.10   Concept of law of intermediate temperatures.

With $\mathrm{emf}_3 = \mathrm{emf}_1 + \mathrm{emf}_2$, the temperature differences are also additive: $(T_1 - T_3) = (T_1 - T_2) + (T_2 - T_3)$. This concept can be used to reduce the number of connecting circuits to a single circuit. In addition, this allows for a thermocouple calibrated at one reference condition to be corrected and used at another reference condition.

Disadvantages and Drawbacks of Thermocouples: While thermocouples are the most common sensor for laboratory temperature measurement, they are not without their drawbacks. Depending on the specific application, researchers should be aware of potential pitfalls of thermocouples that can lead to errors in the temperature measurements. The transient response of the thermocouple depends on junction size: a smaller size provides a faster response. Therefore, in transient experiments, the junction should be on the order of 0.5 mm in diameter to minimize time lag in the experiment.

The insertion of the thermocouple junction into a solid material or fluid can also lead to errors (the measured temperature is not the actual temperature of the substance). The most significant insertion error is due to conduction through the thermocouple wire itself. The wire can act as a fin from a surface and conduct heat along the length of the wire. As the wire is conducting heat, the temperature of the junction will artificially change. This conduction error can be minimized by using small diameter thermocouple wire. In high temperature applications, radiation to the thermocouple from the surrounding environment can also affect the temperature of the junction. In high-speed flows, the thermocouple junction can see a temperature rise due to the adiabatic heating associated with the high-speed fluid stagnating on the junction. Again, as the size the thermocouple junction increases, the error becomes more significant.

Inaccurate thermocouple readings can also be produced when junctions are not isothermal. If intermediate materials are used in the circuit, these materials, and their corresponding junctions, must be kept at uniform temperatures. For the most accurate use of thermocouples, reference junctions should be used. Integrating a reference junction (usually an ice bath at 0°C) into the thermoelectric circuit gives a known reference point for the thermocouple (Law of Intermediate Temperatures). If the experimental setup does not readily lend itself to an ice bath, electronic, cold junction compensators are available for modern data acquisition systems.

3.2.2   TYPICAL THERMOCOUPLE FABRICATION AND CALIBRATION

Standardized Materials: There are many standard materials, A and B, available for thermocouple fabrication. The particular choice of material depends on the range of temperature for a given application. For example, Type T (composed of Copper + and Constantan −) can be used for low temperature measurements, Type K (Chrome + and Aluminum −) can be used for high temperature measurements. The different types of thermocouple wires, with coated insulation, can be ordered as rolls (one roll can be

100–500 feet) from various companies. The following tables show common thermocouple materials, their typical temperature ranges, and the associated error. The thermocouple "types" are identified according to ANSI standards.

Type	Material Combination		Applications
	Positive	Negative	
E	Chromel (+)	Constantan (−)	Highest sensitivity (<1000°C)
J	Iron (+)	Constantan (−)	Nonoxidizing environment (<760°C)
K	Chrome (+)	Alumel (−)	High temperature (<1372°C)
S	Platinum/ 10% rhodium	Platinum (−)	Long-term stability high temperature (<1768°C)
T	Copper (+)	Constantan (−)	Reducing or vacuum environments (<400°C)

Type	Wire		Expected Bias Error[a]
	Positive	Negative	
S	Platinum	Platinum/10% rhodium	±1.5°C or 0.25%
R	Platinum	Platinum/13% rhodium	±1.5°C
B	Platinum/30% rhodium	Platinum/6% rhodium	±0.5%
T	Copper	Constantan	±1.0°C or 0.75%
J	Iron	Constantan	±2.2°C or 0.75%
K	Chromel	Alumel	±2.2°C or 0.75%
E	Chromel	Constantan	±1.7°C or 0.5%

Alloy Designations
Constantan: 55% copper with 45% nickel
Chromel: 90% nickel with 10% chromium
Alumel: 94% nickel with 3% manganese, 2% aluminum, and 1% silicon

From Temperature Measurements ANSI PTC 19-3-1974.
[a] Use greater value; these limits of error do not include installation errors

Thermocouple Fabrication: Thermocouples can be made using commercially available electric welders, as shown in Figure 3.11. When welding large steel components for structural applications, a large current is sent through the work material and filler material. An electrical arc is formed, and the arc melts the work material and filler material forming a bond between the components. A thermocouple welder is similar, but with small diameter wires, the power required to arc the wire, melt the tips, and form a junction is relatively small. Most electric welders have a variable power supply as the power required to create a suitable junction increases as the wire diameter increases.

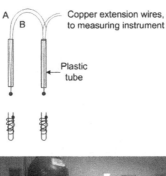

FIGURE 3.11 Thermocouple wires and electric welder to make thermocouple bead.

When purchasing a spool of thermocouple wire commercially, the two conductors are individually insulated (so they are isolated from one another), and an outer sheath of insulation is used to conveniently couple the wires together. The insulation colors are selected according to ANSI standards for each thermocouple type. To fabricate a thermocouple, a section of wire is cut to the desired length. All layers of insulation should be stripped from both ends of the wire, so bare wire is exposed on both ends. Depending on the size of the thermocouple and the amount of exposed wire near the junction, the thermocouple can be held with special pliers at either end of the wire. The end where the junction will be made, is then touched to the carbon electrode on the welder; both wires should touch the electrode simultaneously. Many welding machines are equipped with either a nitrogen or argon purge line to improve the weld integrity of the junction (shielding gas). After forming the bead (junction) between the two wires, the continuity of the thermocouple can be checked using a standard digital multimeter. The junction will close the circuit between the two wires, and by attaching the free ends of the wires (opposite of the junction) to the multimeter, the resistance of the thermocouple can be measured. If the resistance is infinity, an open circuit, the bead joining the two wires was not properly created. When making many thermocouples, of the same type, wire diameter, and length, the electric resistance of the thermocouples will be approximately equal.

Thermocouple Calibration: The newly welded thermocouple beads must be calibrated before for they are used for temperature measurements. Thermometers (mercury or alcohol in glass) can be used for the

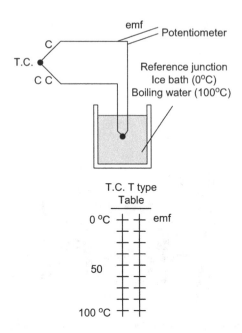

FIGURE 3.12   Concept of thermocouple calibration.

thermocouple calibration. For example, one can use an ice bath at $T = 0°C$ and boiling water at $T = 100°C$. There is a linear relationship between the emf produced in a thermocouple circuit and the temperature producing the emf. Therefore, a two-point calibration is adequate to fully define this relationship. Figure 3.12 shows a thermocouple calibration.

Thermocouple Measurement Principle: The thermocouple reference junction can be an ice bath or the ambient temperature. Cold junction compensators are also available to electronically simulate the voltage produced in an ice bath for a given thermocouple type. These reference junctions can be applied as an external signal to the data acquisition hardware or they can be a built-in function for many modern instruments.

Thermocouple Temperature Measurement Devices: Many devices are available with built-in conversion utilities for direct temperature measurements. From handheld devices for a single thermocouple to large scale, expandable systems for hundreds of thermocouples, researchers have a vast array of possibilities. In addition to selecting hardware based on the number of thermocouples required for an application, the sampling frequency may also be a concern for transient tests or turbulent temperature fluctuations. Common hardware configurations are listed below:

- Potentiometer, 1 channel
- Temperature indicator switch, 2 ~ 12 channels
- Data logger, ~100 channels
- National Instruments–1 module ~ 20 channels; 5 modules ~ 100 channels

Temperature Response: For steady-state temperature measurements, more accurate results are obtained if the time constant of the thermocouple (τ) is as small as possible (i.e., smaller junction/bead 0.5 ~ 1 mm). Thermocouples with small time constants provide faster responses. For example, for a time constant of approximately one second, the thermocouple junction bead should be approximately 1 mm in diameter. Thermocouples most often are used in fluids for time averaged temperature measurements. For high frequency, temperature measurements, i.e., time-dependent flow temperature measurements, very thin film gauges (Chapter 6) or cold wire sensors (Chapter 11) should be used.

Theory of Thermocouple Response Time: A thermocouple bead and wire is inserted into a flow at a given temperature with a given convective heat transfer coefficient (Figure 3.13). How long will it take for the thermocouple junction to come into thermal equilibrium with the fluid (the bead is at the same temperature as the fluid)? From the energy balance equation, the thermocouple bead gains energy by convection from the surrounding fluid. The resultant thermocouple bead temperature can be determined from the following equations. The response time decreases with the so-called time constant, the smaller time constant provides a faster response time; the smaller time constant implies a smaller thermocouple bead diameter [5].

$$\rho V c_p \frac{dT_w}{dt} = -hA_s\left(T_w - T_\infty\right), \text{ if } Bi = \frac{hL_c}{k} < 0.1,\ L_c = \frac{V}{A_s}$$

$$\theta = \frac{T_\infty - T_w}{T_\infty - T_i} = e^{-at} = e^{\frac{-t}{\tau_t}},\text{ when } a = \frac{hA_s}{\rho V c_p}$$

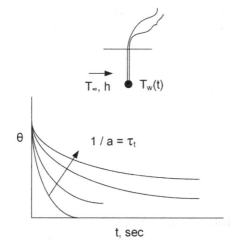

FIGURE 3.13   Theory of thermocouple response time.

$$\downarrow \frac{\rho V c_p}{A_s h} = \frac{1}{a} = \text{Time constant} = \tau_t$$

Fast response for $\dfrac{1}{a} \downarrow$

Example

T.C. bead size $= d = 1\,\text{mm}$

$$L_c = \frac{V}{A_s} = \frac{\dfrac{4}{3}\pi\left(\dfrac{d}{2}\right)^3}{4\pi\left(\dfrac{d}{2}\right)^2}$$

$$t = 1\,\text{second for } T_w = T_\infty$$

For given values of T_∞, h, c_p, ρ, error estimation can be based on the lead wire model.

3.2.3   Error Estimate in Temperature Measurements

Physical Errors in Temperature Sensors: As mentioned previously, there are errors due to the thermocouple insertion and junction heating or cooling. Insertion errors are due to conduction, convection, radiation, and recovery effects. Junction point heating or cooling errors are due to non-isothermal connections and the reference junction. In addition, errors can be due to aging/annealing/work hardening, magnetic field effects, and reference junction errors. Radiation errors need to be considered for high temperature measurements ($T > 500°C$) and recovery effects if Ma > 0.2.

The following lead wire model (Figure 3.14) can be used to estimate thermocouple reading errors for measuring temperatures in fluids and solids. The model is based on the concept of heat conduction through a fin with proper boundary conditions [1].

3.2.3.1   Lead Wire Model

Apply the concept of heat conduction through a fin.

Axial heat conduction,

$$Q_s = -k_w A_w \frac{dT}{dx} \left(\text{underestimation}\right);$$

if $kA \equiv k_w A_w + k_i A_i$ (overestimation)

$$\text{Radial heat conduction,} \quad \frac{dQ_r}{dr} \cong \frac{2\pi k_i \left(T - T_{\text{surf}}\right)}{\ln\left(\dfrac{r_i}{r_w}\right)},$$

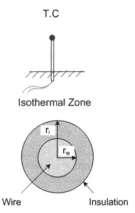

FIGURE 3.14   Concept of single lead wire with insulation.

where $dQ_r \equiv h2\pi r dr \left(T_{surf} - T_f \right)$

$$\frac{dQ_r}{dr} = \frac{T_{ref} - T_f}{\dfrac{1}{2\pi r_i h}} \equiv \frac{\left(T - T_f \right)}{R},$$

where $R = \frac{1}{2\pi r_i h} + \dfrac{\ln\left(\dfrac{r_i}{r_w} \right)}{2\pi k_i}$

Similarly, use the concept of heat conduction through a fin for the double wire with inner and outer insulations (Figure 3.15).

$$r_1 = \sqrt{2} r_w$$

$$r_2 = \frac{L_1 + L_2}{4}$$

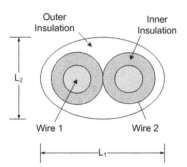

FIGURE 3.15   Concept of double lead wire with insulation.

$$Q_x = -\left(k_{w1} + k_{w2}\right) A_w \frac{dT}{dx} = -\overline{kA}\frac{dT}{dx}$$

$$\frac{dQ_r}{dr} = \frac{\left(T - T_f\right)}{R}, \text{ where } R = \frac{1}{2\pi r_2 h} + \frac{\ln\left(\dfrac{r_2}{r_1}\right)}{2\pi k_i}$$

Examples

1. A thermocouple temperature sensor is inserted in a gas flow (Figure 3.16). What is an estimate of the temperature error? [1]

From fin equation solution:

$$\frac{(T_{tc} - T_r)}{(T_w - T_r)} = \frac{1}{\cosh\left(\dfrac{L}{\sqrt{\tilde{k}AR}}\right)}$$

Temperature error: The temperature difference between the thermocouple and fluid $(T_{tc} - T_r)$ decreases with increasing the values of L, and $\dfrac{L}{\sqrt{\tilde{k}AR}}$,

$$\left(T_{tc} - T_r\right)\downarrow, \text{ if } \frac{L}{\sqrt{\tilde{k}AR}}\uparrow$$

If $M < 0.2$, $T_r = T_f$

If $M > 0.2$, $T_r = T_f\left(1 + r\frac{k-1}{2}M^2\right)$

where:

T_r is the recovery temperature,

M is the Mach number,

r is the Recovery Factor, it depends on the following conditions:

$\sqrt{Pr}$	Laminar
$Pr^{1/3}$	Turbulent
0.68 ± 0.07	Wire normal to flow
0.86 ± 0.09	Wire parallel to flow

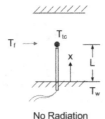

No Radiation

FIGURE 3.16   A thermocouple temperature sensor is inserted in a gas flow.

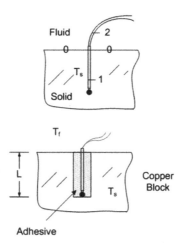

FIGURE 3.17 A thermocouple temperature sensor is embedded in a massive solid.

2. A thermocouple temperature sensor is embedded in a massive solid (Figure 3.17). What is an estimate of the temperature error? [1]

From the solution of previous case, one gets the similar temperature results as shown in below equation (3.1).

$$\frac{(T_{tc} - T_s)}{(T_{00} - T_s)} = \frac{1}{\cosh\left(\dfrac{L}{\sqrt{\tilde{k}AR_I}}\right)} \tag{3.1}$$

where $R_1 = \dfrac{\ln\dfrac{r_2}{r_1}}{2\pi k_i} + \dfrac{\ln\dfrac{r_3}{r_2}}{2\pi k_a}$,

where r_3 = adhesive radius

Fin theory: Consider a fin in solid 1 and extended into fluid 2.

$$Q_I = \sqrt{\frac{\tilde{k}A}{R_I}}\,(T_{00} - T_s)\,\tanh\left(\frac{L}{\sqrt{\tilde{k}AR_1}}\right)$$

$$Q_{II} = \sqrt{\frac{\tilde{k}A}{R_{II}}}\,(T_f - T_{00}) \times 1 (\because L \to \infty)$$

since $Q_I = Q_{II}$:

$$(T_{00} - T_s) = \frac{(T_f - T_s)}{1 + \sqrt{R_{II}/R_I}\,\tanh\left(\dfrac{L}{\sqrt{\tilde{k}AR_1}}\right)} \tag{3.2}$$

From (3.1) and (3.2):

$$\frac{(T_{tc} - T_s)}{(T_w - T_s)} = \frac{1}{\cosh\left(\dfrac{L}{\sqrt{\tilde{k}AR_I}}\right)} \left[\frac{1}{1 + \sqrt{R_{II}/R_I}\,\tanh\left(\dfrac{L}{\sqrt{\tilde{k}AR_I}}\right)}\right]$$

Temperature error: The temperature difference between the thermocouple and solid $(T_{tc} - T_s)$ decreases with increasing the values of L, and R_{II},

$$(T_{tc} - T_r)\downarrow, \quad \text{if } L\uparrow, \ \tilde{k}AR_I \downarrow, \ R_{II} \uparrow$$

3. In a surface-mounted thermocouple temperature sensor (surface temperature measurement of a massive solid, Figure 3.18), what is an estimate of the temperature error? [1]
From the fin equation solution:

$$\frac{(T_{surf} - T_{tc})}{(T_{surf} - T_f)} \sim \frac{\sqrt{\tilde{k}A/R}}{\pi r_1 k_s}\tanh\left(\frac{L}{\sqrt{\tilde{k}AR_I}}\right), \quad \tanh\left(\frac{L}{\sqrt{\tilde{k}AR_I}}\right) \approx 1 \text{ if } L \to \infty$$

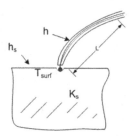

T.C seated in a shallow
hole drilled into the surface

T.C affixed to the surface,
well bonded to the surface

FIGURE 3.18   A surface-mounted thermocouple temperature sensor.

$$\frac{T_{surf} - T_{tc}}{T_{surf} - T_i} \sim \frac{1}{\left(h_s R_1 / k_s\right)} \sim \frac{1}{B_i}$$

Temperature error:

The temperature difference between the thermocouple and solid $(T_{surf} - T_{tc})$ will reduce, if the Biot number increases and if the thermocouple conductance/solid conductance decreases.

$$\left(T_{surf} - T_{tc}\right) \downarrow, \text{ if } \frac{\sqrt{\tilde{k}A/R}}{\pi r_1 k_s} \frac{\text{conductance of T.C.}}{\text{conductance of solid}} \downarrow, \text{ and } \frac{h_s r_1}{k_s} \uparrow$$

3.2.4 HIGH VELOCITY GAS TEMPERATURE MEASUREMENTS

Stagnation Temperature, T_0, or Total Temperature, T_T: When a high-speed flow approaches a sphere, or thermocouple bead, flow dynamic energy converts into thermal energy which increases its temperature. This is referred to as the stagnation temperature or total temperature (Figure 3.19). If a gas flow is brought to rest adiabatically, the stagnation temperature can be determined as shown below. The stagnation temperature is the same as the static temperature for low-speed flows and in liquids [6]. The total temperature, T_T, (or stagnation temperature, T_0) = static temperature T_s (or T_∞) + dynamic temperature $\left(\frac{v^2}{2c}\right)$:

$$T_0 = T_\infty + \frac{v^2}{2c} = T_\infty \left[1 + \left(\frac{\gamma - 1}{2}\right) M^2\right] = T_\infty \left(1 + 0.2 M^2\right)$$

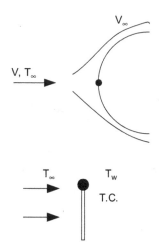

FIGURE 3.19 Concept of stagnation temperature.

If M = 1, deviate 20%

      = 0.5, deviate 5%

where:

$\gamma = \frac{c_p}{c_v} = 1.4$ for air

M is the Mach number $= \frac{V}{\sqrt{\gamma RT}}$

$\sqrt{\gamma RT}$ is the speed of sound

~ 1000 fps at room temperature

Adiabatic Wall Temperature, T_{aw}, or Recovery Temperature, T_r: When a high-speed flow passes over an adiabatic surface, frictional heat $\left(\sim v \left(\frac{\partial u}{\partial y} \right)^2 \right)$ is generated inside the boundary layer due to viscosity and the velocity gradient. This heat increases the surface temperature, and this elevated temperature is referred to as the adiabatic wall temperature, T_{aw}, or recovery temperature, T_r (Figure 3.20). Note that the recovery temperature (T_r) is the stagnation temperature (T_0), if the recovery factor, $r = 1$ [6]. The recovery temperature is strongly dependent upon the probe design. Adiabatic wall temperature T_{aw} (or recovery temperature, T_r) = static temperature T_s (or T_∞) + dynamic temperature $\frac{U_\infty^2}{2c}$:

$$T_{aw} = T_\infty + \frac{U_\infty^2}{2c}, \text{ if } Pr = 1,$$

$$T_r = T_\infty + r_c \frac{U_\infty^2}{2c}, \text{ if } Pr \neq 1$$

where:

r_c = recovery factor

= $Pr^{1/2}$ Laminar flow

= $Pr^{1/3}$ turbulent flow

= $0.75 \sim 0.99$

= 1 if $Pr = 1$ (assumption, ideal)

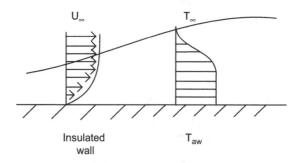

FIGURE 3.20   Concept of adiabatic wall temperature.

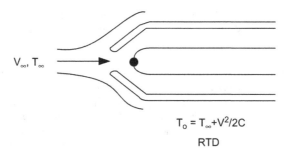

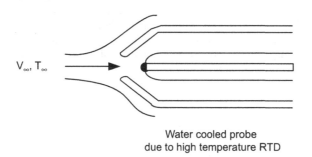

FIGURE 3.21   Concept of high velocity and high temperature probe.

$$T_{aw} = T_r = r_c \left(T_0 - T_\infty\right) + T_\infty$$

$$= T_0, \ \text{if} \ r_c = 1$$

In general $T_{aw} < T_0$, $T_r < T_T$

High Velocity/High Temperature Probe for Gas Temperature Measurements: Temperature measurements in high-speed gas flows, such as a gas turbine combustor, ramjet, or scramjet are difficult and require special measurement probes. The gas velocity must be reduced to zero in order to measure the true temperature (Figure 3.21). High-velocity gas temperature probes (HVGT) must be calibrated.

Example

The space shuttle Columbia broke apart upon re-entry into the Earth's atmosphere at approximately 8:00 a.m. on Saturday, February 1, 2003. As the space shuttle broke apart, debris fell to the ground across East Texas. During this tragedy, all seven crew members on Columbia were killed.

The space shuttle broke apart at approximately 200,000 feet ~ 39 miles above the state of Texas. The shuttle was traveling at approximately 12,500 mph (or 18 times the speed of sound). The shuttle had just reached the hottest point of its descent into the Earth's atmosphere. NASA officials estimated the external temperature generated by friction (so-called aerodynamic heating) as the shuttle travelled into the compressed atmosphere (so-called the stagnation temperature) was 3,000°F.

The heat shield is comprised of approximately 20,000 tiles, each 6″ × 6″. A section of tile was damaged during the shuttle launch as a block of foam insulation (20″ × 16″ × 5″) fell from the fuel tank and struck the heat shield beneath the left wing of the shuttle. The damaged tiles on the shuttle were not detected before re-entry of the shuttle. Note that 1 mile = 5128 feet, speed of sound = $(\gamma.R.T_{a1})^{1/2}$ = 1000 fps, γ = 1.4, M_1 = Mach number = shuttle velocity/speed of sound = 18, T_{a1} = air temperature = 45°F = 280°K at 39 miles sky. Based on the compressible flow equations with a normal shock wave [7], the estimated Mach number and air temperature after the shock M_2 = 0.38 and T_{a2} = 63.8 T_{a1} = 17,885°K, respectively. The following is the estimated shuttle leading edge temperature (the hottest temperature) T_0.

$$T_0 = T_{a2}\left\{1+[(\gamma-1)/2]M_2{}^2\right\}$$

$$= T_{a2}\left\{1+0.2\,M_2{}^2\right\}$$

$$= 1.029\,T_{a2}$$

$$= 18,402\ K\,(\text{stagnation temperature})$$

3.3   HEAT FLUX MEASUREMENTS

All heat transfer occurs because of a driving temperature difference. Rather a temperature gradient through a solid (conduction), a temperature difference between a surface and an adjacent fluid (convection), or a temperature difference between multiple surfaces (radiation), a temperature difference must exist for energy transfer to occur in the form of heat. In laboratory experiments, it is often necessary to use an external heating/cooling source to create the desired temperature difference. A variety of heaters can be used to heat a mass or surface; in addition, electric heaters can be used to raise the temperature of a fluid. While more difficult, it is also possible to incorporate a cooling system to lower the temperature of a surface or fluid.

3.3.1   Power Sources with Meters

The net heat transfer providing a measured temperature difference is readily obtained using electric resistance heaters. These heaters are generally comprised of high resistance, heating elements and are connected to a power source. Using a digital multimeter, the voltage drop across the heater, the current flowing through the heating element, and the resistance of the heater can be measured. Using Ohm's Law, the power supplied to the heater can be calculated from these three measurements:

$$q = VI = I^2R$$

Rather made in-house using heating wire or purchased commercially, the resistance of an electrical heater is fixed. The overall resistance of the heater is proportional to the length of heating wire. Small changes in resistance may occur with temperature changes, but the overall construction of the heater is fixed. Therefore, to control the amount of power supplied to the heater, variable voltage supplies are generally used. Rather with DC or AC power supplies, the power supplied to the heater can be controlled by changing the voltage source. Variable transformers, or variacs, are cost-effective ways to regulate heater voltage, and they can be purchased for a wide range of input voltages and are available for both single-phase and three-phase power sources.

3.3.2 TYPICAL HEATERS FOR VARIOUS APPLICATIONS

Many types of heaters can be integrated into a system to heat a surface. Rod, or cartridge, heaters are fabricated as a rigid diameter. These heaters can be inserted into a surface and provide uniform heating along the length of the rod. Depending on the size of the rod, the power density of these heaters can exceed 200 W/in^2 (32 W/cm^2) and can easily heat to over 1000°F (540°C). The lead wires generally exit the heater at one end, and can be easily routed through the system as needed. Other heaters are generally attached directly to a surface.

Depending on the type of heater, it can be placed on the backside of the surface, away from the fluid flow, or it can be attached on the front side of the heated surface, in direct contact with the fluid. Like rod heaters, plate, or strip, heaters are industrial-grade heaters that can be commercially purchased to meet a variety of needs. The heaters are generally fastened to the back of a surface and are used to heat an entire volume. A stainless steel outer sheath protects resistance wire coils housed within the strip. Like the rod heaters, the rigidity of these metal strips increases the life of the strip heaters. When purchasing strip heaters, the mounting options and lead wire placement must be considered. Depending on the overall size of the heaters they can supply up to 55 W/in^2 (8 W/cm^2) with outer sheath temperatures reaching 1200°F (650°C).

Thin foil and wire heaters can be fabricated in-house or purchased commercially. Thin foil heaters are often fabricated of low-conductivity, high-resistance foil (shim) material. Stainless steel and Inconel are commonly used for this application. The thickness of the foil is generally on the order of 0.5 mm (0.018 in). Increasing the foil thickness increases the rigidity of the heater but decreases the resistance. However, if the foil is placed in direct contact with the fluid, and thermocouples are attached to the back of the foil, having a thicker foil may be undesirable due to the temperature gradient through the thickness of the foil. Foil strips can also be used and connected via "bus bars" rather than using a single foil to cover an entire surface. Heating wire can also be attached to the backside of a surface. The wire is generally coated with an insulating material to allow the wire to be wound and coiled together, without creating a short circuit. The length of wire can be set to provide the desired electrical resistance. Both the foil and wire elements are incorporated into commercially available heater mats (flexible heaters). With the foil or wire protected by silicone rubber, these commercial heaters are very robust. The heaters can be purchased in standard sizes, or they can be custom made for a specific application. Standard silicone rubber heaters are rated to 450°F (232°C) and 5 W/in² (0.78 W/cm²). High-power-density rubber heaters are also commercially available.

Other heaters are required to heat fluids. Pipe, or circulation, heaters are the most common type of process heater for both liquids and gases. Much like a standard hot water heater, a heating element is enclosed within a pipe, and the fluid is forced over the heating element. Baffles are used within the pipe to increase the residence time of the fluid within the pipe. For many process applications, a thermocouple is placed directly at the outlet of the heater, and a feedback controller adjusts the power to the heater to maintain a set outlet temperature. Circulation heaters come in a wide range of sizes and can supply up to 120 W/in² (19 W/cm²) with a maximum operating temperature of 1600°F (870°F). Mesh heaters have also been used to heat gases. These heaters are generally used for short duration tests. A high-resistance mesh (stainless steel) is attached to two bus bars, and a large power is supplied to the bus bars. The power travels through the wire mesh, and provides a step change in the air temperature. As a large current source is needed to create a measurable temperature difference across the mesh, a welding machine is often used as the power source of this heater. The welder supplies a large current for a short duration, making it an ideal power source for the mesh heater.

3.3.3 Error Estimate in Heat Flux Measurements

For steady state experiments, heat flux from a test surface can be estimated from power input through heaters divided by surface area. The uncertainty is directly related to the errors from power measurements (watts, current, voltage, resistance) and surface dimension measurements. More accurate heat flux measurements can be obtained for larger power inputs on a larger test surface. More uncertainty will be involved with smaller power inputs on small test surfaces. In some applications, based on one-dimensional heat conduction, the heat flux can be calculated by a measured temperature difference across a material thickness with a known thermal conductivity for the material.

For unsteady, surface heat flux measurements, i.e., time-dependent temperature and heat flux measurements, a thin film heat flux gauge should be used, as discussed in Chapter 6.

Examples

Referring to Chapter 1, Figures 1.3 and 1.4 show how to measure temperatures in a solid rod by thermocouples and heater power to determine the thermal conductivity of the rod or contact resistance between two different rod materials. Refer to Chapter 1, Figure 1.8, to see how to measure temperatures in wind tunnel air flow and on a cylindrical tube by thermocouples, as well as how to apply heater power to the tube to determine its convection heat transfer coefficient. Refer to Chapter 1, Figure 1.9. Thermocouples attached to a foil heater are used to measure temperatures as a function of time under jet impingement flow. Refer to Figure 1.10

in Chapter 1. A square channel is heated by attaching four flexible heaters to the back-side walls of the test section. Thermocouples are used to measure the wall temperature and the temperature of the air flowing through the square channel.

PROBLEMS

1. As shown in the chapter, in order for a thermocouple to be used for temperature measurements, the voltage difference generated between the junctions of the thermocouple circuit, at the given temperature, must be measured and interpreted. Consider a T-type thermocouple with the thermoelectric behaviors (voltage difference to temperature) described as a NIST ITS-90 Thermocouple by the following relationship (reference junction temperature is 0°C)

$$T = C_0 + C_1 E + C_2 E^2 + \ldots + C_n E^n$$

Coefficient	Value
C_0	0.000000E+00
C_1	2.592800E+01
C_2	−7.602961E-01
C_3	4.637791E-02
C_4	−2.165394E-03
C_5	6.048144E-05
C_6	−7.293422E-07
C_7	0.000000E+00

where E is in mV and T is in °C.

Connect this thermocouple to LabVIEW to monitor the voltage difference of the thermocouple junction. If at a given time, the voltage reading is 0.991977 mV, calculate the corresponding temperature value of this thermocouple junction.

2. The Seebeck effect is the operating principle for a thermocouple. In addition, the Seebeck coefficient quantifies the sensitivity of the thermoelectric voltage change under a small variation of temperature at the junction. Referring to problem 1 (T-type TC and LabVIEW), if the voltage reading is 1.738072 mV.

    a. Calculate the temperature and the Seebeck coefficient by using the relationship given in Problem 1.

    b. Knowing a voltage value of 1.822687 mV is 2°C higher than part (a), assume the Seebeck coefficient is constant for a small temperature increment. Check the temperature difference between (a) and (b) to see if this assumption is valid.

3. The time constant of the thermocouple bead is defined as

$$\tau_t = \frac{\rho V c_p}{h A_s}$$

Using a T-type thermocouple with the bead (junction) diameter of 0.5 mm to obtain the mainstream temperature in a flat plate wind tunnel to be 25°C with a mainstream velocity of 21.6 m/s. The thermocouple is inserted into the flow and the bead is exposed to the mainstream. Assume that before turning on the mainstream of the wind tunnel, the temperature of the thermocouple bead is at 20°C and the bead is a perfect sphere. The density and the specific heat of the thermocouple bead is assumed to be the average value of the two wires comprising the thermocouple.

Information:

a. The heat transfer for flow over a sphere is modeled by Whitaker as,

$$Nu_{sph} = \frac{hD}{k} = 2 + \left[0.4 Re^{1/2} + 0.06 Re^{2/3} \right] Pr^{0.4} \left(\frac{\mu_\infty}{\mu_s} \right)^{1/4}$$

b. The Reynolds number in the above correlation is given to be 691.42.
c. The properties of air at one atmosphere are

Temperature (°C)	Dynamic Viscosity (kg/m-s)	Thermal Conductivity (W/m-K)
20	1.825×10^{-5}	0.02514
25	1.849×10^{-5}	0.02551

d. The properties of the material in T-type thermocouple are

	Material	
Property	Copper	Constantan
Density (kg/m³)	8960	8900
Specific heat (KJ/kg-K)	0.385	0.39

Use the above information to calculate the time constant of the thermocouple.

4. From the theory of thermocouple response time, look at the equation and interpret the physical meaning of time constant.
5. When a temperature sensor (e.g., thermocouple) is inserted inside a hot gas flow, convection occurs from the mainstream to the thermocouple bead and conduction occurs through the thermocouple wire. This conduction loss from the thermocouple bead through the wire can be modeled as a fin with

finite length and fixed-tip temperature; this is so-called stem conduction. A rectangular duct consists of copper plates with heaters attached to them to increase the temperature of air flow through the duct. A thermocouple is used at the exit of the duct to measure the outlet temperature of the air. Estimate the temperature error through the thermocouple wire under the following conditions.

Information:

Thermocouple insertion length: 4.85 inch

Thermocouple wire diameter: 0.01 inch (AWG 30)

Thermal conductivity of thermocouple insulation (PFA): 0.19 W/m-K

Overall thermocouple insulation diameter

   ($L_1 = 0.0236$ inch, $L_2 = 0.0118$ inch): 0.0177 inch

Local duct wall temperature: 45°C

Local flow temperature (not high-speed flow): 29°C

Averaged thermal conductivity of T-type thermocouple

   (Copper: 385; Constantan: 19.5): 202.25 W/m-K

Local heat transfer coefficient: 199.1 W/m²-K

6. For the same duct as Problem 5, it is desired to measure the copper duct wall temperature by drilling a 1/32″ diameter hole to embed the thermocouple 1.92″ deep. Later thermally conductive epoxy is used to fill the hole and secure the thermocouple. The remaining wire is exposed inside the duct (as shown in Figure 3.17). Using the same thermocouple as Problem 5, calculate the temperature error.

   Additional information:

   Thermal conductivity of conductive epoxy: 1.038 W/m-K

7. A thermocouple is needed to measure a surface temperature; the same thermocouple from Problem 5 is used for this purpose. A small, shallow hole is drilled on the target surface (duct wall), and the thermocouple bead is inserted into the hole. The remaining insulated wire remains exposed (Figure 3.18). Assume the flow condition inside the duct to be same as Problem 5, calculate the temperature error if the duct wall material is copper.

8. The Hyperloop is a future high-speed transportation solution capable of traveling between Los Angeles and San Francisco in approximately 35 minutes. To reach this target traveling time, the original proposal is to have the pod of the Hyperloop to cruise between 300 and 760 mph along the 354.4-mile route. As it will be extremely difficult to always maintain the tube along the entire route as a vacuum, the team decides to let the channel possess very low pressure. The research team plans to build a wind tunnel with the mainstream velocity of 760 mph to study the maximum thermal effect of stagnation temperature on the front of the pod. To simplify the analysis, assume the incoming freestream air temperature is 20°C. Calculate the stagnation temperature on the tip of the pod.

9. Researchers want to study heat transfer of turbulent flow through a duct. A channel consisting of 15 sections of copper plates, as shown in the

following figure, is fabricated. The copper plates are embedded with thermocouples with the same properties provided in Problem 5. Flexible, rubber heaters are attached to the backside of the copper plates. The heaters are powered by an AC power supply (0–240 V) through a variable transformer (variac) to regulate the voltage; they are also connected to the multimeter to monitor the voltage and current across the heaters. The heater's resistance is measured to be 51 ohms before supplying the power. When the desired experimental flow condition is reached, the heater voltage and current are measured to be 100.7 V and 1.9 A, respectively. Calculate the heater power output.

10. The external heat transfer coefficients on a turbine blade leading edge need to be measured. Since the turbine leading edge can be approximated as a semicylinder, the problem can be treated as the flow across a heated semicylinder. In order to do so, the leading edge is instrumented with copper plates and heaters and put inside a wind tunnel. Consider the cross section of the instrumented leading edge model, as shown in the following figure. Assume a single, flexible heater is attached to all copper plates that can provide uniform heat flux to the target area and the heater voltage and current applied to the heater is found to be 100.7 V and 1.9 A, respectively. Calculate the heat flux for each section.
    Additional information:
    Semicylinder diameter: 1.5 inches
    Semicylinder height: 10 inches
    Insulation thickness: 0.0625 inch
    Copper plate thickness: 0.125 inch

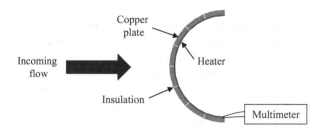

REFERENCES

1. Eckert, E.R.G. and Goldstein, R.J., *Measurements in Heat Transfer*, 2nd ed., Hemisphere Publishing Corporation, Washington, DC, 1976.
2. Lee, T.W., *Thermal and Flow Measurements*, CRC Press, Taylor & Francis Group, NewYork, 2008.
3. Figliola, R.S. and Beasley, D.E., *Theory and Design for Mechanical Measurements*, 4th ed., John Wiley & Sons, New York, 2000.
4. Holman, J.P., *Experimental Methods for Engineers*, 6th ed., McGraw-Hill, New York, 1994.
5. Incropera, F. and Dewitt, D., *Fundamentals of Heat and Mass Transfer*, 5th ed., John Wiley & Sons, New York, 2002.
6. Kays, W.M., *Convective Heat and Mass Transfer*, McGraw-Hill, New York, 1966.
7. Fox, R.W. and McDonald, A.T., *Introduction to Fluid Mechanics*, 3rd ed., John Wiley & Sons, New York, 1985.

4 Experimental Planning and Analysis of Results

4.1 EXPERIMENTAL PLANNING

Before any experiment can be performed, a brief study of the logistics involved must be completed. The materials and resources required to accomplish the objectives of the research should be identified in advance. The researcher must identify what parameters will be studied, complete a detailed design of the experimental setup, and provide a means to incorporate the test section into the existing infrastructure of the laboratory. This experimental planning, simply known as the design procedure, will be described in this chapter. The information presented will be useful for beginning researchers as a reference guide in their attempts to build a worthwhile experiment.

4.1.1 LITERATURE SURVEY AND EQUIPMENT IDENTIFICATION

Research involves the discovery and application of new information. Therefore, before making a substantial investment into a new project, researchers must search open literature to learn if their questions have already been answered.

Before planning any experiment, the first step is to do a detailed literature survey. From the literature survey, it may be found there is no need to design a new experiment because the needed information is already available. It is difficult to know what information is missing in your technical community if you are unaware of what has already been completed. If the required information is not available in the open domain, the literature survey will provide a variety of experimental methods used to obtain similar data sets.

Following the detailed review of open literature, the key components to conduct a successful experimental program must be identified. In some cases, a completely new facility will be required, but often you simply need to modify or fabricate a new test section to incorporate into the existing laboratory infrastructure. For example, you may only need to fabricate a new test model for an existing wind tunnel facility. On the other hand, you may need to build a transonic turbine cascade facility for a new turbine blade heat transfer study. Similarly, you may use the existing equipment for flow and temperature measurements, or you need to purchase the new equipment for the same purpose. Of course, more resources (money and time) are required to build new facilities with new measurement hardware.

4.1.2 TEST SECTION DESIGN, FABRICATION, AND INSTRUMENTATION

The objectives of any experimental program must be clearly identified before investing in the study. The objectives must clearly state the parameters to be studied and

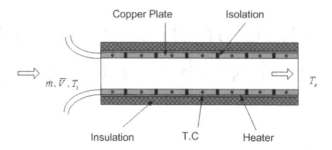

FIGURE 4.1   Sketch of forced convection heat transfer experiment.

under what range of conditions. For example, in an experiment for forced convection in a duct (Figure 4.1), the primary flow parameter is defined by the non-dimensional Reynolds number. A range of Reynolds numbers should be chosen that are relevant to the desired application (HVAC, electronic cooling, gas turbine cooling, etc.). The heat transfer will be measured over this range of flow conditions.

$$\mathrm{Re}_{D_h} = \frac{\rho V D_h}{\mu}$$

By assuming the appropriate conditions for density and dynamic viscosity, the velocity and hydraulic diameter can be manipulated. Extremely high velocities should be avoided unless the experiment demands it. By limiting the velocity for the highest Reynolds number, the hydraulic diameter of the test section can be determined. Once this parameter is known, the size of the test section can be determined. The desired maximum volume flow rate of the working fluid can thus be calculated for this size and the highest Reynolds number. This flow rate should match the maximum flow that can be obtained from the available compressor or blower. Of course, mass flow rate can be related to flow velocity, density and hydraulic diameter.

It should be noted that when designing an experiment to be performed in a wind tunnel, such as flow across a circular cylinder (Figure 4.2), flow over a flat plate

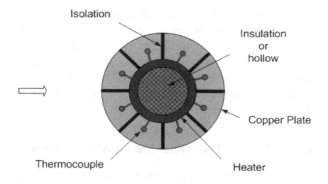

FIGURE 4.2   Sketch of flow across a cylinder heat transfer experiment.

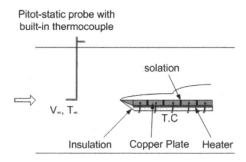

FIGURE 4.3 Sketch of flow over a flat plate heat transfer experiment.

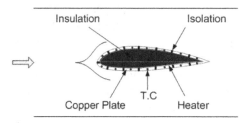

FIGURE 4.4 Sketch of flow over an airfoil heat transfer experiment.

(Figure 4.3), and flow over an airfoil (Figure 4.4), the flow rate depends on the cross-sectional area of the wind tunnel. In this case, the wind tunnel blower should be rated to handle this maximum flow rate. The blower power can also be calculated by estimating the total pressure drop (wind tunnel wall and test section surface) and multiplying it with the volumetric flow rate. The pressure drop in the test section can be estimated using an empirical correlation for the friction factor for a smooth test surface. If experiments are to be performed on a rough surface, an augmented friction factor can be assumed.

4.1.2.1 Flow Measurement

An accurate measurement of this flow rate is required when conducting experiments for the predetermined range of Reynolds numbers. Several flow measurement techniques are available, which can be applied to an experiment and are described in more detail in Chapter 2. For internal flow experiments, orifice flow meters and rotameter flow meters are widely used upstream of the test section. Venturi flow meters may be used when there are limitations on the maximum pressure drop allowed in the test section.

A Pitot probe with static pressure taps can also be used to determine the velocity. When using Pitot probes, to obtain an accurate flow rate in the test section, the Pitot-static probe should be traversed along a predefined grid to account for the velocity change due to the boundary layer development. The velocity profile can be averaged to get the total flow. The flow measurement techniques discussed measure a differential pressure to calculate the fluid velocity or flow rate. Suitable pressure measurement devices must be used to measure these pressures, as described in Chapter 2. A manometer is an inexpensive and reliable instrument for measuring relatively small differential pressures. When dealing with low flow rates, inclined and micro manometers are often used to accurately measure these small pressure differences. The primary advantage of using manometers is that they do not require calibration. Pressure transducers, which produce a measurable voltage or current due to change in pressure, can also be used. They are available for varying ranges and need calibration against a dead-weight tester or another accurate instrument before they can be used.

4.1.2.2 Experimental Methods and Measurement Techniques

Prior to designing the test section, the measurement technique that will be used to measure heat transfer coefficients must to be selected. What types of measurements are required? For example, are you looking for local or detailed velocity or temperature measurements? The required spatial and temporal resolution will dictate which experimental technique is needed, and thus the required instrumentation. A detailed description of several measurement techniques is described in the following chapters. If an optical technique is to be used, space should be allotted for a window to allow the camera to capture images of the test surface. In general, windows can be made from any transparent material such as Plexiglas but for certain measurement techniques such as infrared thermography, infrared transmissible windows must be used.

4.1.2.3   Heater Power Specifications

All heat transfer experiments require some form of heat addition (or removal) to obtain a temperature difference for convection to occur. Either the test surface can be heated via a surface heater or the mainstream air can be heated with a pipe heater. The amount of heat supplied depends on the desired temperature difference between the working fluid and test surface. There are several types of heaters available for heating a surface. The type of heater used is strongly dependent on the measurement technique that will be employed. The most commonly used surface heater is a flexible wire heater when only average heat transfer results are needed, such as the steady-state heat transfer measurement techniques with copper plates (presented in Chapter 5). This heater consists of a thin, high-resistance wire wound over the entire shape of the test surface. They are available in standard sizes and can be custom ordered to fit the test geometry (Figure 4.5). Slight non-uniformity in the heat flux distribution exists due to the gap between adjacent wire strands. Hence, these heaters are typically mounted on the underside of a high conductivity copper plate which ensures that the power emitted by the heater is distributed uniformly over the entire copper surface resulting in a constant wall temperature. However, this non-uniformity in heat flux distribution makes these heaters unsuitable for local heat transfer measurements.

Foil heaters are preferred for heating the test surface when local heat transfer coefficients are needed (Figure 4.6). A more uniform local temperature distribution can be obtained as the gap between wires in flexible wire heaters is avoided. The foil

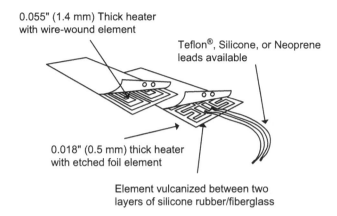

0.055" (1.4 mm) Thick heater
with wire-wound element

Teflon®, Silicone, or Neoprene
leads available

0.018" (0.5 mm) thick heater
with etched foil element

Element vulcanized between two
layers of silicone rubber/fiberglass

FIGURE 4.5   Flexible wire heater for heat transfer experiment.

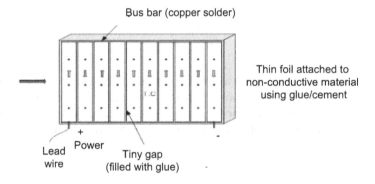

Bus bar (copper solder)

Thin foil attached to
non-conductive material
using glue/cement

Lead
wire

+
Power

Tiny gap
(filled with glue)

-

FIGURE 4.6   Sketch of thin foil heater for heat transfer experiment.

heaters can be directly exposed to the fluid without a copper plate interface, which is required for flexible wire heaters. They can be used with steady-state as well as transient measurement methods as discussed in Chapters 5 and 6. Average heat transfer measurements can also be obtained using thin foil heaters.

Steam condensation can be used as a heat source for heat transfer experiments at a constant surface temperature (Figure 4.7). A typical steam jacket at 100°C and one atmosphere pressure can be placed inside or outside of the test tube to provide uniform wall temperature boundary conditions.

4.1.2.4   Heater Power Estimations
In order to calculate the power input to the heater, the Nusselt number enhancement when compared to a smooth surface for the same geometry should be estimated prior to fabricating the experimental test section. For example, for flow in a duct

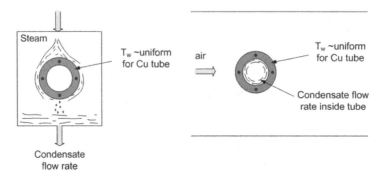

FIGURE 4.7   Sketch of steam condensation for heat transfer experiment.

with turbulators, used to augment heat transfer, a maximum enhancement of 3 can be estimated. This assumes the heat transfer in your turbulated test section will be three times greater than the heat transfer in a standard duct with smooth walls.

$$\frac{\text{Nu}}{\text{Nu}_0} = 3$$

A higher estimated enhancement will result in a more conservative design, as the heater will be rated for a higher power input than needed. The Nusselt number for a smooth surface can be determined by using an empirical correlation available in open literature such as the Dittus-Boelter correlation for turbulent flow in a smooth pipe.

$$\text{Nu}_0 = 0.023 \cdot \text{Re}^{0.8}\, \text{Pr}^{0.4}$$

From this enhancement factor, the highest heat transfer coefficient can be calculated from the estimated Nusselt number.

$$\text{Nu} = \frac{hD_h}{k} = 3\text{Nu}_0 = 3(0.023 \cdot \text{Re}^{0.8}\, \text{Pr}^{0.4})$$

This value for the heat transfer coefficient can be introduced in Newton's law for convective heat transfer and the resulting heat input, Q_{input} can be calculated. This heat input and the extraneous heat losses, Q_{loss}, from the heater will contribute to the total heat supplied by the heater.

$$Q_{\text{heater}} = Q_{\text{input}} + Q_{\text{loss}} = hA_{\text{heated}}\left(T_w - T_m\right) + Q_{\text{loss}} = \dot{m}C_p\left(T_{m,\text{inlet}} - T_{m,\text{exit}}\right) + Q_{\text{loss}}$$

The extraneous heat losses are proportional to the temperature difference between the heated surface and the mainstream fluid. This is typically a small percentage of the power input if the backside of the heater is well-insulated. For a higher temperature difference, heat losses are higher while for low values, the relative uncertainties for the heat transfer coefficients increase rapidly. A temperature difference of approximately 20°C–30°C is optimal in reducing the experimental uncertainty while keeping the magnitude of the heat loss in check. By using a heater and power source with this minimal power rating, the desired heat transfer rates and temperature differences can be achieved. The heat output Q from a heater can be calculated by measuring the voltage, V and current supplied, I to the heater. A multimeter can be used to record these parameters.

$$Q_{\text{heater}} = VI = \frac{V^2}{R} = I^2 R$$

To regulate the voltage supplied to the heater, a variable auto-transformer should be used between the heater and the input power line. If the heat flux needed from the heater is very high, a three-phase voltage source may be used. This will require a three-phase heater and variable autotransformers.

When experiments are being performed for internal heat transfer measurements in ducts, an alternative method of measuring the heat input is by determining the sensible heat gained by the fluid from the entrance to the exit of the test section. From the energy balance, the heat emitted by the heater is equivalent to the heat gained by the fluid plus the extraneous heat losses. Thus, by measuring the inlet and outlet temperatures of the fluid, the total heat input can be determined.

$$Q_{\text{heater}} = Q_{\text{fluid}} + Q_{\text{loss}} = \dot{m} C_p \left(T_{m,\text{inlet}} - T_{m,\text{exit}} \right) + Q_{\text{loss}}$$

If the test section is exposed to high pressures and temperatures or involves moving parts, a safety analysis should be performed to ensure that the design is fail-safe. Stress analysis can be performed by using a reasonable factor of safety. Sufficient safety precautions should be taken when using heaters. Hazardous materials such as wood, which can catch fire, should be avoided when building experimental test sections.

4.1.3 TEST FACILITY CALIBRATIONS AND MEASUREMENTS

Any new experimental setup, facility, or user, should validate the setup against a known standard. For the above-mentioned channel flow test loop design, one needs to calibrate the mass flow rates from orifice or rotameter flow meters, which relate to the channel Reynolds numbers. A heat loss calibration for the test section is also required to determine the net heat flux to the fluid. Thermocouples, and the corresponding data acquisition hardware, should be properly calibrated to minimize error in the temperature measurements. If pressure transducers are involved for flow rate or friction measurements, they should also be calibrated before they are installed into the test section.

For wind tunnel tests, the Pitot-static probe (and corresponding differential pressure transducer) should be calibrated to ensure the correct freestream velocities (Reynolds numbers) are calculated and tested. The heaters and heat loss (along with the multimeters) should be calibrated as needed. Thermocouples and pressure transducers also require calibration before implementation into a research facility. These calibrations ensure the measurement devices are working properly, and they can be safely used over the entire range of flow and heat transfer parameters.

4.2 ANALYSIS OF RESULTS

Any measurement has error or uncertainty. All measured values, instruments, and handbook values have errors. The error of a measured or calculated value can be defined as the difference between the measured value and the true value. However, the true value is generally unknown. Therefore, the researcher must estimate the error contained in their experiment to estimate a range in which the true value might fall (based on the measured value). Often the term "uncertainty" is used to express the error of a measurement or experiment. Uncertainty is the "degree of accuracy with which the experimenter believes the measurement has been made."

Errors find their way into all experiments, regardless of how carefully the researcher performs the experiments. The errors can be completely random, or the researcher can commit a blatant error leading to inaccurate results. Researchers must take the necessary steps to eliminate as many errors as possible. However, real uncertainty will still exist in the experimental data. The factors leading to these errors are often vague and difficult to quantify.

Errors are classified into two basic types: systematic and random. Systematic errors are due to imperfect equipment, bias, and physical effects. These errors usually shift measured values in one direction. Systematic errors can be minimized (eliminated to the extent possible) through experiment design and calibration. Random errors are differences in measurements due to normal variations. These errors can be positive or negative; therefore, random error can be reduced (minimized) by taking multiple measurements.

All measured and calculated quantities have an associated error: mass, length, diameter, time, pressure, pressure difference, velocity, mass flow rate, temperature, temperature gradient, heat transfer rate, Reynolds number, heat transfer coefficient, etc. Researchers must estimate and report the error, or uncertainty, of each directly measured (pressure, temperature, etc.) or indirectly derived quantity (Reynolds number, heat transfer coefficient, etc.). Because the error varies between all sensors and transducers, it is necessary to estimate the experimental uncertainty prior to purchasing new instrumentation. High-accuracy transducers can be expensive, but in some cases, this expense may be necessary to significantly reduce the error of your experiment.

Methods have been proposed to estimate the uncertainty of individual measurements and calculated quantities. The following sections discuss how to report the errors, or uncertainties, for derived values such as velocity, flow rate, Reynolds number, friction factor, heat transfer rate, heat transfer coefficient, etc.

4.2.1 Data Reduction and Uncertainty Analysis

There are several ways to estimate the error, or uncertainty, from an individual measured quantity. The straightforward way is to repeat the measurement several times and calculate the mean value of the measurements. For example, the diameter of a circular tube is measured 10 times, and each measurement is reported in the following table.

Measurement Number	1	2	3	4	5	6	7	8	9	10
Diameter (cm)	2.16	2.20	2.10	2.11	1.87	1.89	1.89	2.01	1.80	1.98

Based on the 10 measurements, the measurement mean for the cylinder diameter is 2.00 cm. The maximum deviation from the mean value is 0.2 (or 10% of the mean value). Therefore, the researcher could report the diameter of the tube to be 2.00 cm ± 10%. Rather than basing the uncertainty interval off the maximum deviation, the uncertainty can also be presented in terms of the standard deviation of the measurements. The standard deviation of the ten measurements shown in the table is 0.13 cm (or 6.7% of the mean value). Therefore, the diameter of the cylinder could also be reported as 2.00 cm ± 6.7% (or 2.00 ± 0.13 cm).

4.2.1.1 Overall Uncertainty Based on the Second Power Equation in a Single-Sample Experiment

In most experiments, a final, derived quantity is required (friction factor, heat transfer coefficient, etc.). This dependent variable is a function of several independent variables, which are directly measured (temperature, pressure, etc.). Each measurement contributes to the final, total uncertainty of the derived quantity. This combination is referred to as the propagation of the uncertainty. There are several ways to calculate and report the propagation of errors leading to the final uncertainty of the derived variable. The following equations are an abstract from the original paper of Kline and McClintock [1]. The interested readers can also refer to several papers published by Moffat [2,3].

For one given measurement (pressure or temperature), the arithmetic mean of the specific measurement is "m." The uncertainty interval for this measurement, defined by the researcher or manufacturer of the instrument, will be represented as "w." Kline and McClintock proposed presenting this measurement range with a specified interval of confidence, "b." In other words, how confident is the researcher that the true value of the measurement will fall into the range given by the uncertainty interval, "w"?

For example, a pressure measurement may be reported as 50.2 ± 0.5 psia (20 to 1). The experimentalist is willing to bet with 20 to 1 odds that the pressure measurement is within 0.5 psia of the mean value. The 20-to-1 odds is often used as this represents the 95% confidence interval (or ~2 standard deviations). Higher odds, of say 100 to 1, broaden the range to 99% confidence.

Researchers need to estimate the uncertainty of the derived quantity (dependent variable) based on the uncertainties of the primary measurements (independent variables). How does the uncertainty in each individual measurement affect the uncertainty of the derived quantity? Assume "R" to be a derived quantity that is a function of multiple independent variables ($v_1, v_2, v_3, \ldots, v_n$).

$$R = R\left(v_1, v_2, v_3, \ldots, v_n\right)$$

If the uncertainties of the independent variables are small, the error propagation to the final result can be represented in terms of these small variations in each variable.

$$\delta R = \frac{\partial R}{\partial v_1}\delta v_1 + \frac{\partial R}{\partial v_2}\delta v_2 + \ldots + \frac{\partial R}{\partial v_n}\delta v_n$$

Theorem 4.1

Instead of calculating the uncertainty based on variations of the measured values, it is more desirable to describe the uncertainty distributions in terms of uncertainty intervals, w, with stated odds (or confidence). Therefore, it is necessary to find the overall uncertainty interval, w_r, for the resultant quantity.

Kline and McClintock proposed, "If the maximum deviation of the ith variable from its mean is $(\pm\delta v_i)_{max}$, then the maximum deviation of R from its mean value is given by":

$$\delta R_{max} = \left|\frac{\partial R}{\partial v_1}\delta v_{1max}\right| + \left|\frac{\partial R}{\partial v_2}\delta v_{2max}\right| + \ldots + \left|\frac{\partial R}{\partial v_n}\delta v_{nmax}\right|$$

$$\text{Or, } w_R = \left|\frac{\partial R}{\partial v_1}w_1\right| + \left|\frac{\partial R}{\partial v_2}w_2\right| + \ldots + \left|\frac{\partial R}{\partial v_n}w_n\right|$$

This equation is referred to as the linear equation.

Theorem 4.2

If a sufficiently large sample size is available for each measured quantity, the uncertainty can be presented in terms of the standard deviation for each independent variable. More data is required for this approach, so the statistical analysis can be completed. According to Kline and McClintock, "If the standard deviation of each variable is given by σ_i, then the standard deviation of R is given as":

$$\sigma_R = \left[\left(\frac{\partial R}{\partial v_1}\right)^2\sigma_1^2 + \left(\frac{\partial R}{\partial v_2}\right)^2\sigma_2^2 + \ldots + \left(\frac{\partial R}{\partial v_n}\right)^2\sigma_n^2\right]^{1/2}$$

The derivation of the above expression (the square root of the sum of the squares) is based on two standard deviations. Therefore, the standard deviation of R, σ_R, is being reported with 95% confidence (20-to-1 odds).

Theorem 4.3

Experimentalists typically use a combination of Theorems 4.1 and 4.2 to present the accuracy of their results. Using the second-power equation (Theorem 4.2), and the stated uncertainty intervals (similar to Theorem 4.1), Kline and McClintock propose, "If each of the variables are normally distributed, then the relation between w_i and w_R, which gives the same odds for each of the variables and for the result, is":

$$w_R = \left[\left(\frac{\partial R}{\partial v_1} w_1 \right)^2 + \left(\frac{\partial R}{\partial v_2} w_2 \right)^2 + \ldots + \left(\frac{\partial R}{\partial v_n} w_n \right)^2 \right]^{1/2}$$

The result, R, can be reported with 95% confidence as $R \pm w_R$ (20-to-1 odds).

Example:

The velocity of air, c, can be determined from Bernoulli's equation using a measured pressure difference, ΔP, the static pressure of the air, P_a, and the static temperature of the air, T_a.
  Bernoulli's equation: $P_1 + \frac{1}{2}\rho V_1^2 + \rho g z_1 = P_2 + \frac{1}{2}\rho V_2^2 + \rho g z_2$
  With P_2 being the stagnation pressure ($V_2 = 0$), and $z_1 = z_2$

$$P_1 + \frac{1}{2}\rho V_1^2 = P_2$$

Rearrange to solve for V_1

$$V_1 = c = \sqrt{\frac{2 \cdot \Delta P}{\rho}}$$

Perfect gas equation: $P = \rho RT$ or $\rho = \frac{P_a}{R \cdot T_a}$
  Combine the perfect gas equation with the velocity from Bernoulli's equation:

$$c = \sqrt{\frac{2 \cdot \Delta P \cdot T_a \cdot R}{P_a}}$$

The following data was measured

$\Delta P = 8.0 \pm 0.1$ in. H_2O (20 to 1)
$T_a = 67.4°F = 527.1 \pm 0.2$ °R (20 to 1)
$P_a = 14.7 \pm 0.3$ psia (20 to 1)

Using the second-power equation:

$$w_R = \left[\frac{1}{4} \frac{2RT_a g_o}{(\Delta p) p_a} \left(w_{\Delta p} \right)^2 + \frac{1}{4} \frac{2(\Delta p) RT_a g_o}{p_a^3} \left(w_{p_a} \right)^2 + \frac{1}{4} \frac{2(\Delta p) R g_o}{p_a T_a} \left(w T_a \right)^2 \right]^{1/2}$$

Dividing the uncertainty equation by the original velocity calculation provides a more compact expression for the uncertainty. The expression yields the uncertainty as a percentage of the derived value.

$$\frac{w_c}{c} = \frac{w_R}{R} = \left[\left(\frac{1}{2} \frac{w_{\Delta p}}{\Delta p} \right)^2 + \left(\frac{1}{2} \frac{w_{p_a}}{p_a} \right)^2 + \left(\frac{1}{2} \frac{w_{T_a}}{T_a} \right)^2 \right]^{1/2}$$

Substituting in the given values:

$$\frac{w_c}{c} = \frac{1}{2} \left[1.56 \times 10^{-4} + 4.15 \times 10^{-4} + 1.44 \times 10^{-7} \right]^{1/2} = 1.1\%$$

4.2.1.2 Generalization

Many forms of similar equations commonly appear across many experiments. The following shows several simplified uncertainty expressions.

---

1. $R = \dfrac{AB^m}{C^n}$ $\dfrac{w_R}{R} = \left[\left(\dfrac{w_A}{A} \right)^2 + \left(m \dfrac{w_B}{B} \right)^2 + \left(n \dfrac{w_C}{C} \right)^2 \right]^{1/2}$

2. $Y = a X_1^b X_2^c X_3^{-d}$ $\dfrac{w_Y}{Y} = \left[\left(b \dfrac{w_{X_1}}{X_1} \right)^2 + \left(c \dfrac{w_{X_2}}{X_2} \right)^2 + \left(-d \dfrac{w_{X_3}}{X_3} \right)^2 \right]^{1/2}$

3. $Y = X_1^a \left(X_2 - X_3 \right)^b$ $\dfrac{w_Y}{Y} = \left[\left(a \dfrac{w_{X_1}}{X_1} \right)^2 + \left(b \dfrac{w_{X_2}}{(X_2 - X_3)} \right)^2 + \left(b \dfrac{w_{X_3}}{(X_2 - X_3)} \right)^2 \right]^{1/2}$

4. $Y = X_1^a \left(X_2^b - X_3^c \right)$ $\dfrac{w_Y}{Y} = \left[\left(a \dfrac{w_{X_1}}{X_1} \right)^2 + \left(\dfrac{X_2^b}{(X_2^b - X_3^c)} . b \dfrac{w_{X_2}}{X_2} \right)^2 + \left(\dfrac{X_3^c}{(X_2^b - X_3^c)} . c \dfrac{w_{X_3}}{X_3} \right)^2 \right]^{1/2}$

Examples:

Estimate the uncertainties for the following derived parameters (dependent variables): velocity; mass flow rate; Reynolds number; friction factor; heat transfer rate; heat transfer coefficient; Nusselt number. You need to provide the measured data (independent variables) such as $\Delta P, P, T, D, L, T_w, T_b, VI$, etc., and their uncertainty values from the real experiments, for the derived parameter uncertainty calculations.

$$V = \sqrt{\frac{2(\Delta P)RTg}{P}}$$

$$\dot{m} = \rho A_c V = \frac{P}{RT} A_c V$$

$$Re_D = \frac{\rho V D}{\mu} = \frac{V D}{v} = \frac{4\dot{m}}{\pi D \mu}$$

$$Re_x = \frac{\rho V x}{\mu} = \frac{V x}{v}$$

$$C_f = \frac{\tau_w}{\frac{1}{2}\rho V^2} = \frac{\Delta P}{4\frac{L}{D}\frac{1}{2}\rho V^2}$$

$$Q = \dot{m}c_P\left(T_0 - T_i\right) = hA_s\left(T_w - T_b\right)$$

$$h = \frac{q - q_{loss}}{A_s\left(T_w - T_b\right)} = \frac{VI - q_{loss}}{A_s\left(T_w - T_b\right)}$$

$$Nu = \frac{hD}{k}$$

4.2.1.3   The Uncertainty Based on the Precision Limit and the Bias Limit

The *ASME Journal of Heat Transfer* [4] suggests that all uncertainty evaluation be performed in accordance with the 95% confidence interval and the presentation of results be made with the following information:

1. The precision limit, P: The precision limit is an estimate of the lack of repeatability caused by random errors and unsteadiness.
2. The bias limit, B: The bias limit is an estimate of the magnitude of the fixed, constant error.
3. The uncertainty, U: The $\pm U$ interval about the nominal result is the band within which the experimenter is 95% confident that the true value of the result lies. The uncertainty is calculated from

$$U = [B^2 + P^2]^{1/2}$$

The precision limit represents random errors (normal variations of a measurement). Variations of a reading from an analog instrument, ambient changes during the course of an experiment, variations within the instrument's accuracy specifications are all random errors. Bias (or systematic) errors can be caused from incorrect calibrations, electrical charge "leak" from transducers, radiation error in thermocouple measurements, or bias error from the experimentalist (consistently incorrect

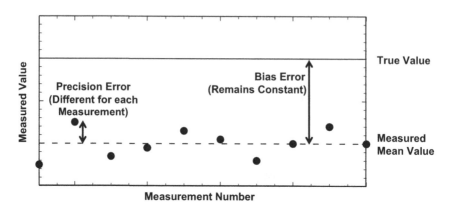

FIGURE 4.8 Conceptual representation of bias and precision errors.

readings). As mentioned earlier, bias errors should be mitigated as much as possible through careful design and experimentation. Figure 4.8 graphically shows bias and precision errors. While the "true value" is shown as a reference in this figure, for most experiments, this "true value" is not known.

Let's take the example of a heat transfer experiment

$$q = mc(T_o - T_i) \tag{4.1}$$

Following the second-power equation of Kline and McClintock:

$$P_q^2 = \left(\frac{\partial q}{\partial m}\right)^2 P_m^2 + \left(\frac{\partial q}{\partial c}\right)^2 P_c^2 + \left(\frac{\partial q}{\partial T_o}\right)^2 P_{T_o}^2 + \left(\frac{\partial q}{\partial T_i}\right)^2 P_{T_i}^2$$

$$B_q^2 = \left(\frac{\partial q}{\partial m}\right)^2 B_m^2 + \left(\frac{\partial q}{\partial c}\right)^2 B_c^2 + \left(\frac{\partial q}{\partial T_o}\right)^2 B_{T_o}^2 + \left(\frac{\partial q}{\partial T_i}\right)^2 B_{T_i}^2 + 2\left(\frac{\partial q}{\partial T_o}\right)\left(\frac{\partial q}{\partial T_i}\right) B'_{T_o} B'_{T_i}$$

The last term is neglected if the bias limits in the temperature measurements are uncorrelated.

Using Equation (4.1) and defining $\Delta T = T_o - T_i$:

$$\left(\frac{P_q}{q}\right)^2 = \left(\frac{P_m}{m}\right)^2 + \left(\frac{P_c}{c}\right)^2 + \left(\frac{P_{T_o}}{\Delta T}\right)^2 + \left(\frac{P_{T_i}}{\Delta T}\right)^2 \tag{4.2}$$

$$\left(\frac{B_q}{q}\right)^2 = \left(\frac{B_m}{m}\right)^2 + \left(\frac{B_c}{c}\right)^2 + \left(\frac{B_{T_o}}{\Delta T}\right)^2 + \left(\frac{B_{T_i}}{\Delta T}\right)^2 \tag{4.3}$$

If bias limits in the temperature measurements are estimated as 0.5°C, the bias limit on the specific heat property value is 0.5%, the bias error of the mass flow meter system is 0.25%, and $\Delta T = 20$°C

$$\left(\frac{B_q}{q}\right) = \sqrt{(0.0025)^2 + (0.005)^2 + \left(\frac{0.5°C}{20°C}\right)^2 + \left(\frac{0.5°C}{20°C}\right)^2} = 0.036 \ (= 3.6\%)$$

If the random error is such that the precision limit for q calculated from Equation (4.2) was 2.7%, the overall uncertainty in the determination of q, U_q would be:

$$\left(\frac{U_q}{q}\right) = \sqrt{\left(\frac{B_q}{q}\right)^2 + \left(\frac{P_q}{q}\right)^2} = \sqrt{(0.036)^2 + (0.027)^2} = 0.045 = 4.5\%$$

Theoretically, both the bias limit, B, and the precision limit, P, need to be estimated and reported. That means the researcher should estimate the resultant uncertainty due to the bias limit and the precision limit for the individual parameters using the second-power equation of Kline and McClintock. However, many researchers choose to report overall uncertainty based on the individual parameters that combine the bias and precision errors together before using the second-power equation of Kline and McClintock. For example, a thermocouple company says 0.5°C error in temperature measurement; this is the bias uncertainty, B. However, there is another measured 0.5°C uncertainty during a certain experiment; this is the precision uncertainty, P. You may estimate the B and P separately based on 0.5°C error (and other individual parameters) by using the second-power equation and then calculate the overall uncertainty, U, by combining B and P. Often researchers choose to calculate the overall uncertainty, U, based on 1.0°C error (and other individual parameters) by using the second-power equation. The key is to identify which parameters are the primary contributors driving the uncertainty of the experiment. With these measurements identified, the experimentalists should identify methods to reduce the error, or uncertainty, to improve the accuracy of the experimental results.

4.2.1.4 Test Uncertainty Performance Test Code
ASME published the Test Uncertainty- Performance Test Code. The code or standard was developed under procedures accredited as meeting the criteria for American National Standards. The interested readers can consult with the ASME Performance Test Code [5].

4.2.2 RESULTS COMPARISONS AND CONCLUSIONS

To ensure the measured data is reliable, the results should first be compared to existing results or correlations. For internal flow experiments, one can compare the measured friction factors and Nusselt numbers with the existing correlations for a smooth channel at similar Reynolds numbers and Prandtl numbers. For external flow experiments, one can compare the measured drag coefficients and Nusselt numbers with the existing correlations for a flat plate or circular cylinder at similar Reynolds numbers and Prandtl numbers. These favorable comparisons with existing correlations guarantee proper experimental techniques and reliable measurement devices. Therefore, the test sections are ready for modification for the proposed rough channel experiments or for the proposed rough surface, flat plate or circular cylinder experiments at similar flow conditions. The presentation of the results should be followed by a discussion and physical explanation of the new test results. All papers and reports should include concluding remarks to summarize the major findings of the work.

PROBLEMS

1. A rectangular smooth channel is being investigated. The flowrate is metered using an orifice plate where the pipe diameter is 4 inch and the bore diameter is 3 inches. The pressure drop across the meter is 0.57 in WC with an upstream pressure of 407.73 in WC. The temperature before the orifice is 26°C. Calculate the mass flow rate using the appropriate equation for the orifice plate and the corresponding Reynolds number in the rectangular smooth channel. What is the uncertainty in the mass flow rate (MFR) and the Reynolds number?

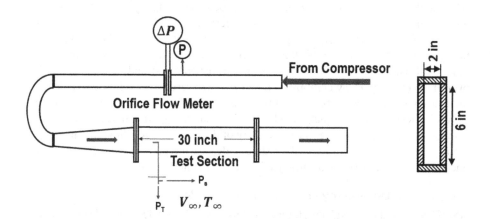

2. In Problem 4.1, the length of the test section is 30 inches. A heater is attached to the outer surface of each longer side (made of several copper segments with thermocouples and insulation material). The supplied voltage from a variac to the heater is 45 V with a current rating of 2.67 A on each longer side. Calculate the Nusselt number at the end of longer side (based on rectangular channel hydraulic diameter) and corresponding uncertainty for the following three cases:
   a. If the temperature difference $(T_w - T_{bulk})$ at the channel last region is 10°C.
   b. If the temperature difference $(T_w - T_{bulk})$ at the channel last region is 20°C.
   c. If the temperature difference $(T_w - T_{bulk})$ at the channel last region is 30°C.

3. An inlet cross section of a suction-type wind tunnel is 30×10 in where an airfoil model (10-inch long and 10-inch wide) will be studied. A Pitot-static tube is placed in the mid-plane of the channel indicating a pressure difference of 1.25 in WC where the static pressure is 405.75 in WC and the corresponding temperature is 20°C. Estimate the mass flow rate through the wind tunnel (assume a uniform velocity) and the corresponding Reynolds number over the airfoil model based on its length. What are the uncertainties?

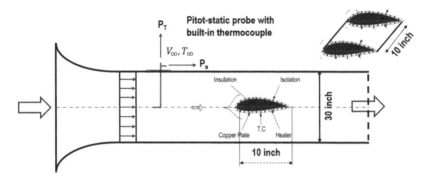

4. In the Problem 4.3, if a variac supplies 85 V with a current rating of 1.4 A to a heater that is attached to the inner surface covering a total length of 22 inch around the airfoil and the temperature difference between the end of the airfoil surface and the flow is found to be 20°C, calculate the Nusselt number at the end of the airfoil surface (based on its length). What is the uncertainty of Nusselt number?

5. An experiment is designed to study the flat plate (19.68-inch long and 10-inch wide) heat transfer at a moderate Reynolds number where the Pitot-static probe measured a differential pressure of 0.65 in WC and static pressure of 406.35 in WC. The temperature is recorded as 25°C. The inlet area of the test section is 30×10 inch. Calculate the mass flow rate (MFR) through the wind tunnel (assume a uniform velocity) and the corresponding Reynolds number over the flat plate based on its length. Also, estimate the corresponding uncertainties.

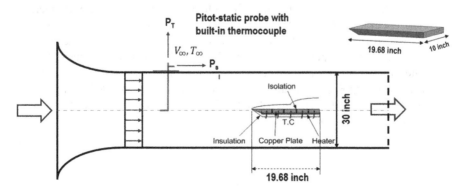

6. In Problem 4.5, a heater is used to heat the flat plate and make a temperature difference (T_w–T_∞) of 24°C at the end of the flat plate. To do so, a variac is used to supply 78 V with a current rating of 1.6 A. Calculate the Nusselt number at the end of the flat plate and the corresponding uncertainty of Nusselt number.

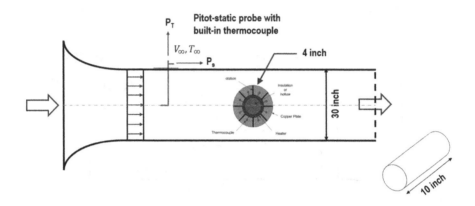

7. A test section is required to study the heat transfer from a 4-inch diameter circular cylinder under cross flow. The cross section of the wind tunnel is 30 × 10 inches and powered by a suction blower. The Pitot-static probe indicates a differential pressure of 0.76-inch WC with a static pressure of 394.5 in WC. The temperature reading is 21°C. Calculate the mass flow rate through the wind tunnel (assume a uniform velocity) and the corresponding Reynolds number over the cylinder based on its diameter. What are the uncertainties?

8. In Problem 4.7, a heater is used to heat the cylinder inner wall and produce a temperature difference (T_w–T_∞) of 25°C at the cylinder leading edge region. If the power input to the heater is 110 V with corresponding 2.2 A, calculate the Nusselt number at the leading-edge region of the cylinder based on its diameter. What is the uncertainty of Nusselt number?

9. A turbine vane cascade test section of 20.7 × 4.8 inch provides a high-speed flow. The pressure range is relatively high and requires a pressure-scanner. The accuracy level is ±0.05 in WC. The Pitot-static probe measurement shows a differential pressure of 3.46 in WC where 501.5 in WC is the static pressure. The temperature is recorded as 21°C. Calculate the mass flow rate through the turbine vane cascade (assume a uniform velocity) and the corresponding Reynolds number over the turbine vane cascade based on its inlet velocity and axial chord length (C_{ax} = 3.26 inch). What are the corresponding uncertainties?

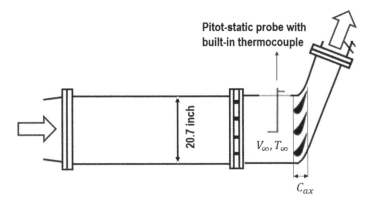

10. In Problem 4.9, the middle vane is instrumented with a series of thin foil heaters that use a power supply of 110 V and 2.1 A current. The total length covered by the thin foil heater is 12 inches with a span of 4.8 inch. The temperature difference between the thin foil heater at a measured location and flow is 27°C. Calculate the Nusselt number at the measured location based on turbine vane axial chord length. What is the corresponding uncertainty of Nusselt number?

REFERENCES

1. Kline, S.J. and McClintock, F.A., "Describing Uncertainties in Single-Sample Experiments," *Mechanical Engineering*, Vol. 75, No.1, 1953, pp. 3–8.
2. Moffat, R.J., "Contributions to the Theory of Single-Sample Uncertainty Analysis," *ASME Journal of Fluids Engineering*, Vol. 104, 1982, pp. 250–260.
3. Moffat, R.J., "Describing the Uncertainties in Experimental Results," *Experimental Thermal and Fluid Science*, Vol. 1, 1988, pp. 3–17.
4. Kim, J.H., Simon, T.W., and Viskanta, R., "Journal of Heat Transfer Policy on Reporting Uncertainties in Experimental Measurements and Results," *ASME Journal of Heat Transfer*, Vol. 115, No. 1, 1993, pp. 5–6.
5. An American National Standard, *Test Uncertainty Performance Test Code*, ASME PTC 19.1-2013 (Revision of ASME PTC 19.1-2005), ASME, New York, 2013.

5 Steady-State Heat Transfer Measurement Techniques

5.1 INTRODUCTION AND MEASUREMENT THEORY

There are several techniques employed by researchers to measure convective heat transfer from a surface under steady-state conditions. Steady-state implies that the heat input to the surface, the wall temperature, and the fluid temperature are under equilibrium conditions (i.e., they do not change with time). Local as well as averaged (both overall and regional) measurements can be performed using the steady-state method. This chapter discusses techniques for measuring average (overall and regional) heat transfer coefficients employing copper plates fitted with a heater and thermocouples and steady-state, local heat transfer coefficients using foil heaters instrumented with thermocouples. Detailed, local steady-state measurements can also be obtained using optical techniques such as liquid crystal thermography, infrared thermography, and temperature sensitive paint and are discussed in Chapters 7–9, respectively. All these techniques differ only in the method for measuring the surface temperature. In this chapter, techniques which use thermocouples for wall temperature measurements are described. The basic underlying principle for measuring the heat transfer coefficients remains the same. The test surface is heated by means of a suitable heater. The power input to this heater should be controllable through a device such as a variable autotransformer. The heater is assumed to supply a constant heat input to the test surface resulting in a constant heat flux boundary condition. Circuit current and voltage measurement, from a digital multimeter, provides the total heat input. By maintaining a temperature difference between the wall and the fluid, a thermal boundary layer is developed resulting in heat exchange between the surface and fluid. The primary disadvantage of any steady-state, heat transfer experiment is the measurement of the miscellaneous heat lost from the test section. The consideration of the heat losses will be repeatedly discussed throughout the chapter.

While the measurement technique can be applied to both internal and external flows, there are subtle differences that arise for the respective flow conditions. Therefore, both internal and external flows will be considered separately in the upcoming sections.

5.1.1 Fundamental Measurement Theory for Internal Flows

The heat transfer coefficient is dictated by Newton's Law for convective heat transfer and is given by

$$h = \frac{q''_{net}}{T_w - T_b} = \frac{q''_{gen} - q''_{loss}}{T_w - T_b} \qquad (5.1)$$

where q''_{net} is the net convective heat flux from the surface, q''_{gen} is the power output per unit area from the heater, as obtained from voltage-current measurement, and q''_{loss} is the heat flux lost from the test surface and is a function of the wall temperature. The steady-state wall temperature, T_w, is measured by thermocouples and is generally maintained approximately 20°C higher than the fluid temperature. A higher temperature difference ensures that relative uncertainty in the heat transfer coefficients due to temperature is low. At the same time, heat losses, which are proportional to the wall temperature, increase with increasing wall temperature. A 20°C temperature difference ensures an optimum balance with low relative uncertainties as well as low heat losses. For internal flows, the fluid temperature of interest is the fluid bulk temperature, T_b. It can be either measured using thermocouples inserted in the mainstream or can be determined by applying an energy balance for the test section. The bulk temperature at the exit of the test section, $T_{b,outlet}$ can be calculated using

$$T_{b,exit} = T_{b,inlet} + \frac{\dot{Q}_{net}}{\dot{m} C_p} \qquad (5.2)$$

where $\dot{Q}_{net}$ is the net rate of heat supplied to the fluid $\left(\dot{Q}_{net} = \dot{Q}_{gen} - \dot{Q}_{loss} \right)$, $\dot{m}$ is the mass flow rate in the test section, and C_p is the specific heat of the fluid. $T_{b,inlet}$ is the bulk temperature of the fluid at the test section inlet and is typically measured using thermocouples or RTD's placed at the entrance. The mean bulk temperature can then be calculated as

$$T_b = \frac{T_{b,inlet} + T_{b,outlet}}{2} \qquad (5.3)$$

It should be noted that the measured and calculated temperatures at the exit should ideally be the same. This phenomenon is frequently used as a "sanity check" to ensure that the measured and calculated magnitudes are comparable. Matching values provide confidence that the calculations and measurements performed are reliable. This equation is valid only for duct flow with no mass addition. For experiments with mass addition or removal, such as impingement flows, the energy balance will result in a different equation.

The outlet bulk temperature measurement/calculation comparison is also a simple method to validate the estimation of the miscellaneous heat losses. While test sections should be constructed to minimize the amount of heat that is conducted through the support structure, it is impossible to construct a perfectly insulated test section. Therefore, any heat that is not convected away from the surface must also be taken into account. Generally, the heat loss through the test section is taken into account by one of two methods. The first method is a "real-time" heat loss calibration, and the second method requires a separate "heat loss calibration" test.

The "real-time" heat loss calibration is advantageous because the heat losses can be determined during the actual channel flow experiment. However, additional instrumentation is required to complete the calibration without sacrificing the primary instrumentation for the wall and fluid temperature measurements within the channel. During this "real-time" calibration, additional thermocouples are placed throughout the insulating test section material. With knowledge of the material thermal conductivity, the thickness of the material (distance between thermocouples), and the temperature measurements, the rate of heat transfer through the material can be determined from a simple one-dimensional thermal circuit, as shown in Figure 5.1. The heat generated by the heater is either convected away from the wall or is conducted through the test section. Increasing the resistance of the conduction resistance of the circuit will minimize the amount of heat loss.

As a temperature difference is the driving force for the rate of heat transfer, it is necessary to provide an adequate number of thermocouples to determine the rate of heat conduction through the material. Ideally, a matrix of "heat loss" thermocouples will be available, so the "local" rate of heat loss can be determined for each local wall temperature measurement.

When the measurement demands exceed the availability of the data acquisition hardware, it is necessary to use the existing wall instrumentation to determine the miscellaneous heat losses. When running a separate "heat loss calibration," the heat losses are determined in a series of "no flow" tests. The channel is traditionally filled with an insulating material to eliminate the occurrence of natural convection within the channel. The insulating material (i.e., fiberglass insulation) should have a lower thermal conductivity than the test section material itself. Power is supplied to the

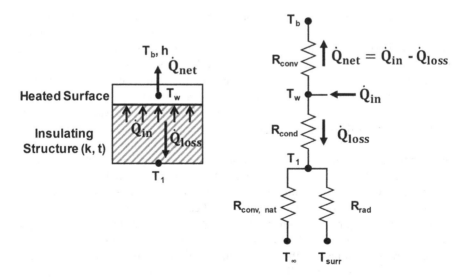

FIGURE 5.1 Schematic and one-dimensional circuit for heat flow through a test section.

heaters, and the channel is allowed to reach steady-state. At least two power settings should be used: steady-state temperatures should be reached that are (a) lower than the expected wall temperatures and (b) higher than the expected wall temperatures. The steady-state wall temperature distributions are recorded, as well as the power supplied to the heaters for both the high and low tests. As the inside of the channel is filled with the low conductivity insulation, it is assumed that all the power supplied to the heaters is being conducted through the test section support. In Figure 5.1, all power supplied to the heater is assumed to conduct downward, through the insulating support material. With this heat loss calibration, it is important to measure the room temperature during both the heat loss calibration and the actual tests, as the driving temperature difference is that between the test section wall and the ambient room. Using linear interpolation, the actual rate of heat loss can be determined for the actual experiment. While this method is more time consuming, it does not require additional instrumentation (as the real-time calibration does). This method is shown in greater detail in the upcoming examples.

5.1.2 Fundamental Measurement Theory for External Flows

The primary difference between the internal and external convective heat transfer experiments lies with the definition of the heat transfer coefficient. Replacing the bulk temperature in Equation (5.1) with the freestream (or adiabatic wall) temperature, the convective heat transfer coefficient for external flows can be calculated in Equation (5.4).

$$h = \frac{q''_{net}}{T_w - T_\infty} = \frac{q''_{gen} - q''_{loss}}{T_w - T_{aw}} \tag{5.4}$$

The adiabatic wall temperature definition is typically used for external flows as well as internal flows when heat generation due to viscous dissipation can be significant. The adiabatic wall temperature represents the condition when there is no heat exchange between the test surface and the fluid. It can be measured using the same measurement techniques employed for measuring the wall temperature. Thus, two tests need to be performed when this definition is used to determine the heat transfer coefficient. In the first test, the test section is unheated and the wall temperatures are recorded under the predetermined flow conditions. The wall temperatures then correspond to the adiabatic wall temperatures. It should be noted that for the adiabatic condition to be true, the test section should be well-insulated such that no heat exchange occurs between the test section and the ambient surroundings. This is particularly true, when the inlet fluid to the test section and the ambient surroundings are at different temperatures. In the second test, heat is applied to the surface and the wall temperatures are recorded after a steady-state has been reached at the same flow conditions. At high mach numbers, under compressible flow conditions, the adiabatic wall temperature will include the effect of heat generation from viscous dissipation. For external flows, when the adiabatic wall temperature is not known, the recovery factor can be used to calculate it using the equation

$$T_{aw} = T_\infty + r \frac{V^2}{2C_p} \tag{5.5}$$

where r is the recovery factor. For laminar flow, $r = Pr^{1/2}$ while for turbulent flow $r = Pr^{1/3}$. For gases, fluid properties can be calculated using the Eckert reference temperature [1], T_r, which is given by Equation (5.6).

$$T_r = T_\infty + \frac{1}{2}\left(T_w - T_\infty\right) + 0.22\left(T_{aw} - T_\infty\right) \tag{5.6}$$

The above equation is valid for both laminar and turbulent flows.

However, for internal flow experiments, the adiabatic wall temperature has to be corrected for the changing fluid temperature due to heat addition, as the fluid flows in the test section. The changing fluid temperature can be determined from the energy balance, and it is the same as the mean bulk temperature in Equation (5.3). Assuming the inlet fluid temperatures for both tests to be same, the corrected average fluid temperature can then be given as

$$T_b = T_{aw} + \Delta T_{corr} = T_{aw} + \left(\frac{T_{b,inlet} + T_{b,exit}}{2} - T_{b,inlet}\right) = T_{aw} + \left(\frac{T_{b,exit} - T_{b,inlet}}{2}\right) \tag{5.7}$$

As with internal flows, the miscellaneous heat losses must be determined for the steady-state external experiments. With the external flows, the "real-time" heat loss calibration method is most often preferred. However, either method can be modified for determined of the heat losses.

The test surface is typically instrumented with several thermocouples connected to a data acquisition system. Thermocouples are chosen mainly due to their flexibility and availability for large temperature ranges. The thermocouple type depends on the required accuracy and temperature range. Chapter 3 provides more details for choosing the appropriate thermocouples. Several vendors manufacture data acquisition systems which can be programmed to suit the application. During the experiment, before the application of heat, the wall temperature is usually equal to the fluid temperature. Once heat is applied by regulating power to the heater through an autotransformer, the test surface initially rises quickly within a few minutes and then the wall temperature slowly asymptotes to a steady-state value. It can take up to 2 hours to achieve steady-state, depending on the test section design. However, true steady-state will never be obtained as the wall temperature is sensitive to the inlet fluid temperature and to the ambient temperature to a smaller extent. Changes in the inlet temperature with time will affect the wall temperature even if a constant heat flux is applied to the test surface. Steady-state can be assumed when the temperature change is less than 0.1°C in 15 minutes. The power supplied to the heater should be sufficient to maintain an average temperature difference of about 20°C between the surface and the fluid.

In the above discussion, the driving temperature potential for convective heat transfer to take place is supplied by the wall heater. The use of a wall heater can be avoided if the mainstream temperature is hot and a heat extraction mechanism is available to continually remove heat from the test surface such that a temperature difference exists between the mainstream fluid and the test surface. Setting up this heat extraction mechanism, such as a refrigeration cycle, along with the added drawback of heating the inlet fluid, is more cumbersome than installing a wall heater with its associated accessories. Thus, the use of this inverse method is unpopular and is not discussed in this text.

5.2   STEADY-STATE, OVERALL AVERAGE HEAT TRANSFER MEASUREMENT TECHNIQUE

As the name suggests, this technique gives the overall, average heat transfer over the entire test surface. This technique gives a representative idea of the overall heat transfer from a test surface and was frequently used about one to two decades ago. The technique's inability to get the local heat transfer coefficient is a major drawback. A modified version of this technique gives the regionally averaged heat transfer over the test surface and will be described in the next section.

5.2.1   SINGLE COPPER PLATE WITH HEATER AND THERMOCOUPLES

A single metal plate with high thermal conductivity is used as the test wall for which heat transfer is to be studied. Copper, which has a very high thermal conductivity of ~400 W/mK, is frequently used. Aluminum, which has a thermal conductivity of ~177 W/mK, can also be used. By using a material which has a high conductivity, the entire test surface attains the same temperature at steady-state conditions. If the Biot number is less than 0.1, the entire test surface will give the same wall temperature. The Biot number is given in Equation (5.8)

$$\text{Bi} = \frac{hL_c}{k} = \frac{h}{k}\left(\frac{V}{A_s}\right) \tag{5.8}$$

where h is the heat transfer coefficient on the surface, k is the thermal conductivity of the metal, and L_c is a characteristic length equal to the ratio of the volume of the plate, V, and the surface area, A_s, available for convection. If aluminum is used as the test wall material, the researchers should ensure the convective heat transfer coefficients are sufficiently low enough to ensure the Biot number remains less than 0.1. The test wall is instrumented with thermocouples for temperature measurement and fitted with a heater on the underside to provide a uniform heat flux.

5.2.2   EXPERIMENT EXAMPLE

Internal cooling passages likely to be found in the trailing region of gas turbine blades were experimentally investigated by Lau et al. [2]. Trailing edge cooling passages are generally relatively narrow channels, as required by the tapered shape of the turbine airfoils. To increase the structural rigidity within these narrow passages and increase the convective surface area, these channels are commonly lined with pin-fins. These pin-fins are most often an array of short cylinders spanning from the pressure side to the suction side of the channel.

Lau et al. [2] acquired "overall" heat transfer coefficient under a variety of flow conditions typically found in trailing edge cooling channels. In addition to varying the Reynolds number of the coolant through the channels, they also considered multiple pin-fin arrays, and they quantified the effect of trailing edge ejection (the lateral expulsion of coolant from the channel for trailing edge film cooling).

While two separate test sections were used by Lau et al., only one of those channels is described in this example. Figure 5.2 shows the experimental setup for measurement of overall heat transfer coefficients [2]. As explained previously, for the measurement of "overall" heat transfer coefficients, it is necessary to construct the cooling channel from a material with a high thermal conductivity. For the present example, all walls of the channel (as well as the circular pin-fins) were fabricated from copper. For their baseline case (referred to a "Configuration A" in the referenced paper), the rectangular channel had a cross-sectional area of 95.3 × 6.4 mm, and the length of the channel was 95.3 mm. A staggered array of pin-fins was used in the channel, and the pins were arranged such that streamwise and spanwise spacings of the pins were 2.5 times the pin diameter ($D = 6.4$ mm).

The copper walls of the channel were constructed of 6.4 mm thick copper, and strip heaters were placed on the backside of the copper to heat the walls of the channel. As shown in Figure 5.2, nine thermocouples were imbedded in the top wall of the channel and four thermocouples were used on the bottom wall. The thermocouple outputs were recorded by a thermocouple data logger, and the power supplied to the strip heaters was measured using a digital multimeter (the power to the heaters was varied using a variable autotransformer). For this stationary, rectangular channel, the power to the heaters was adjusted so the wall temperatures were approximately 19.4°C greater than the inlet air temperature.

For the current experiment, the following raw data were recorded for each cooling configuration: (a) ambient room temperature (needed for heat loss calculation), (b) inlet bulk temperature, (c) wall temperature from each thermocouple, and (d) the power supplied to the heater. In addition, a separate heat loss calibration was

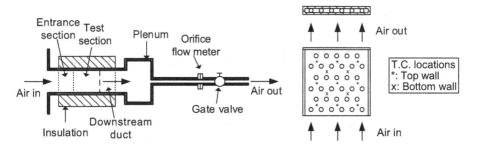

FIGURE 5.2 Experimental setup for measurement of overall heat transfer coefficients.

required to determine the magnitude of the energy that was lost through the support structure of the test section (not convection away from the wall by the cooling air). For this heat loss calibration, two sets of data (wall temperatures and heater powers) were obtained, so the heat loss during the actual test could be approximated using linear interpolation.

5.2.2.1 Mass Flow Rate Calculation

For the present example, care must be taken with the calculation of the Reynolds number for the coolant. While the Reynolds number of the coolant is varied, the Reynolds number is specific to flow through a pin-fin array. As shown in Equation 5.9, the pin diameter is used as the characteristic length and the coolant velocity is taken as the maximum velocity of the coolant occurring as the coolant travels through the minimum cross-sectional area (between two pins).

$$\text{Re} = \frac{\rho u_{max} D_{pin}}{\mu} \tag{5.9}$$

At a specific Reynolds number, the maximum velocity can be calculated by rearranging Equation (5.9). However, an average channel velocity is needed to determine the required mass flow rate through the channel. Based on the pin-fin spacing and mass conservation, the average velocity through the channel can be determined.

$$\dot{m} = \rho A_{c,min} u_{max} = \rho A_c V_{avg} \tag{5.10}$$

Rearranging the Equation 5.10 provides the average coolant velocity through the channel.

$$V_{avg} = \frac{A_{c,min} u_{max}}{A_c} \tag{5.11}$$

With the given pin-fin arrangement, the cross-sectional area can be calculated as $A_c = h(2.5D)$, and the minimum area is $A_{c,min} = h(1.5D)$.

Combining Equations 5.9 through 5.11 the mass flow rate desired for a specific Reynolds number can be calculated, and this is shown in Equation (5.12).

$$\dot{m} = \rho A_c V_{avg}$$

$$= \left(1.23 \frac{kg}{m^3}\right)(95.3 \text{ mm})(6.4 \text{ mm})\left(\frac{1 \text{ m}}{1000 \text{ mm}}\right)^2 \left(\frac{1.5}{2.5}\right)$$

$$\text{Re}\left[\frac{1.79 \times 10^{-5} \dfrac{Ns}{m^2}}{\left(1.23 \dfrac{kg}{m^3}\right)(6.4 \text{ mm})\left(\dfrac{1 \text{ m}}{1000 \text{ mm}}\right)}\left(\frac{kgm}{Ns^2}\right)\right] \tag{5.12}$$

TABLE 5.1

Calculated Mass Flow Rates over Range of Reynolds Numbers

Nominal Re	Actual Re	$\dot{m}\left(\dfrac{kg}{s}\right)$
6000	5779	0.00592
10,000	10,088	0.0103
15,000	14,127	0.0145
20,000	20,160	0.0206
25,000	25,558	0.0262

Table 5.1 summarizes the coolant mass flow rates over the range of Reynolds numbers considered in this example.

5.2.2.2 Heat Loss Calibration (Raw Data)

The copper test section was insulated to maximize the resistance to heat transfer through the support material. Using a low conductivity, insulating material will deter the transfer of heat through the support material by conduction. However, the test section cannot be perfectly insulated, and therefore, the amount of heat transferred by modes other than convection must be quantified. For the present experiment, supplementary steady-state tests were performed to establish a matrix of heat loss data. Two heat loss tests were completed in which the wall temperature within the duct was heated to 11°C and 50°C above the ambient room temperature. It is necessary to record the ambient room temperature during both the actual tests and the heat loss tests because the temperature difference between the test section and the room is the driving temperature difference for heat transfer away from the test section. In addition to the wall and ambient temperatures, the power supplied to the heater during these tests must be recorded. The raw data (including the average wall temperature taken from the array of thermocouples) acquired during the heat loss tests is shown in Table 5.2.

5.2.2.3 Heat Transfer Enhancement (Raw Data)

During each heat transfer experiment, it is necessary to measure the wall temperature, inlet air temperature, ambient room temperature, and the power supplied to the

TABLE 5.2

Raw Data from Supplementary Heat Loss Tests

	$\overline{T_w}$ (°C)	T_{amb} (°C)	Heater Voltage (V)	Heater Current (Amp)
Low data	32.1	21.0	4.5	0.256
High data	71.1	21.0	15.2	0.865

TABLE 5.3

Raw Data Acquired over Range of Reynolds Numbers for Overall Heat Transfer Experiment

	Re				
	5778	10,088	14,127	20,160	25,558
T_{amb} (°C)	21.1	21.1	21.1	21.1	21.1
T_{inlet} (°C)	21.1	21.2	21.0	20.7	21.0
$T_{w,1}$ (°C)	41.2	40.4	40.6	38.4	40.3
$T_{w,2}$ (°C)	41.0	40.1	40.3	38.1	39.8
$T_{w,3}$ (°C)	41.3	40.6	40.8	38.7	40.6
$T_{w,4}$ (°C)	40.7	39.7	39.8	37.5	39.2
$T_{w,5}$ (°C)	41.6	40.8	41.2	39.1	41.3
$T_{w,6}$ (°C)	40.8	40.0	40.6	37.9	39.7
$T_{w,7}$ (°C)	41.1	40.2	40.4	38.2	40.1
$T_{w,8}$ (°C)	41.1	40.2	40.4	38.2	40.1
$T_{w,9}$ (°C)	41.2	40.4	40.7	38.6	40.4
Heater voltage (V)	31.9	38.3	43.3	46.6	52.9
Heater current (Amp)	1.81	2.17	2.46	2.65	3.00

heater. For this overall heat transfer experiment, multiple thermocouples are embedded into the copper surface. However, as one can see in Table 5.3, the variation in these thermocouples is minimal.

Comparing Tables 5.2 and 5.3 it is obvious the wall temperatures maintained during the actual test fall between those of the "high" and "low" heat loss tests. It is a good practice to have the heat loss data bracket the actual test data, so linear interpolation can be used to estimate the actual heat losses during the steady-state experiment.

5.2.2.4   Heat Transfer Enhancement Data Reduction

The authors of the current experimental study sought to compare overall Nusselt numbers obtained in rectangular cooling channels under variety of flow conditions and channel configurations. Before acquiring the non-dimensional Nusselt number, the convective heat transfer coefficient must first be calculated. As the wall temperature of the channel is constant, it is necessary to modify the traditional form of the convective heat transfer Equation to incorporate the constant wall temperature boundary condition. Equation (5.13) shows the heat transfer coefficient definition with the driving temperature difference taken as the log mean temperature difference (which is defined in Equation (5.14)).

$$h = \frac{\left(\dot{Q}_{in} - \dot{Q}_{loss}\right)/A_s}{\Delta T_{lm}}$$

(5.13)

$$\Delta T_{\text{lm}} = \frac{T_{\text{inlet}} - T_{\text{outlet}}}{\ln\left(\dfrac{\overline{T_w} - T_{\text{outlet}}}{\overline{T_w} - T_{\text{inlet}}}\right)} \tag{5.14}$$

In order to calculate the log mean temperature difference, and thus the convective heat transfer coefficient, it is necessary to know the outlet bulk temperature. While the outlet temperature could be measured with thermocouples, the outlet temperature can also be calculated using the first law, energy balance. The calculation of the outlet temperature serves as a check for the estimation of the heat losses; if the heat losses have been properly taken into the account, the calculated outlet temperature should be close to the measured outlet temperature. The outlet temperature can be calculated using Equation (5.15).

$$T_{\text{outlet}} = T_{\text{inlet}} + \frac{\left(\dot{Q}_{\text{in}} - \dot{Q}_{\text{loss}}\right)}{\dot{m}c_p} \tag{5.15}$$

As shown above in Equation (5.15), the rate of heat loss, $\dot{Q}_{\text{loss}}$, must be known prior to calculating the outlet temperature. Linear interpolation can be applied to determine the actual rate of heat loss during the test using the supplementary heat loss calibration data. Equation (5.16) shows the basic linear interpolation, and this can be rearranged to explicitly solve for the rate of heat loss, as shown in Equation (5.17).

$$\frac{\dot{Q}_H - \dot{Q}_L}{\left(\overline{T_{H,w}} - T_{H,amb}\right) - \left(\overline{T_{L,w}} - T_{L,amb}\right)} = \frac{\dot{Q}_{loss} - \dot{Q}_L}{\left(\overline{T_w} - T_{amb}\right) - \left(\overline{T_{L,w}} - T_{L,amb}\right)} \tag{5.16}$$

$$\dot{Q}_{loss} = \left[\frac{\dot{Q}_H - \dot{Q}_L}{\left(\overline{T_{H,w}} - T_{H,amb}\right) - \left(\overline{T_{L,w}} - T_{L,amb}\right)}\right]\left[\left(\overline{T_w} - T_{amb}\right) - \left(\overline{T_{L,w}} - T_{L,amb}\right)\right] + \dot{Q}_L \tag{5.17}$$

As shown in Equation (5.17), the ambient, room temperature is needed to determine the rate of heat loss, as the heat is being lost to the room. Also, as this is an overall heat transfer experiment, the wall temperature always represents the average of the nine thermocouples embedded in the channel wall. The rates of heat transfer can be calculated using the measured voltages and currents supplied to the heaters $\left(\dot{Q} = VI\right)$. Using this equation, and Equation (5.17), the power input and heat loss for each Reynolds number is summarized in Table 5.4.

With the calculated heat loss, and the known power input, the outlet temperature of the air can be calculated using Equation (5.15). The mass flow rate was previously determined, and is shown in Table 5.1 for each Reynolds number. Taking the specific heat of air to be $c_p = 1005$ J/kgK, the outlet air temperature was calculated and shown in Table 5.5.

TABLE 5.4

Power Input and Lost over Reynolds Number Range

Actual Re	$\dot{Q}_{input}$ (W)	$\dot{Q}_{loss}$ (W)	$\dot{Q}_{net} = \dot{Q}_{input} - \dot{Q}_{loss}$ (W)
5779	57.79	3.93	53.87
10,088	83.18	3.66	79.52
14,127	106.3	3.74	102.6
20,160	123.3	3.05	120.2
25,558	158.6	3.62	155.0

TABLE 5.5

Calculated Outlet Temperatures

Re	T_{outlet} (°C)
5779	30.2
10,088	28.8
14,127	28.1
20,160	26.5
25,558	26.9

TABLE 5.6

Calculated Log Mean Temperature Differences

Re	ΔT_{lm} (°C)
5779	15.0
10,088	14.9
14,127	15.8
20,160	14.5
25,558	16.0

With knowledge of the outlet bulk temperatures, the log mean temperature difference for each Reynolds number can be calculated using Equation (5.14), and these values are shown for the current case in Table 5.6.

Finally, the overall Nusselt number for the channel can be determined. For the calculation of the Nusselt number, the heat transfer coefficient shown in Equation (5.13), should be non-dimensional. For the present experiment, the characteristic length is taken as the pin diameter. With this knowledge, Equation (5.18) can be used to determine the channel Nusselt numbers with the thermal conductivity of air, $k = 0.0254$ W/mK.

$$\text{Nu} = h \cdot \frac{D_{\text{pin}}}{k} = \frac{\left(\dot{Q}_{\text{in}} - \dot{Q}_{\text{loss}}\right) / A_s}{\Delta T_{\text{lm}}} \cdot \frac{D_{\text{pin}}}{k} \tag{5.18}$$

The surface area shown in Equation (5.18) is the total area of copper exposed to the cooling air. For the current case where air only travels streamwise through the channel (no lateral ejection), the total surface area includes the endwalls on the top and bottom walls, the two sidewalls, and the area of the pins. Given the channel and pin-fin dimensions, the total area of exposed copper is $A_s = 0.0215 \text{ m}^2$. With this information, the channel Nusselt numbers can be calculated, and for the given range of Reynolds numbers, they are shown in Table 5.7.

Plotting the data in Table 5.7 on a log-log plot replicates a single set of data shown in Figure 5.3.

TABLE 5.7
Channel Averaged (Overall)
Nusselt Numbers

Re	Nu
5779	42.0
10,088	62.3
14,127	76.3
20,160	96.9
25,558	113.3

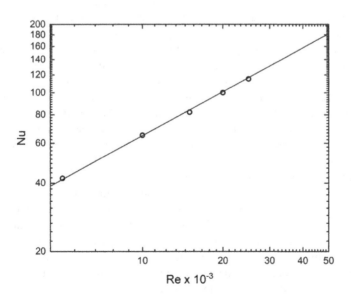

FIGURE 5.3 Overall Nusselt number as a function of Re.

5.3   STEADY-STATE, REGIONAL AVERAGE HEAT TRANSFER MEASUREMENT TECHNIQUE

This technique is one of the most widely employed measurement methods, for convective heat transfer, used by researchers. The main advantage of this technique is that it is a reliable method which can give repeatable results. The relative uncertainty of this method, if correctly applied, can be controlled to be around 5%. In fact, data obtained by using this technique has been used by several researchers as a basis for comparing and calibrating more advanced techniques such as liquid crystal thermography, mass transfer techniques, and infrared thermography.

The primary disadvantage of this measurement technique is its inability to obtain a true local distribution of the heat transfer coefficient on the test surface. This disadvantage can be partly overcome by resolving the test surface into several smaller regions and using a copper plate for each region. Either a single heater, supplying a uniform heat flux over the heated area, or individual heaters for each region can be used to heat the test surface. By using individual heaters, the copper plate temperature, or the wall temperature, can be maintained equal for each region resulting in a uniform wall temperature boundary condition. A disadvantage of using many heaters is that variable transformers equal to the number of regions are needed to control the power input to each heater. This can be operationally inconvenient for the experimentalist, as well as expensive. It should be noted that when using a copper plate grid as a test surface, each region should be isolated from the adjacent regions using an insulating material such as rubber or balsa wood to prevent conduction losses.

5.3.1   Multiple Copper Plates with Heaters and Thermocouples

This method involves using copper plates with embedded thermocouples as the test wall, with one side of the plate in contact with a heater while the other side is exposed to the convective fluid. The reason copper is used for the test surface is due to its very high thermal conductivity (~400 W/mK) which results in low Biot number. If the Biot number for the copper plate is less than 0.1, the wall temperature can be assumed to be uniform in the plate volume due to high conduction within the copper plate. Thus, accurate, average heat transfer coefficients can be measured with this method. Aluminum can be used for the surface if the Biot numbers are less than 0.1.

Ideally, several thermocouples should be embedded in each copper plate so that an accurate average wall temperature can be obtained. If the Biot number is less than 0.1, all thermocouples should read the same temperature. Tiny holes can be drilled in the copper plate symmetrically for the thermocouples. An adhesive, such as a thermally conductive epoxy, can be used to attach the thermocouples to the copper plate. It should be noted that the holes should be drilled in such a manner as to leave the heat transfer surface smooth. If the thermocouples are to be inserted from the heater side, grooves may be cut in the copper plate to allow passage for the thermocouple wires. This is much more convenient than cutting the heater to allow the wires to pass through it. A data acquisition system can monitor the temperatures of the copper plates as measured by the thermocouples.

Power supplied to the heater raises the temperature of the copper plate and is generally controlled through a variable transformer which can regulate the applied voltage. A resistance wire, flexible heater or a foil heater can be used. If the test surface is circular, an axially placed rod heater can be used to heat the wall. If one side of the heater is exposed, it should be covered with a suitable insulation material (e.g., fiberglass, wood, etc.) to prevent any extraneous heat losses. The minimum insulation thickness needed can be approximated by applying heat conduction principles between the heated wall and the ambient surroundings.

The heater power input can be predetermined before the experiment by estimating the heat transfer coefficients on the surface using a suitable empirical correlation. For example, the Dittus-Boelter correlation can be used for turbulent flow in a duct to determine the heat transfer coefficients for a specified Reynolds number. If turbulators are to be studied, a suitable enhancement factor can also be employed on the heat transfer coefficient. By using this magnitude for the heat transfer coefficient, the power input can be estimated for the given test surface area and temperature difference between the wall and fluid.

The heated wall exposed to the flowing fluid will attain a steady-state (unchanging wall temperature for a specified heat flux) after some time has elapsed. The wall temperature should be maintained approximately 20°C higher than the fluid temperature. By measuring the power input to the heater and the related fluid and wall temperatures, the heat transfer coefficients can be calculated. The heater power input should be corrected for any extraneous heat losses in the system.

5.3.2   EXPERIMENT EXAMPLE

5.3.2.1   Background and Introduction to Experiment

Heat transfer enhancement due to various rib configurations was considered in a square channel by Han et al. [3]. The objective of this study was to evaluate the thermal performance of various rib turbulators in a square channel. The thermal performance of each rib configuration was evaluated through the measurement of both the heat transfer enhancement and the associated frictional losses. Regional heat transfer coefficients, and thus regional heat transfer enhancement, were obtained using copper plates and heaters.

Figure 5.4 shows the square test channel used in this study by Han et al. [3]. As the figure indicates the length of the square channel was divided into 10 isolated sections. The copper plates were in direct contact with the cooling air passing through the channel. Foil heaters were placed on the back of the copper plates. Finally, the copper plates and foil heaters were encased in wooden plates (to minimize the stray heat loss from the heaters away from the copper).

The length of the heated test section was 101.6 cm, and the cross section was 5.08 cm by 5.08 cm. Each of the forty copper plates was 5.08 cm wide by 10.16 cm long, and they were separated by 0.159 cm thick wooden strips. A blind hole is drilled on the back of each copper plate, and a single 24 AWG, type-T thermocouple was glued in each hole using high temperature, high conductivity epoxy. These 40 thermocouples measuring the regional wall temperatures were connected

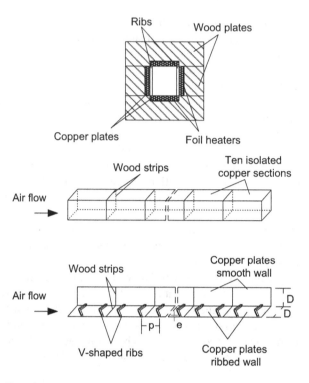

FIGURE 5.4 Experimental setup for measurement of regional heat transfer coefficients.

to a Fluke 2285 B data logger for real-time monitoring of the channel temperatures. Thermocouples were also placed in the mainstream flow at the inlet and the outlet of the channel to measure the bulk air temperature at these locations.

Four foil heaters (one for each wall) were constructed from a 0.001-inch-thick stainless steel foil. Each heater was 5.08 cm wide by 101.6 cm long to cover one wall of the square channel. Lead wires were soldered to opposite ends of the heater, and power to each of the four heaters was controlled independently through four separate variable transformers. The foil heaters were attached between the copper plates and wood using double sided tape.

While this channel was used to study many turbulator configurations over a wide range of Reynolds numbers, only one case is considered in this example: the channel with 45° V-shaped ribs. Data collected from a representative Reynolds number (approximately 30,000) is shown. Furthermore, the required steps to properly reduce the raw data for this case are provided.

5.3.2.2   Mass Flow Rate Calculation

Prior to beginning the experiment, it is necessary to calculate the required coolant mass flow rate. With the desired Reynolds number of 30,000, it is possible to determine the required flow rate.

$$\text{Re} = \frac{\rho V D_h}{\mu} = \frac{D_h \dot{m}}{\mu A_c} \tag{5.19}$$

After evaluating the viscosity of air at the channel's bulk inlet temperature, the required mass flow rate, $\dot{m}$, of air can be calculated from Equation (5.19). Care must be taken by the experimentalist to ensure that correct characteristic length is chosen for the experiment. In this case, the Reynolds number is defined based on the hydraulic diameter of the channel.

$$D_h = \frac{4 A_c}{P} \tag{5.20}$$

With the square channel described above, the required mass flow rate was determined by combining Equations (5.19) and (5.20) and rearranging the terms.

$$\dot{m} = \text{Re} \times \frac{\mu P}{4} \tag{5.21}$$

$$\dot{m} = \frac{30{,}000}{4} \times \left(1.79 \times 10^{-5} \frac{\text{Ns}}{\text{m}^2}\right) \times (4 \times 5.08 \text{ cm}) \left(\frac{1 \text{ m}}{100 \text{ cm}}\right) \left(\frac{\text{kgm}}{\text{Ns}^2}\right) = 0.0273 \frac{\text{kg}}{\text{s}} \tag{5.22}$$

While 0.0273 kg/s is the target mass flow rate for Re = 30,000, it is difficult to meter the flow rate precisely to obtain flow; therefore, it is expected that the actual Reynolds number will vary slightly from the target number.

5.3.2.3 Heat Loss Calibration (Raw Data)

Although the test section has been constructed, so the structural components have relatively low thermal conductivities, stray heat losses have not been completely mitigated. Therefore, it is necessary to determine the magnitude of the heat loss. The heat losses were determined for the present channel from a "no flow" experiment. During this no flow experiment, the square channel was filled with a low conductivity, fiberglass insulation. Power was supplied to the foil heaters, and the walls were heated. Two different power settings were applied to the heaters to generate two sets of wall temperatures. These two sets of temperatures were chosen so one set was below the anticipated wall temperature during the actual test, and the second set was above the expected wall temperature. The two sets of temperatures bracketing the actual wall temperature make it possible to interpolate the heat loss during the actual test.

Due to the symmetry of the current channel, heat loss data is provided for one wall of the duct, and this data is assumed to be applicable for all four walls (Table 5.8). Upon reaching steady-state at the two different power settings, the resistance of the heaters was measured and combined with the measured voltage. These resistance and voltage measurements are shown in Table 5.9.

TABLE 5.8

Temperature Distributions Acquired through the Heat Loss Calibration

x (cm)	Low–T_w (°C)	High–T_w (°C)
5.16	48.1	78.3
15.5	49.1	77.7
25.5	49.3	78.3
35.8	48.9	77.1
45.8	49.8	78.9
56.1	49.7	80.7
66.3	49.6	78.5
76.4	49.5	79.3
86.6	49.4	79.1
96.6	49.3	78.9
Room	22.3	23.1

TABLE 5.9

Voltage and Resistance Measurements from Heat Loss Calibration

	Voltage (V)	Resistance (Ω)
Low	2.15	6.0
High	4.0	6.0

5.3.2.4 Heat Transfer Enhancement (Raw Data)

Tables 5.10 and 5.11 summarize the data collected within the rib roughened channel at an approximate Reynolds number of 30,000. As the tables indicate, data was collected on two walls within the square channel: a smooth wall and a wall roughened with V-shaped ribs.

It is convenient to compare the wall temperatures measured in the actual test to those obtained during the heat loss calibration. Figure 5.5 compares the wall temperature distributions on the ribbed and smooth walls with those obtained from the high and low sets of heat loss data. As the figure indicates the temperature distributions measured during the heat loss calibration are more uniform through the channel than distributions obtained during the actual heat transfer enhancement test. These actual distributions can vary depending on the type of heater used to heat the test section walls. Figure 5.5 also shows how the "high" and "low" sets of heat loss data bracket the temperatures measured on both the smooth and ribbed walls.

TABLE 5.10

Temperature Distributions Acquired at Re = 32,050

x (cm)	Smooth–T_w (°C)	Roughened–T_w (°C)
5.16	58.3	57.5
15.5	64.1	59.8
25.5	61.7	59.5
35.8	61.8	61.7
45.8	65.8	65.4
56.1	69.8	66.9
66.3	70.2	69.5
76.4	74.5	72.7
86.6	75.5	76.8
96.6	75.4	76.5
Inlet Bulk	28.1	
Outlet Bulk	41.4	
Room	22.5	

TABLE 5.11

Voltage and Resistance Measurements at Re = 32,050

	Voltage (V)	Resistance (Ω)
Smooth wall	28.9	6.0
Roughened wall	33.1	6.0

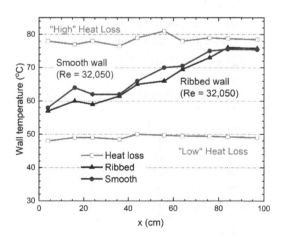

FIGURE 5.5 Comparison of wall temperature distributions in actual test and heat loss calibration.

5.3.2.5 Heat Transfer Enhancement Data Reduction

Recalling that the goal is to determine the heat transfer enhancement in the rib roughened channel, it is necessary to recognize how the heat transfer enhancement will be evaluated. In the study shown with this example, the heat transfer enhancement is presented in terms of the Nusselt number ratio (Nu/Nu_o). This is a convenient method for comparing the increased heat transfer in the rib roughened channel (Nu) to that of a smooth channel (Nu_o). In order to determine the Nusselt number, the heat transfer coefficient must first be calculated, and the heat transfer coefficient can be determined the convective heat transfer equation.

$$h_x = \frac{\dot{Q}_{net,x}/A_s}{T_{w,x} - T_{b,x}} \qquad (5.23)$$

The local wall temperatures have been measured directly with the thermocouples mounted in the copper plates. The local bulk temperature, $T_{b,x}$, must be determined from linear interpolation, and the local, convective heat flux is the difference of the power input to the heaters and the heat loss determined from the calibration.

Using the measured inlet and outlet temperatures, the local bulk temperature can be determined at each location in the channel. Table 5.12 presents the bulk temperature distribution through the channel. As this is the average coolant temperature through the channel, this bulk temperature distribution is valid for both the ribbed wall and the smooth wall of the square channel.

The measured wall temperatures are compared to the interpolated bulk fluid temperature in Figure 5.6. As shown in the figure, a moderate temperature difference (between the wall and fluid) is maintained through the length of the channel. One would expect the wall temperature distribution and the bulk fluid temperature

TABLE 5.12

Bulk Temperature Distribution through Rib Roughened Channel

x (cm)	x/L	$T_{b,x}$ (°C)
0	0.0	28.1 (inlet)
5.16	0.0508	28.88
15.5	0.152	30.43
25.5	0.251	31.94
35.8	0.352	33.49
45.8	0.451	35.0
56.1	0.552	36.55
66.3	0.652	38.08
76.4	0.752	39.61
86.6	0.852	41.14
96.6	0.951	42.65
101.6	1.0	43.4 (outlet)

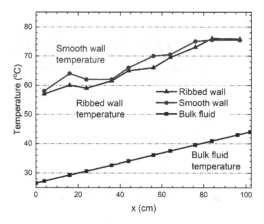

FIGURE 5.6 Comparison of wall temperature and bulk fluid temperature distributions through the square channel.

distribution to be parallel curves in the fully developed region of the channel. While both the ribbed and smooth walls maintain approximately the same slope from $x = 36.8$–86.6 cm, they are not changing at the exact rate as the fluid temperature. While each wall in the square channel is exposed to the same heat flux, circumferentially the heat flux is not uniform. Therefore, the temperature distributions slightly vary from those predicted from internal heat transfer theory.

With the regional wall and air (bulk) temperatures known, it is now necessary to determine the regional rate of convective heat transfer $\left(\dot{Q}_{net,x} \right)$. The net rate of heat transfer is the difference between the heat transfer supplied to the channel and the miscellaneous heat losses, as shown by Equation (5.24). Therefore, both the regional heat applied to the channel and the regional heat losses must be determined.

$$\dot{Q}_{net,x} = \dot{Q}_{in,x} - \dot{Q}_{loss,x} \tag{5.24}$$

Table 5.11 shows the measured resistance for the two foil heaters and the voltage supplied (via variable transformers) to these heaters. With these measurements, the total power (heat) supplied to the channel can be determined with Equation (5.25).

$$\dot{Q}_{in} = \frac{V^2}{R} \tag{5.25}$$

While the total power supplied to the channel is useful, power supplied to each copper plate is needed for the calculation of the regional heat transfer coefficient. For the present case, the total power supplied to the channel is evenly distributed to the 10 copper plates on each wall. The total power supplied to each channel as well as the fraction supplied to each copper plate on the surface are summarized in Table 5.13.

For the determination of the heat losses, it is necessary to determine the amount of power supplied to each copper plate during the "low" and "high" sets of heat loss

TABLE 5.13

Power Input to Each Copper Plate within the Channel

	Total $\dot{Q}_{in}$ (watts)	Regional $\dot{Q}_{in,x}$ (watts)
Ribbed wall	182.6	18.26
Smooth wall	139.2	13.92

TABLE 5.14

Power Input to Each Copper Plate during the Heat Loss Calibration

"Low" Total $\dot{Q}_L$ (watts)	"Low" Regional $\dot{Q}_{L,x}$ (watts)	"High" Total $\dot{Q}_H$ (watts)	"High" Regional $Q_{H,x}$ (watts)
0.770	0.0770	2.67	0.267

data. Using the power supplied (through voltage and resistance measurements), the total power and regional power distributions can be determined. The power supplied to the heaters during the heat loss calibration is shown in Table 5.14.

Linear interpolation can be used to determine the actual heat loss during the heat transfer enhancement test. Using the measured wall temperatures and power inputs during the calibration and the measured wall temperature during the actual test, the heat losses can be approximated. As the driving temperature difference for the heat losses is the temperature difference between the heated wall and the ambient room, the ambient room temperature must be considered in the calculation of the heat losses. As shown in Equation (5.26), the room temperature must be recorded during for each set of calibration data, as well as during the Reynolds number test.

$$\frac{\dot{Q}_{H,x} - \dot{Q}_{L,x}}{\left(T_{H,wx} - T_{H,\text{room}}\right) - \left(T_{L,wx} - T_{L,\text{room}}\right)} = \frac{\dot{Q}_{\text{loss},x} - \dot{Q}_{L,x}}{\left(T_{wx} - T_{\text{room}}\right) - \left(T_{L,wx} - T_{L,\text{room}}\right)} \qquad (5.26)$$

Rearranging Equation (5.26) to solve for the actual, regional heat loss, $\dot{Q}_{\text{loss},x}$, yields Equation (5.27). It should be noted that the expression for the regional heat loss shown in Equation (5.27) could take several forms depending on how linear interpolation was used with the known data.

$$\dot{Q}_{\text{loss},x} = \left[\frac{\dot{Q}_{H,x} - \dot{Q}_{L,x}}{\left(T_{H,wx} - T_{H,\text{room}}\right) - \left(T_{L,wx} - T_{L,\text{room}}\right)}\right]\left[\left(T_{wx} - T_{\text{room}}\right) - \left(T_{L,wx} - T_{L,\text{room}}\right)\right] + \dot{Q}_{L,x}$$

$$(5.27)$$

Combining the heat supplied to each copper plate with the heat losses, the net heat transfer can be determined as shown in Equation (5.24). The heat flux distributions for each copper plate are summarized in Tables 5.15 and 5.16.

TABLE 5.15

Heat Flux Distributions for the Ribbed Wall

x (cm)	$\dot{Q}_{in,x}$ (W)	$\dot{Q}_{loss,x}$ (W)	$\dot{Q}_{net,x}$ (W)
5.16	18.26	0.136	18.12
15.5	18.26	0.148	18.11
25.5	18.26	0.144	18.12
35.8	18.26	0.164	18.10
45.8	18.26	0.180	18.08
56.1	18.26	0.184	18.08
66.3	18.26	0.210	18.05
76.4	18.26	0.227	18.03
86.6	18.26	0.256	18.00
96.6	18.26	0.255	18.01

TABLE 5.16

Heat Flux Distributions for the Smooth Wall

x (cm)	$\dot{Q}_{in,x}$ (W)	$\dot{Q}_{loss,x}$ (W)	$\dot{Q}_{net,x}$ (W)
5.16	13.92	0.142	13.78
15.5	13.92	0.178	13.74
25.5	13.92	0.159	13.76
35.8	13.92	0.165	13.76
45.8	13.92	0.183	13.74
56.1	13.92	0.202	13.72
66.3	13.92	0.215	13.71
76.4	13.92	0.239	13.68
86.6	13.92	0.247	13.67
96.6	13.92	0.248	13.67

With the net rate of heat transfer $\left(\dot{Q}_{net,x}\right)$ and the difference between the regional wall and fluid temperatures $(T_{w,x} - T_{b,x})$, it is possible to determine the regional heat transfer coefficients using Equation (5.23). The surface area (A_s) used in this calculation is the area of each individual copper plate. In the study considered in this example, the surface area of each copper plate is:

$$A_s = (5.08 \text{ cm}) \times (10.16 \text{ cm}) = 51.61 \text{ cm}^2 \tag{5.28}$$

The calculated heat transfer coefficients are shown in Table 5.17. These values were arrived at using Equation (5.23), and the data presented in Tables 5.10, 5.12, 5.15, and 5.16 (wall temperatures, bulk temperatures, ribbed wall net rate of heat transfer, smooth wall net rate of heat transfer).

TABLE 5.17

Regional Heat Transfer Coefficients on the Ribbed and Smooth Walls

x (cm)	h_x (W/m²K) Ribbed Wall	h_x (W/m²K) Smooth Wall
5.16	122.7	90.7
15.5	119.5	79.1
25.5	127.6	89.6
35.8	124.3	94.1
45.8	115.2	86.4
56.1	115.4	79.9
66.3	111.3	82.7
76.4	105.6	76.0
86.6	97.8	77.1
96.6	103.1	80.9

The heat transfer coefficients can be non-dimensionalized, and presented in terms of the Nusselt number. The non-dimensional Nusselt number is defined in Equation (5.29).

$$\text{Nu}_x = \frac{h_x D_h}{k} \tag{5.29}$$

The hydraulic diameter was previously defined in Equation (5.20); for the square channel considered in this example, the hydraulic diameter is equal to the length of side of the channel ($D_h = 5.08$ cm). Taking the thermal conductivity of air at the inlet bulk temperature, the regional Nusselt numbers can be calculated, and are presented in Table 5.18. Also, the streamwise dimension (x) has been non-dimensionalized with the channel hydraulic diameter.

Finally, when investigating roughness schemes which will enhance heat transfer, it is desirable to quantify the amount of heat transfer enhancement. While several options are available to express the level of heat transfer enhancement, the most widely used method is through the Nusselt number ratio. This ratio compares the measured Nusselt number to that expected in a simple smooth tube with fully developed, turbulent flow. The Dittus-Boelter/McAdams correlation is commonly used to approximate the Nusselt number for fully developed, turbulent flow in a smooth tube, and this correlation is shown in Equation (5.30).

$$\text{Nu}_o = 0.023 \text{Re}^{0.8} Pr^{0.4} \tag{5.30}$$

Using the Reynolds number presented in the paper (Re = 32,050), and the Prandtl number evaluated at the inlet bulk temperature, the Nusselt number expected in a smooth tube is 80.25. Using this value, and the data presented in Table 5.18, the

TABLE 5.18

Regional Nusselt Numbers on the Ribbed and Smooth Walls

x/D_h	Nu_x Ribbed Wall	Nu_x Smooth Wall
1.01	240.6	178.0
3.05	234.4	155.1
5.02	250.2	175.7
7.05	243.8	184.7
9.02	226.0	169.5
11.05	226.4	156.8
13.05	218.3	162.2
15.05	207.1	149.0
17.05	191.9	151.2
19.01	202.2	158.7

TABLE 5.19

Regional Nusselt Number Ratios on the Ribbed and Smooth Walls

x/D_h	$\left(\dfrac{Nu}{Nu_o}\right)_x$ Ribbed Wall	$\left(\dfrac{Nu}{Nu_o}\right)_x$ Smooth Wall
1.01	3.00	2.22
3.05	2.92	1.93
5.02	3.12	2.19
7.05	3.04	2.30
9.02	2.82	2.11
11.05	2.82	1.95
13.05	2.72	2.02
15.05	2.58	1.86
17.05	2.39	1.88
19.01	2.52	1.98

regional Nusselt number ratios can be calculated using Equation (5.31). These ratios expressing the level of heat transfer enhancement over a smooth tube are shown in Table 5.19.

$$\left(\frac{Nu}{Nu_o}\right)_x = \frac{Nu_x}{0.023Re^{0.8}Pr^{0.4}} \tag{5.31}$$

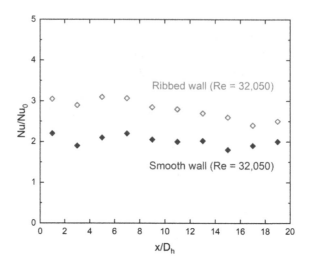

FIGURE 5.7   Regional Nusselt number ratio distributions in the square channel.

The final results shown in Table 5.19 can be plotted to show the heat transfer enhancement through the channel with two walls lined with V-shaped ribs. Figure 5.7 can be compared to Figure 7 of the referenced paper at Re = 32,050 [3].

5.3.2.6   Frictional Losses (Raw Data)

To fully evaluate heat transfer enhancement mechanisms, the cost of the heat transfer enhancement must also be considered. Internal heat transfer enhancement comes at the expense of an increased pressure drop through the duct. In the current experiment from Han et al., the pressure drop through the channel was measured directly with static pressure taps located at the inlet and the outlet of the test duct ($L = 101.6$ cm). As the pressure drop through these cooling passages is typically less than 1 psi (depending on rib geometry and channel length), a suitable pressure transducer/gauge should be selected. For the current experiment, a U-tube "microtector" manometer from Dwyer Instruments was used to measure the pressure drop in the ribbed channel. The manometer has a range of 2 inches of water, and is accurate to ± 0.00025 inches of water.

For the selected case of V-shaped ribs at the nominal Reynolds number of 30,000, Table 5.20 shows the difference between the inlet and outlet pressures within the channel.

TABLE 5.20

Pressure Drop Acquired at Re = 32,050

$P_{in} - P_{out}$ (in H$_2$O)
0.769

5.3.2.7 Frictional Losses Data Reduction

Just as the measured Nusselt number is normalized by the Nusselt number expected in a smooth channel, the pressure penalty is presented in terms of the friction factor ratio. Therefore, the measured pressure drop will be used to calculate the friction factor for the rib roughened channel. Equation (5.32) defines the friction factor in terms of the measured pressure drop.

$$f = \frac{\Delta P}{4\left(\dfrac{L}{D_h}\right)\left(\dfrac{1}{2}\rho V^2\right)} \tag{5.32}$$

Maintaining unit homogeneity, the friction factor can be calculated.

$$V = \mathrm{Re}\,\frac{\mu}{\rho D_h} = 9.18\,\frac{\mathrm{m}}{\mathrm{s}} \tag{5.33}$$

$$f = \frac{\left(0.769\ \mathrm{in\ H_2O}\right)\left(\dfrac{249.1\,\mathrm{Pa}}{1\ \mathrm{in\ H_2O}}\right)\left(\dfrac{1\,\mathrm{N}/\mathrm{m^2}}{1\,\mathrm{Pa}}\right)\left(\dfrac{\mathrm{kg\cdot m}}{\mathrm{N\cdot s^2}}\right)}{4\left(\dfrac{101.6\,\mathrm{cm}}{5.08\,\mathrm{cm}}\right)\left[\dfrac{1}{2}\left(1.23\,\dfrac{\mathrm{kg}}{\mathrm{m^3}}\right)\left(9.18\,\dfrac{\mathrm{m}}{\mathrm{s}}\right)^2\right]} = 0.0462 \tag{5.34}$$

The measured friction factor should be normalized by that predicted for a smooth duct (thus indicating the frictional losses incurred beyond those of a smooth duct). The friction factor for fully developed turbulent flow in a smooth, circular cylinder has been proposed by Blasius and is shown in Equation (5.35).

$$f_o = 0.046\,\mathrm{Re}^{-0.2} \qquad \left(10^4 < \mathrm{Re} < 10^6\right) \tag{5.35}$$

Combining the measured friction factor with that for flow through a smooth tube will yield the friction factor ratio.

$$\frac{f}{f_o} = \frac{f}{0.046\,\mathrm{Re}^{-0.2}} = \frac{0.0462}{0.046(32050)^{-0.2}} = 7.99 \tag{5.36}$$

Therefore, at the nominal Reynolds number of 30,000, the proposed V-shaped configuration incurs a pressure drop 7.99 times greater than that of a smooth tube.

Example: Square Channel with Transverse Ribs

The following data is available for an experimental test run in a square channel with transverse ribs along its length. Air at 1.2 bar, 20°C is used as the working fluid. All four walls of the channel are heated with four separate heaters. The channel is divided into 10 regions with equal lengths and with four thermocouples in each region for measuring the wall temperature.

Desired Reynolds number = 50,000
Width of each side of the square channel = 5 cm
Length of channel = 1 m
Length of each region = 0.1 m
Average wall temperature at region 9 = 40°C
Voltage applied to heater = 30 V
Resistance of heater = 20 Ω

Assuming heat loss to be 3% of the heat supplied to the heater, calculate the heat transfer coefficient and Nusselt number for the 9th region.

The hydraulic diameter of the channel is given by

$$D_h = \frac{4 \cdot A_{cs}}{P} = \frac{4(0.05)^2}{4 \cdot 0.05} = 0.05 \, \text{m}$$

Air properties at 1.2 bar and 20°C is:

$\rho = 1.427 \, \text{kg/m}^3$
$k = 0.02592 \, \text{W/mK}$
$\mu = 0.000012497 \, \text{kg/ms}$
$C_p = 1007 \, \text{J/kgK}$

The Reynolds number is given by

$$\text{Re} = \frac{\rho V D_h}{\mu} = \frac{4\dot{m}}{\mu P} = 50,000$$

The mass flow rate can be calculated from the Reynolds number

$$\dot{m} = \frac{\text{Re}\,\mu P}{4} = \frac{50,000 \cdot 0.000012497 \cdot (4 \cdot 0.05)}{4} = 0.03124 \, \text{kg/s}$$

Heat supplied from one heater to all regions

$$Q_{\text{input}} = Q_{\text{heater}} - Q_{\text{loss}} = \frac{V^2}{R} - 0.03 \cdot \frac{V^2}{R}$$

$$= \frac{30^2}{20} - 0.03 \cdot \frac{30^2}{20} = 43.65 \, \text{W}$$

Heat input per region can be expressed as

$$Q_{\text{input/region}} = \frac{Q_{\text{input}}}{N} = \frac{43.65}{10} = 4.365 \, \text{W}$$

For the ninth region, the bulk temperature at the inlet can be calculated using the energy balance in the first eight regions (recall, all four walls are heated, so the heat input around the entire perimeter must be included).

$$T_{b,9,inlet} = T_{b,inlet} + n \sum_{1}^{8} \frac{Q_{input/region}}{\dot{m}C_p} = 20 + 4 \cdot 8 \cdot \frac{4.365}{0.03124 \cdot 1007}$$

$$= 24.44\,°C$$

The bulk temperature at the exit of the ninth region is also given by

$$T_{b,9,exit} = T_{b,9,inlet} + n \frac{Q_{input/region}}{\dot{m}C_p} = 24.44 + 4 \cdot \frac{4.365}{0.03124 \cdot 1007}$$

$$= 24.995°C$$

The average bulk temperature of air is

$$T_{b,9} = \frac{T_{b,9,inlet} + T_{b,9,exit}}{2} = \frac{24.44 + 24.995}{2}$$

$$= 24.717\,°C$$

The heat transfer coefficient for the ninth region is given by

$$h_9 = \frac{Q_{input/region}}{A_9\left(T_{w,9} - T_{b,9}\right)} = \frac{4.365}{(0.05 \cdot 0.1)(40 - 24.717)}$$

$$= 57.124 \ W/m^2K$$

The Nusselt number is given by

$$Nu = \frac{h_9 D_h}{k} = \frac{57.124 \cdot 0.05}{0.02592} = 110.19$$

Example: Rectangular Channel with Hemispherical Dimples

A rectangular test section (3.0625″ × 1.0625″) is augmented with hemispherical dimples, and the heat transfer from the dimpled surface is to be evaluated. The dimples are machined into small, square copper plates (0.9375″ × 0.9375″ × 0.125″ thick). On the wide side of the channel, the copper plates are arranged in a 3 × 15 grid (3 copper plates across the width of the channel and 15 in the streamwise direction). A flexible, rubber heater is applied to the backside of the copper plates to heat the surface. The test section support structure is fabricated from Garolite, G-10. Each of the 45 copper plates is equipped with a thermocouple, and the thermocouples are spaced 1″ apart in both the spanwise and streamwise directions. To evaluate the heat transfer enhancement in this channel, the following data must be obtained.

Steady-state wall temperature, $T_{w,x}$
Inlet bulk air temperature, T_{inlet}
Outlet bulk air temperature, T_{outlet}
Power supplied to the heater, Q_{in} (voltage, current, resistance)
Room temperature, T_{room}

Pressure drop, static pressure, and temperature at the orifice meter, ΔP, P_1, T_1

High and low steady-state wall temperatures from the heat loss calibration, $T_{H,x}$ and $T_{L,x}$

Room temperatures during the heat loss calibration, $T_{H,room}$, $T_{L,room}$

Power supplied to the heater during high and low heat loss calibrations, Q_H and Q_L (voltage, current, resistance)

From these measured quantities, the thermal performance of the dimples can be evaluated. Reflecting on the previous examples and envisioning how this type of experiment should be fabricated, instrumented, and conducted, the following ideas should be considered:

1. How can the Reynolds number and air mass flow rate be calculated based on the supplied instrumentation?
2. How can the experimental uncertainty of the Reynolds number be calculated? How do you determine dominant factor driving the magnitude of the uncertainty?
3. What would be the baseline for comparison of this type of experiment? How can the Nusselt number in a smooth channel with turbulent flow be calculated?
4. What is the expected trend for the wall temperature distribution through the channel?
5. What is the expected trend for the bulk temperature through the channel? How does this trend compare to that for the wall temperature distribution?
6. How can the heat loss from the channel be estimated?
7. What method is available to calculate the outlet bulk temperature? Why is it useful to calculate the outlet temperature and directly compare it to the measured value?
8. Based on the measured quantities, how can the regionally averaged heat transfer coefficients be calculated? Based on the streamwise distribution, what is required for the flow to be considered fully developed?
9. How can the heat transfer enhancement afforded by the dimples be evaluated (relative to a smooth channel)?
10. How can the experimental uncertainty of the measured Nusselt number be evaluated?

Being able to understand the type of results provided by an experiment is critical to designing and executing a successful experimental program. While no specific numbers are provided with this example, it gives the researcher an opportunity to reflect on the expected outcomes for the described test section and instrumentation.

Additional References: The aforementioned copper-heater-thermocouple measurements have been applied to determine the regional-average heat transfer coefficient distributions for many enhanced channel flow applications. Several selected papers have been published using this methodology for cylindrical leading-edge impingement cooling by Chupp et al. [4], impingement cooling on a flat-wall rectangular channel by Kercher and Tabakoff [5] and Florschuetz et al. [6], rectangular cooling channel with pin fins by Metzger et al. [7], and circular tube with repeated ribs by Webb et al. [8]. This regional-average heat transfer measurements have also been applied to determine the regional-average heat transfer coefficient distributions for non-rotating cooling channels with various turbulence promoters by Han et al. [9,10] as well as for rotating two-passage cooling channels by Wagner et al. [11] and Han et al. [12]. Interested readers can read these papers for their detailed flow loop design, test geometry, experimental procedure, and data analysis and results presentation.

5.4 STEADY-STATE, LOCAL HEAT TRANSFER MEASUREMENT TECHNIQUE

This technique can be used to obtain the local temperature distribution on the surface as opposed to the regionally averaged heat transfer distributions. Typically, the test surface is covered with a thin metal foil which itself acts as a heater. By applying a voltage across the foil, heat is generated by the foil proportional to its electrical resistance. The temperature of the foil is measured using thermocouples attached to the underside of the foil heater. The measured temperature corresponds to the wall temperature of the test surface. As the foil heater is thin, lateral heat conduction through its cross section is small due to its small cross-sectional area even though large temperature gradients might exist. Thus, it can be assumed that the temperature distribution measured is local and hence, local heat transfer coefficients can be calculated. Foil heaters are also the most popular choice for heating a surface when used in combination with optical techniques such as liquid crystal thermography, temperature sensitive paint (TSP), and infrared thermography under steady-state conditions.

5.4.1 THIN FOIL HEATERS WITH THERMOCOUPLES

Along with the copper plate method discussed earlier, this technique is one of the simplest and more inexpensive methods to measure heat transfer coefficients. This technique has been available in literature for several decades with studies performed to measure both heat transfer coefficient and film cooling effectiveness distributions. Typical uncertainty with this technique is approximately 5%–10%. This method has been used extensively by researchers as it can give consistent results. A major disadvantage of this method is the requirement of skilled labor in setting up the foil heaters and thermocouples. This technique demands that the experimentalist is dexterous when attaching the thermocouples to the foil without damaging the foil. Slight damage to the foil can cause hot spots resulting in biased data. Also, in order to have a comprehensive local temperature distribution on the surface, many thermocouples must be used, necessitating a large data acquisition system.

As opposed to the copper plate method where the copper plate separates the fluid and the heater, with this method, the foil acts as a heater and is in direct contact with the fluid. A stainless steel or Inconel thin foil can be used as the heater. The thickness of the foil used is typically about 0.05 mm (0.002″). A thin foil is desired as it reduces lateral heat conduction in the foil. However, thinner foils are difficult to handle and tend to get damaged easily. The foil is cut to the desired shape and mounted on a substrate. The substrate is typically a material with low conductivity such as polystyrene or Plexiglas.

The foil heater can be constructed by the experimentalist by cutting a foil sheet to the desired shape corresponding to the test surface. By applying a voltage across the ends of the foil, heat is generated internally within the foil. The foil heater should be designed such that the resistance across the heater is large. A low heater resistance will require large currents to generate heat. The test surface area can be covered with a single foil or several foil heater strips depending on the electrical resistance of each strip. Each foil strip is relatively small in width compared to its length. A smaller width also ensures a more uniform heat flux from the heater. The resistance of a foil heater can be calculated using

$$R = \frac{\rho L}{A} \tag{5.37}$$

where ρ is the electrical resistivity and depends on the foil material. As resistance is proportional to the length of the heater and inversely proportional to the cross-section area, by maintaining a small width and a large span, the resistance of each foil strip can be manipulated so that the current requirement is manageable and is lower than the maximum current that can be drawn from the available power supply line. The heat generated by the foil can be given as

$$Q_{\text{heater}} = I^2 R = I^2 \frac{\rho L}{A} \tag{5.38}$$

where I is the current flowing through the foil. Thus, the foil heater can be cut into an "S" pattern to adjust the length of the heater and hence the resistance to appropriate limits. The "S" pattern can be avoided by using individual foil strips connected to each other in series through copper bus bars. The heater is then connected to a power supply through a variable auto-transformer to obtain the desired heat output. When all the foil strips are connected appropriately, the entire bank of strips provides a uniform heat flux over the surface. Care should be taken when handling the foil due to its delicate nature. Wrinkling the foil can cause a non-uniform heat distribution resulting in hot spots.

As with the copper plate method, thermocouples are used to record the temperature. Thermocouples can be soldered on the underside of the foil strip along its length. A thermally conductive epoxy can also be used to attach the thermocouples. In some cases, using thermally conductive epoxy is advantageous as it acts as an electrical insulator preventing large currents in the foil heater from flowing through the thermocouples and into the data acquisition system which may damage it. As these heaters can provide local data too, the chosen spatial resolution of the thermocouples should provide a sufficiently detailed temperature distribution on the surface.

The foil strips can be attached to the substrate using double-sided tape or some sprayable adhesive. Manual application of glue may cause bumps in the foil surface due to inconsistency of the glue thickness when it is being applied. Using a double-sided tape or sprayable adhesive ensures that the test surface is smooth. The non-conducting substrate helps to reduce extraneous heat losses. Additional insulation such as fiberglass, wood, etc. can be used to further reduce the heat loss. The foil heaters should be designed such that the requisite heat input for the highest Reynolds number to be investigated can be achieved.

After starting the experiment, the test section will attain steady-state after some time has elapsed. The heater power input should be set so that the wall temperature is about 20°C higher than the fluid temperature. The power input and the temperatures must be recorded along with the other flow measurements. Heat loss information can also be obtained for the test section under no flow conditions. By knowing these parameters, the heat transfer coefficients can be calculated.

5.4.2 EXPERIMENT EXAMPLE

5.4.2.1 Background and Introduction to Experiment

Han et al. in 1993 [13] investigated local heat transfer distributions on a turbine blade airfoil using a linear cascade. The linear cascade consisted of five blades, and each blade was a two-dimensional representation of a traditional high-pressure turbine blade. With the selected blade profile, the air turns 107° while accelerating 2.5 times from the inlet of the cascade to the exit. Heat transfer coefficient (Nusselt number) distributions were obtained with varying flow conditions (Re = 100,000–300,000) and wake conditions. A spoked wheel was used to simulate the presence of the upstream stator, and the wheel was rotated to accurately model the passing of the rotating blade relative to the stationary vane.

As indicated by Han et al. [13], the center blade was instrumented with thin foil heater strips and thermocouples to provide a local Nusselt number distribution on the blade from the leading edge to the trailing edge. A total of 26 heaters strips were used on the blade surface (as shown in Figure 5.8), and the heater strips were connected in series. Each heater strip was 25.4 cm long, 2 cm wide, and 0.00378 cm thick. Thermocouples were attached to the backside of the foils to provide a local wall temperature distribution. The instrumented blade was manufactured from wood to minimize heat loss by conduction through the blade substrate.

While many cases were considered in the experimental study, only one case will be considered in the present example. The baseline case of "no wake" will be considered at the inlet velocity of 21 m/s (corresponding to Reynolds number of 300,000). The characteristic length in both the Reynolds number and Nusselt number calculation is the chord length of the airfoil, which is given as $C = 22.68$ cm.

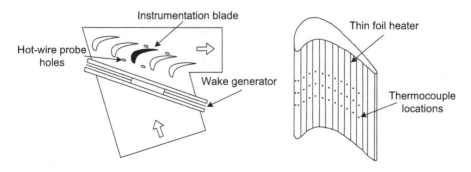

FIGURE 5.8 Overview of experimental facilities for measurement of local heat transfer coefficients using thin foil heaters and thermocouples.

5.4.2.2 Freestream Velocity Calculation

With wind tunnel studies of this nature, the freestream velocity is often monitored using a Pitot tube. As Pitot tubes measure both the static and dynamic pressures, the velocity inside the wind tunnel can easily be calculated from Bernoulli's equation. Therefore, it is necessary to determine the flow velocity (and differential pressure) that is needed to provide the desired Reynolds number. In the current example, the desired Reynolds number is Re = 300,000. For the given experimental setup, the Reynolds number can be calculated in Equation (5.39).

$$\text{Re} = \frac{\rho V C}{\mu} \tag{5.39}$$

Equation (5.39) can be rearranged to solve for the required freestream velocity at a desired Reynolds number.

$$V = \text{Re}\,\frac{\mu}{\rho C} \tag{5.40}$$

For the given geometry and flow through the open loop wind tunnel, the required velocity can be calculated as shown in Equation (5.41).

$$V = 300{,}000 \times \left[\frac{1.85 \times 10^{-5}\,\frac{\text{Ns}}{\text{m}^2}}{\left(1.184\,\frac{\text{kg}}{\text{m}^3}\right)(22.28\,\text{cm})}\right]\left(\frac{100\,\text{cm}}{1\,\text{m}}\right)\left(\frac{\text{kgm}}{\text{Ns}^2}\right) = 21.0\,\frac{\text{m}}{\text{s}} \tag{5.41}$$

As a Pitot tube is a logical choice to monitor the freestream velocity, it is necessary to use Bernoulli's equation to determine the differential pressure that corresponds to the desired velocity of 21.0 m/s. A simplified form of Bernoulli's equation is provided in Equation (5.42).

$$\Delta P = \frac{1}{2}\rho V^2 = \frac{1}{2}\left(1.184\,\frac{\text{kg}}{\text{m}^3}\right)\left(21.0\,\frac{\text{m}}{2}\right)^2\left(\frac{\text{Ns}^2}{\text{kgm}}\right)\left(\frac{\text{Pa}}{\text{N/m}^2}\right) = 261\,\text{Pa} \tag{5.42}$$

With pressures of this low magnitude it is often desired to convert the pressure into a column of water or mercury. For the present experiment, the Pitot tube is attached to a digital manometer with the pressure is given in inches of water. Therefore, the differential pressure reading can be converted to inches of water as shown in Equations (5.43) and (5.44).

$$\Delta P = \rho_{H_2O}gh \tag{5.43}$$

$$h = \frac{\Delta P}{\rho_{H_2O}g} = \frac{261\,\text{Pa}}{\left(998.2\,\frac{\text{kg}}{\text{m}^3}\right)\left(9.81\,\frac{\text{m}}{\text{s}^2}\right)}\left(\frac{\text{N/m}^2}{\text{Pa}}\right)\left(\frac{\text{kgm}}{\text{Ns}^2}\right)\left(\frac{1\,\text{in}}{0.0254\,\text{m}}\right) = 1.05\,\text{in H}_2\text{O}$$

$$\tag{5.44}$$

5.4.2.3 Acquired Experimental Data (Heat Loss and Convective Heat Transfer Data)

As with other steady-state heat transfer experiments, the bulk of the test section should be composed of a low conductivity material to minimize the conduction of heat through the material (and maximize the heat convected away from the surface). However, even with the use of low conductivity materials, stray heat losses exist, and they must be taken into account.

To calculate the heat lost during the actual experiments, two sets of calibration data were obtained. In a "no flow" scenario, the instrumented blade was heated to a "low" temperature (lower wall temperature than expected during actual experiment) and to a "high" temperature. The wall temperature distributions were obtained at both of these conditions (low and high temperature) under a steady-state condition. In addition, the room temperature and the power supplied to the heaters were recorded. Under this no flow condition, the power supplied to the heaters is equivalent to the heat lost during the actual test (which is dominated by forced convection). Table 5.21 presents the wall temperature distributions obtained during the heat loss calibration.

The origin ($x = 0$) corresponds to the leading edge of the airfoil. The x-direction is defined in the streamline direction, with the positive direction being along the suction surface, and the negative direction along the pressure surface.

TABLE 5.21

Temperature Distributions Acquired through the Heat Loss Calibration

x (cm)	Low–T_w (°C)	High–T_w (°C)	x (cm)	Low–T_w (°C)	High–T_w (°C)
	Suction Surface			Pressure Surface	
0.71	29.5	49.5	−0.425	34.9	55.3
2.34	35.2	55.2	−2.06	34.1	54.1
4.25	34.7	54.7	−3.97	35.4	55.4
6.02	34.9	54.9	−5.81	34.7	54.7
8.08	35.3	55.3	−7.73	34.8	54.8
9.92	35.0	55.0	−9.57	35.0	55.0
11.8	34.8	54.8	−11.5	35.3	55.3
13.7	34.7	54.7	−13.3	34.9	54.9
15.6	35.4	55.4	−15.2	34.7	54.7
17.4	34.1	54.1	−17.1	35.2	55.2
19.3	34.9	54.9	−19.1	35.1	55.1
21.2	33.9	53.9			
23.0	34.6	54.6	Room	25.0	25.0
24.9	34.8	54.8			
26.9	35.3	55.3			

TABLE 5.22

Voltage and Resistance Measurements from Heat Loss Calibration

	Voltage (V)	Resistance (Ω)
Low	3.21	7.53
High	12.43	7.53

The input power to the heater (comprised of multiple foil strips connected in series) was measured for each test, and the voltages and resistances are shown in Table 5.22.

The temperature distribution and the heater input power measured during the selected case are shown in Tables 5.23 and 5.24, respectively.

TABLE 5.23

Wall Temperature Distribution Acquired at Re = 300,000

x (cm)	Low–T_w (°C)	x (cm)	Low–T_w (°C)
Suction Surface		Pressure Surface	
0.71	31.4	−0.425	34.4
2.34	36.2	−2.06	44.3
4.25	38.4	−3.97	49.8
6.02	40.1	−5.81	49.5
8.08	40.9	−7.73	49.1
9.92	42.4	−9.57	48.6
11.8	44.0	−11.5	46.9
13.7	45.5	−13.3	46.0
15.6	47.8	−15.2	45.7
17.4	49.5	−17.1	44.9
19.3	51.8	−19.1	44.7
21.2	49.5		
23.0	42.1	Room	25.2
24.9	37.3		
26.9	35.9		

TABLE 5.24

Voltage and Resistance Measurements at Re = 300,000

Voltage (V)	Resistance (Ω)
35.0	7.53

5.4.2.4 Airfoil Heat Transfer Coefficient Distributions

The objective of this experimental study was to obtain local heat transfer coefficient (Nusselt number) distributions on a blade surface under a variety of freestream turbulence conditions. Therefore, it is necessary to utilize the measured quantities (wall temperatures and heater power) to obtain heat transfer coefficient distributions. Rearranging the traditional convective heat transfer equation, equation (5.45) can be used to calculate the local heat transfer coefficient.

$$h = \frac{q''}{T_w - T_\infty} = \frac{q''_{gen} - q''_{loss}}{T_w - T_{aw}} \tag{5.45}$$

As indicated in the work by Han et al. [13], the adiabatic wall temperature (T_{aw}) is taken as room temperature and is approximately 25°C.

Similar to the regional heat transfer enhancement example (from Han et al. [13]), linear interpolation can be used to determine the "net" rate of heat transfer for the heater. Using the "high" and "low" heat loss data along with the actual test data (both temperature and power input), the local distribution for the rate of heat transfer can be determined. The heat transfer rates are summarized in Table 5.25.

Given the physical size of each heater strip (25.4 cm long × 2 cm wide) and the total number of heater strips (26), the total heater area is known ($A = 0.132$ m^2).

TABLE 5.25

Heat Generation, Heat Loss, and Net Power Distribution

x (cm)	$\dot{Q}_{gen}$ (W)	$\dot{Q}_{loss}$ (W)	$\dot{Q}_{net}$ (W)	x (cm)	$\dot{Q}_{gen}$ (W)	$\dot{Q}_{loss}$ (W)	$\dot{Q}_{net}$ (W)
	Suction Surface				Pressure Surface		
0.71	162.7	3.00	159.7	−0.425	162.7	0.715	162.0
2.34	162.7	2.15	160.5	−2.06	162.7	10.9	151.7
4.25	162.7	4.72	158.0	−3.97	162.7	15.0	147.7
6.02	162.7	6.13	156.6	−5.81	162.7	15.3	147.4
8.08	162.7	6.49	156.2	−7.73	162.7	15.9	147.8
9.92	162.7	8.27	154.4	−9.57	162.7	14.2	148.5
11.8	162.7	9.56	152.7	−11.5	162.7	12.3	150.4
13.7	162.7	11.5	151.2	−13.3	162.7	11.8	150.9
15.6	162.7	13.1	149.6	−15.2	162.7	11.7	151.0
17.4	162.7	15.6	146.8	−17.1	162.7	10.4	152.3
19.3	162.7	17.4	145.3	−19.1	162.7	10.4	152.3
21.2	162.7	16.1	146.6				
23.0	162.7	8.31	154.4				
24.9	162.7	3.59	159.1				
26.9	162.7	1.71	161.0				

TABLE 5.26

Local Heat Transfer Coefficient Distribution

x (cm)	h(W/m²K)	x (cm)	h(W/m²K)
Suction Surface		Pressure Surface	
0.71	195.0	−0.425	133.2
2.34	110.3	−2.06	60.2
4.25	90.6	−3.97	45.4
6.02	79.7	−5.81	46.0
8.08	75.6	−7.73	46.8
9.92	67.9	−9.57	48.0
11.8	61.6	−11.5	52.4
13.7	56.5	−13.3	55.0
15.6	50.1	−15.2	55.7
17.4	45.8	−17.1	58.6
19.3	41.4	−19.1	59.0
21.2	45.7		
23.0	69.3		
24.9	99.4		
26.9	114.4		

With the known surface area, the net heat flux $\left(\dot{Q}_{net}/A\right)$ can be determined, and from Equation 5.45, the local heat transfer coefficient can be calculated. Table 5.26 shows the heat transfer coefficient distribution from the selected case.

Finally, both the length scale (x) and the heat transfer coefficients (h) can be non-dimensionalized. The x-coordinate is normalized by the chord length of the blade (22.68 cm), and the Nusselt number is calculated from the heat transfer coefficient (as shown in Equation (5.46)).

$$\mathrm{Nu} = \frac{hC}{k} \tag{5.46}$$

Table 5.27 shows the Nusselt number distributions, and the values are plotted in Figure 5.9. The data calculated in the example and shown in Figure 5.9 represent the "No Wake" case shown in Figure 9 from the work of Han et al. [13].

With the thin foil heater technique, the local Nusselt number distribution clearly captures the separation of the boundary layer near the trailing edge of the suction surface (x/C = 0.9) as indicated in the sudden increase in the local Nusselt numbers. The ability to capture both transition and separation points, is a marked advantage of the thin foil technique.

TABLE 5.27
Local Nusselt Number Distribution

x/C	Nu	x/C	Nu
Suction Surface		Pressure Surface	
0.0313	1682	−0.0188	1149
0.103	951.4	−0.0906	518.8
0.188	781.4	−0.175	391.4
0.266	687.2	−0.256	396.4
0.356	651.5	−0.341	403.3
0.438	586.0	−0.422	414.0
0.519	531.2	−0.506	451.8
0.606	487.1	−0.588	474.2
0.688	431.9	−0.672	480.4
0.769	394.8	−0.753	505.5
0.853	356.6	−0.841	508.6
0.934	394.3		
1.01	598.0		
1.10	856.9		
1.19	986.3		

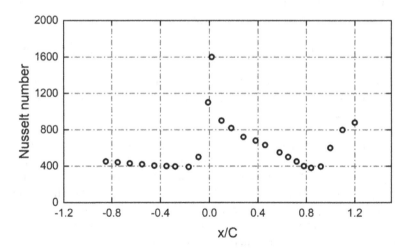

FIGURE 5.9 Local Nusselt number distributions for "No Wake" case at Re = 300,000.

5.4.2.5 Additional References

The aforementioned foil-heater-thermocouple measurements have been applied to determine the local heat transfer coefficient distributions for many film cooling and internal enhanced channel flow applications. Several selected papers have been published using this methodology for flat-plate film cooling by Eriksen and Goldstein [14] and for turbine blade film cooling by Mehendale et al. [15]. This local heat transfer measurement technique has also been applied to determine the local heat transfer coefficient distributions for rectangular cooling channels with various turbulence promoters by Han and Park [16]. Interested readers can read these papers for their detailed flow loop design, test geometry, experimental procedure, and data analysis and results presentation.

PROBLEMS

1. A square channel is designed to measure the convective heat transfer coefficient (HTC) of an internal cooling passage using the steady-state, regional average heat transfer measurement technique. The inner dimensions of the channel are 0.1 m $(W) \times 0.1$ m $(H) \times 1$ m (L). The channel is divided in 10 streamwise regions. The length of each region is 0.1 m. All four walls are heated by four separate heaters. Given the following parameters:

   Re = 30,000
   Heat loss = 5%
   AC voltage applied to the heater = 40 ± 0.1 V
   Current in the heater = 2 ± 0.1 A
   Air properties at inlet: $T_{in} = 25°C$, $\rho = 1.168$ kg/m³, $k = 0.026$ W/(m×K), $\mu = 18.48 \times 10^{-6}$ Pa×s, $Pr = 0.71$, and $C_p = 1004$ J/(kg×K)
   Calculate the hydraulic diameter D_h, the mass flow rate, $\dot{m}$, HTC, Nu, Nu_0, and (Nu/Nu_0) for the seventh region if the surface temperature T_w of that region is $55°C \pm 0.2°C$.
   Calculate the friction factor (f) and the friction factor ratio (f/f_0) if the pressure drops from inlet to exit of the channel is 0.02 in H_2O.
   Estimate the uncertainty of the HTC. Assume the uncertainty of the bulk mean temperature at that region is $\pm 0.5°C$ and there is no dimensional uncertainty.

2. A rectangular channel is designed to measure the convective heat transfer coefficient (HTC) of an internal cooling passage using the steady-state, regional average heat transfer measurement technique. The inner dimensions of the channel are 0.1 m $(W) \times 0.05$ m $(H) \times 1$ m (L). The channel is divided in 10 streamwise regions. The length of each region is 0.1 m. The two wide walls (0.1 m) are heated by two heaters. (The other two walls have no copper plates or heaters). Given the following parameters:

   $\dot{m} = 0.05$ kg/s
   Heat loss = 5%
   AC voltage applied to the heater = 45 ± 0.1 V
   Resistance of heater = 25 ± 0.1 Ω
   Air properties at inlet $T_{in} = 20°C$, $\rho = 1.189$ kg/m³, $k = 0.025$ W/(m×K), $\mu = 18.23 \times 10^{-6}$ Pa×s, $Pr = 0.71$, and $C_p = 1007$ J/(kg×K)

Calculate the hydraulic diameter D_h, the Reynolds number, HTC, Nu, Nu_0, and (Nu/Nu_0) for the eighth region if the surface temperature T_w of that region is $40°C \pm 0.2°C$.

Calculate the friction factor (f) and the friction factor ratio (f/f_0) if the pressure drops from inlet to exit of the channel is 0.06 in H_2O.

Estimate the uncertainty in the HTC. Assume the uncertainty of the bulk mean temperature at that region is $\pm0.5°C$ and there is no dimensional uncertainty.

3. A test section is designed to measure the convective heat transfer coefficient (HTC) of a jet impingement cooling passage using the steady-state, regional average heat transfer measurement technique. The test channel has a dimension of 0.12 m $(W) \times 0.03$ m $(H) \times 1$ m (L). The channel is divided in 10 streamwise regions. The length of each region is 0.1 m. There are four impingement jet holes for each region; therefore, there are 40 jets for the entire test section. The diameter of each jet hole is 0.5 cm. The plenum air temperature is $T_{in} = 20°C \pm 0.2°C$. Given the following parameters:

Heat loss = 3%
Total mass flow rate $\dot{m} = 0.1$ kg/s
AC voltage applied to the heater $= 120 \pm 0.1$ V
Resistance of heater $= 20 \pm 0.1 \Omega$
Air properties $k = 0.025$ W/(m×K), $\mu = 18.23 \times 10^{-6}$ Pa×s

Calculate the average jet Reynolds number Re_{jet}, HTC and Nu for the fifth region, if the surface temperature T_w of that region is $50°C \pm 0.2°C$. (Hint: You will use the jet temperature instead of the bulk mean temperature to calculate the HTC, and use the jet diameter to calculate the Nusselt number.)

Estimate the uncertainty in the HTC. Assume there is no dimensional uncertainty.

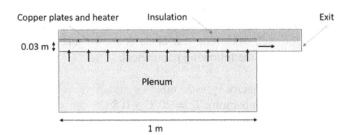

4. The steady-state, regional average heat transfer measurement technique is used for measuring the convective heat transfer coefficient (HTC) of an externally cooled surface. The surface has a dimension of 0.15 m (W) by 0.5 m (L). The test surface is divided in 10 streamwise regions. The length

of each region is 0.05 m. One heater is used as the heat source. The test
plate is shown schematically in the below picture. Given the following
parameters:

Approaching flow velocity $V_\infty = 25$ m/s
Approaching flow temperature $T_\infty = 20°C \pm 0.5°C$
AC voltage applied to the heater $= 60 \pm 0.1$ V
AC current applied to the heater $= 3 \pm 0.1$ A
Heat loss $= 8\%$
Air properties $k = 0.025$ W/(m×K), $\mu = 18.23 \times 10^{-6}$ Pa×s, $\rho = 1.189$ kg/m³
Calculate the HTC of the last (10th) copper plate if the surface temperature
   T_w of that region is $50°C \pm 0.2°C$.
Find the corresponding Re_x, Nu_x, and $Nu_{x, corr.}$ from the correlation.
Estimate the uncertainty in the HTC. Assume there is no dimensional
   uncertainty.

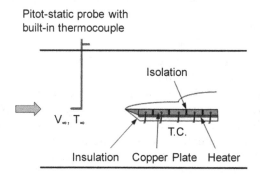

Pitot-static probe with
built-in thermocouple

Isolation

V_∞, T_∞

T.C.

Insulation   Copper Plate   Heater

5. The steady-state, regional average heat transfer measurement technique is
   used for measuring the convective heat transfer coefficient (HTC) of an
   externally cooled turbine blade. The blade has a chord length of $C = 0.2$ m
   and a blade height of 0.2 m. Both the pressure side and suction side surfaces
   are divided in seven stream-wise regions. Two heaters are used as the heat
   source (one for each side). The test blade is shown schematically in the
   below picture. Given the following parameters:

   Approaching flow velocity $V_\infty = 25$ m/s
   Approaching flow temperature $T_\infty = 20°C \pm 0.5°C$
   AC voltage applied to the pressure side heater $= 50 \pm 0.2$ V
   AC current applied to the pressure side heater $= 2 \pm 0.1$ A
   Heat loss $= 2\%$
   Length of the pressure side surface $= 0.224$ m
   Air properties: $k = 0.025$ W/(m×K), $\mu = 18.23 \times 10^{-6}$ Pa×s, $\rho = 1.189$ kg/m³
   Calculate the HTC of the fourth copper plate if the surface temperature T_w
      of that region is $60°C \pm 0.5°C$.

Find the corresponding Re_c and Nu_c.
Estimate the uncertainty in the HTC. Assume there is no dimensional
uncertainty.

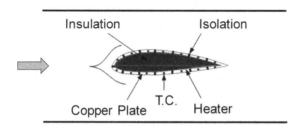

6. The steady-state, regional average heat transfer measurement technique is
used for measuring the convective heat transfer coefficient (HTC) of an
externally cooled cylinder. The cylinder has an outer diameter of $D = 0.1$ m
and the length of the cylinder is 0.3 m. It is divided in 16 circumferen-
tial regions and three longitudinal regions (length of each region is 0.1 m).
There are total 48 copper plates on the cylinder. One heater is used to heat
the entire cylinder. The approaching flow impinges on the leading edge of
the cylinder, and then travels around the circumference, as shown schematically in the below figure. Given the following parameters:

Approaching flow velocity $V_\infty = 50$ m/s
Approaching flow temperature $T_\infty = 20°C \pm 0.2°C$
AC voltage applied to the heater $= 110 \pm 0.2$ V
AC current applied to the heater $= 3 \pm 0.1$ A
Assume heat loss is purely radiative and the surface is a black body. You
   may use the relation $q = \sigma \times (T_w^4 - T_\infty^4) \times A$ (Watts), where A is the surface
   area.
Air properties $k = 0.025$ W/(m×K), $\mu = 18.23 \times 10^{-6}$ Pa×s, $\rho = 1.189$ kg/m³,
   $Pr = 0.7$.
Calculate the HTC of the third copper plate (from the stagnation point) if
   the surface temperature T_w of that region is $60°C \pm 0.2°C$.
Find the corresponding Re_D and Nu_D.
Estimate the HTC of that region by using the following equation:

$$h_\vartheta = 1.14 \left(\frac{k}{D}\right)(Re_D)^{0.5} Pr^{0.4}\left[1-\left(\frac{\vartheta}{90}\right)^3\right], \quad 0° < \vartheta < 80°$$

How does the predicted value from the correlation compare to you calculated
   value from the experiment?
Estimate the uncertainty in the HTC. Assume there is no dimensional
   uncertainty.

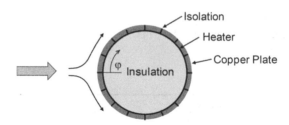

7. The heat transfer coefficient (HTC) inside a circular pipe is to be determined using the steady-state, local heat transfer measurement method. The pipe has an inner diameter of 5 cm. The length of the region of interest (ROI) is 15 cm. Sixteen strips of equally spaced, foil heaters are placed longitudinally inside the pipe, and all heaters are connected in series. The width of each heater is 9 mm. There are four thermocouples along each heater strip. A Variac is used to provide power to the heater. Given the following parameters:

Re = 30000
Assume the heat loss = 5%
AC voltage applied to the heater = 22 V
Resistance of heater = 15 Ω
Air properties at inlet $T_{b, in}$ = 20°C, ρ = 1.189 kg/m³, k = 0.025 W/(m×K), μ = 18.23 × 10⁻⁶ Pa×s, Pr = 0.71, and C_p = 1007 J/(kg×K)
Wall temperatures measured by T.C.s along a heater strip at different locations:

x (m)	T_w (°C)
0.03	51
0.06	53
0.09	55
0.12	57

Calculate the mass flow rate ṁ in kg/s, the velocity V in m/s, the bulk mean temperature at the exit of the ROI, $T_{b, out}$, using the energy balance method.
Plot the HTC, Nu, and (Nu/Nu₀) versus the dimensionless parameter x/D. Nu₀ is calculated from the correlation for fully developed, turbulent flow in a smooth tube.

8. The heat transfer coefficient (HTC) inside a circular pipe is to be determined using the steady-state, local heat transfer measurement method. The pipe has an inner diameter of 5 cm. The length of the region of interest (ROI)

is 15 cm. Sixteen strips of equally spaced foil heaters are placed longitudinally inside the pipe, and all heaters are connected in series. The width of each heater is 9 mm. There are four thermocouples along each heater strip. A Variac is used to provide power to the heater. Given the following parameters:

$Re = 60,000$

Assume the heat loss flux, $Q(T_w) = 3.0 \times (T_w - T_{room})$ W/m^2

Room temperature $T_{room} = 18°C$

Bulk mean temperature at exit of ROI $T_{b, out} = 21.5°C$

AC voltage applied to the heater = 40 V

Resistance of heater = 25 Ω

Air properties at inlet $T_{b, in} = 20°C$, $\rho = 1.189$ kg/m^3, $k = 0.025$ W/(m×K), $\mu = 18.23 \times 10^{-6}$ Pa×s, and $Pr = 0.71$

Wall temperatures measured by T.C.s along a heater strip at different locations:

x (m)	T_w (°C)
0.03	51
0.06	53
0.09	55
0.12	57

Calculate the mass flow rate $\dot{m}$, in kg/s, the velocity V, in m/s, and the bulk mean temperatures using the interpolation method.

Plot HTC, Nu, and (Nu/Nu$_0$) versus the dimensionless parameter x/D. Nu$_0$ is calculated from the correlation for fully developed, turbulent flow in a smooth tube.

9. The heat transfer coefficients (HTC) on a jet cooled surface are to be determined using the steady-state, local heat transfer measurement method. The arrangement of the test is shown schematically in the below figure. There are 10 strips of foil heaters connected in series. Each strip heater is 9 mm wide and 10 cm long. A Variac is used to provide power to the heater. Given the following parameters:

$Re_{jet} = 20,000$

Jet nozzle diameter $D = 0.01$ m

Assume the heat loss is negligible

AC voltage applied to the heater = 40 V

Resistance of heater = 25 Ω

Air properties of jet $T_{jet} = 22°C$, $k = 0.025$ W/(m×K), $\mu = 18.23 \times 10^{-6}$ Pa×s

Wall temperatures measured by T.C.s at different radial locations r:

r (m)	T_w (°C)
0.01	45
0.05	53
0.095	56

Calculate the mass flow rate ṁ, in kg/s.

Plot HTC and Nu versus the dimensionless parameter r/D.

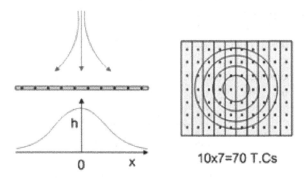

10x7=70 T.Cs

10. The heat transfer coefficients (HTC) on the pressure side surface of a turbine blade are to be determined using the steady-state, local heat transfer measurement method. The test blade is shown schematically in the below figure. There are 10 strips of foil heaters connected in series. Each strip heater is 15 mm wide and 20 cm long. A Variac is used to provide power to the heater. The chord length of the blade is $C = 0.2$ m. There is a Pitot-static probe 0.5C upstream of the blade leading edge to measure the approaching flow total and static pressures. Given the following parameters:

Measured total pressure = 55 in H_2O (gauge)

Measured static pressure = 37 in H_2O (gauge)

Assume the heat loss is negligible

AC voltage applied to the heater = 50 V

Current of heater = 2 A

Air properties at inlet: $T_{in} = 40°C$, $k = 0.027$ W/(m×K), $\mu = 19.2 \times 10^{-6}$ Pa×s, $Pr = 0.71$

Wall temperatures measured by T.C.s at mid-span of the blade at $x = 0.1$ m is $T_w = 45°C$,

(where x is the downstream distance from the leading edge)

Calculate the flow velocity, in m/s, the inlet Reynolds number based on the chord length, the HTC at the above-mentioned point, and Nusselt number based on the chord length.

Estimate the uncertainty in the HTC. Assume there is no dimensional uncertainty.

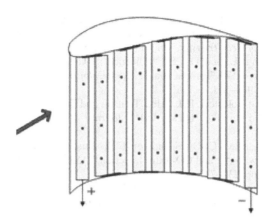

REFERENCES

1. Eckert, E.R.G., "Engineering Relations for Heat Transfer and Friction In High-Velocity Laminar and Turbulent Boundary Layer Flow Over Surfaces with Constant Pressure and Temperature," *Transactions of the ASME*, Vol. 78, No. 6, 1956, pp. 1273–1283.
2. Lau, S.C., Han, J.C. and Kim, Y.S., "Turbulent Heat Transfer and Friction in Pin Fin Channels with Lateral Flow Ejection," *ASME Journal of Heat Transfer*, Vol. 111, No. 1, 1989, pp. 51–58.
3. Han, J.C., Zhang, Y.M. and Lee, C.P., "Augmented Heat Transfer in Square Channels with Parallel, Crossed, and V-Shaped Angled Ribs," *ASME Journal of Heat Transfer*, Vol. 113, 1991, pp. 590–596.
4. Chupp, R.E., Helms, H.E., McFadden, P.W. and Brow, T.R., "Evaluation of Internal Heat Transfer Coefficients for Impingement Cooled Turbine Airfoils," *AIAA Journal of Aircraft*, Vol. 6, 1969, pp. 203–208.
5. Kercher, D.M. and Tabakoff, W., "Heat Transfer by a Square Array of Round Air Jets Impinging Perpendicular to a Flat Surface including the Effect of Spent Air," *ASME Journal of Engineering for Power*, Vol. 92, 1970, pp. 73–82.
6. Florschuetz, L.W., Berry, R.A. and Metzger, D.E., "Periodic Streamwise Variations of Heat Transfer Coefficients for Inline and Staggered Arrays of Circular Jets with Crossflow of Spent Air," *ASME Journal of Heat Transfer*, Vol. 102, 1980, pp. 132–137.
7. Metzger, D.E., Berry, R.A. and Bronson, J.P., "Developing Heat Transfer in Rectangular Ducts with Staggered Arrays of Short Pin Fins," *ASME Journal of Heat Transfer*, Vol. 104, 1982, pp. 700–706.
8. Webb, R.L., Eckert, E.R.G. and Goldstein, R.J., "Heat Transfer and Friction in Tubes with Repeated-Rib Roughness," *International Journal of Heat and Mass Transfer*, Vol. 14, No. 4, 1971, pp. 601–617.
9. Han, J.C., Glicksman, L.R. and Rohsenow, W.M., "An Investigation of Heat Transfer and Friction for Rib-Roughened Surfaces," *International Journal of Heat and Mass Transfer*, Vol. 21, 1978, pp. 1143–1156.
10. Han, J.C. and Zhang, Y.M., "High Performance Heat Transfer Ducts with Parallel Broken and V-Shaped Broken Ribs," *International Journal of Heat and Mass Transfer*, Vol. 35, No. 2, 1992, pp. 513–523.
11. Wagner, J.H., Johnson, B.V. and Kopper, F.C., "Heat Transfer in Rotating Serpentine Passages with Smooth Walls," *ASME Journal of Turbomachinery*, Vol. 113, 1991, pp. 321–330.

12. Han, J.C., Zhang, Y.M. and Kalkuenhler, K., "Uneven Wall Temperature Effect on Local Heat Transfer in a Rotating Two-Pass Square Channel with Smooth Walls," *ASME Journal of Heat Transfer*, Vol. 114, No. 4, pp. 850–858.

13. Han, J.C., Zhang, L. and Ou, S., "Influence of Unsteady Wake on Heat Transfer Coefficient from a Gas Turbine Blade," *ASME Journal of Heat Transfer*, Vol. 115, No. 4, 1993, pp. 904–911.

14. Eriksen, V.L. and Goldstein, R.J., "Heat Transfer and Film Cooling Following Injection through Inclined Circular Holes," *ASME Journal of Heat Transfer*, Vol. 96, 1974, pp. 239–245.

15. Mehendale A.B., Han, J.C., Ou, S. and Lee, C.P., "Unsteady Wake over a Linear Turbine Blade Cascade with Air and CO_2 Film Injection," *ASME Journal of Turbomachinery*, Vol. 116, 1994, pp. 730–737.

16. Han, J.C. and Park, J.S., "Developing Heat Transfer in Rectangular Channels with Rib Turbulators," *International Journal of Heat and Mass Transfer*, Vol. 31, No. 1, 1988, pp. 183–195.

6 Time Dependent Heat Transfer Measurement Techniques

6.1 INTRODUCTION AND MEASUREMENT THEORY

As the name suggests, time dependent techniques involve measuring how a temperature or heat flux changes over a given period of time; this contrasts steady-state experiments where all measured quantities do not change with time. Although the method changes from steady to transient, the goal remains the same with the methods presented in Chapter 4—obtain the convective heat transfer coefficient on a surface. Researchers have used transient techniques for several decades to obtain heat transfer coefficient distributions. These methods have evolved over the years in response to new methods to acquire surface temperature distributions.

With steady-state experiments, only "boundary" conditions were required to yield a heat transfer coefficient. In time dependent tests, the "initial" condition must be well-defined to start the transient process. In addition to recording how temperatures change over time, researchers must also have an accurate way of measuring time, as this becomes another measured quantity. A primary advantage of transient heat transfer tests compared to steady-state tests is that heat loss does not in over the short time of the transient test. Other precautions must be taken, depending on the type of the transient experiment, and these will be discussed in the following sections.

As with steady-state techniques, many options are available for choosing the temperature measurement system. Lumped capacitance techniques, thin film heat flux gauges, optical techniques such as liquid crystal thermography, infrared thermography, etc., are some of the methods successfully used for measuring the surface temperature under transient conditions. Depending on the duration of the experiment, the researcher should choose a temperature measurement system capable of recording at sufficiently high frequencies in order to capture changes in temperature.

Multiple transient techniques will be discussed in detail, but there are general characteristics that are common to most transient tests. For any heat transfer experiment, a temperature differential must exist; for convective heat transfer this involves a temperature difference between the surface and the fluid. With the steady-state methods described in Chapter 5, the surface was heated, and a relatively cold fluid was in contact in with the surface. With the transient tests, it is more common to heat (or cool) the fluid. As you will see in the following sections, this method is often preferred due to the boundary conditions that must be imposed during the experiments. If the surface begins at room temperature, the test begins ($t = 0$) when the hot fluid begins to pass over the surface. During the course of the experiment, the surface temperature will increase, and this transient response of the surface is recorded. If a surface heater is used, it is common to heat the entire volume of the test section, and at $t = 0$, the heater is turned off, a cool fluid passes over the surface, and the cooling of the surface is recorded.

With each transient test, the initial temperature of the test surface must be known ($t = 0$). Therefore, the test begins with the measurement of this initial temperature. Next, the fluid is allowed to pass over the surface, and the temperature changes of the fluid and/or surface are recorded. While this may seem straightforward, there can often be unexpected challenges. For instance, when starting a wind tunnel, it does not instantly come on at the desired velocity; it requires several seconds to achieve the desired speed. How should this time be taken into account? Another potential problem arises with the measurement of time. Is the data acquisition hardware being used as a clock during the experiment? Is there an internal delay with the hardware? How does this effect the measurement of time during the tests?

The combination of experimental test time and working temperatures must be carefully considered. A setup is designed, so the test will begin with a heated surface (hotter than room temperature). When the test starts, the surface begins to cool, and given enough time, the surface will eventually cool to the ambient temperature. If a large initial temperature difference is imposed (T_i-T_∞), the surface temperature will decrease rapidly. However, when a relatively low temperature difference is imposed, the temperature change with respect to time will be much smaller (both cases are moving toward the surface eventually cooling to the ambient temperature). With the large initial temperature, both the temperature and time must be measured very accurately (large temperature changes are occurring quickly). If the initial temperature difference is smaller, the experiment will be run over a longer period of time in order to achieve sufficient drop. However, long duration transient tests run the risk of violating fundamental assumptions for the various transient methods. For this reason, most transient tests are run over 1–2 minutes. As the test length drops below thirty seconds, a small error in time measurement can lead to large uncertainties in the heat transfer coefficients. For longer tests (greater than 3–4 minutes), lateral conduction can occur, which violates the one-dimensional heat transfer assumption used for several transient methods. The time frame of 1–2 minutes allows for accurate time measurement while minimizing lateral conduction through the test surface.

6.2   TRANSIENT LUMPED CAPACITANCE MEASUREMENT TECHNIQUE

6.2.1   Fundamental Principle

This technique relies on the principle of lumped capacitance. Consider an object at a relatively high temperature, which is immersed in a fluid at a lower temperature. Sudden quenching will occur with the solid object's temperature decreasing dramatically until it attains the fluid temperature. This quenching phenomenon occurs due to convection at the solid-fluid interface. If the temperature of the solid is spatially uniform at each time step during the transient process, the lumped capacitance method can be applied. The validity of the spatially uniform temperature assumption can be checked by calculating the Biot number (Bi) of the solid object given by Equation (6.1).

$$Bi = \frac{hV}{A_{surface}k_{solid}} = \frac{hL_c}{k_{solid}} \tag{6.1}$$

The ratio of volume, V to the surface area, $A_{surface}$ is defined as the characteristic length, L_c to simplify calculations. Thus, the Biot number is similar in form to the Nusselt number; however, the characteristic length is defined differently and the thermal conductivity of the fluid is used in the Nusselt number definition. If the Biot number is less than 0.1, the assumption can be considered as valid. The value of $Bi = 0.1$ indicates that only 10% of the total temperature difference that occurs between the inner most region of the solid object and the fluid actually takes place within the solid. With a relatively small temperature change through the volume of the solid, it can be assumed to be at a uniform temperature. Application of the energy balance at the solid-fluid interface results in the following differential equation (assuming the initial temperature of the object is greater than the quenching fluid).

$$-hA_{surface}\left(T_{w,t} - T_\infty\right) = \rho V C_p \frac{dT}{dt} \tag{6.2}$$

The terms ρ, V and C indicate the density, volume and specific heat capacity of the test object. The solution of the above differential equation can be obtained in many heat transfer textbooks [1] and is given in Equation (6.3).

$$\frac{T_\infty - T_{w,t}}{T_\infty - T_{w,initial}} = \exp[-at] \quad \text{where} \quad a = \frac{hA_{surface}}{\rho V C_p} \tag{6.3}$$

Thus, if the Biot number of a quenched object is less than 0.1, the principle of lumped capacitance can be applied to the test section to calculate the heat transfer coefficients. Equation (6.3) can be easily rearranged and written explicitly for the heat transfer coefficient, h.

$$h = -\frac{\rho V C_P}{A_{surface}t} \ln\left(\frac{T_\infty - T_{w,t}}{T_\infty - T_{w,initial}}\right) \tag{6.4}$$

In Equation (6.4), it appears the calculated heat transfer coefficient is a function of time (implying the heat transfer coefficient is time dependent). However, this is not true. The measured heat transfer coefficient is independent of time. To demonstrate how the temperatures change with time, but the heat transfer coefficient remains constant, Equation (6.3) can be written in terms of a time constant, τ.

$$\frac{T_\infty - T_{w,t}}{T_\infty - T_{w,initial}} = \theta = \exp\left[-\frac{t}{\tau}\right] \quad \text{where} \quad \tau = \frac{\rho V C_p}{hA_{surface}} \tag{6.5}$$

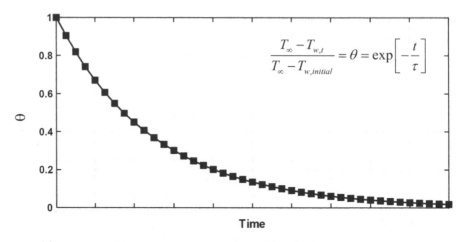

FIGURE 6.1 General temperature response during a lumped capacitance experiment.

Figure 6.1 shows how the non-dimensional temperature, θ, changes as a function of time. Only one value of τ can satisfy the curve. If both time and the time constant are changing, the data will not collapse to the single curve shown in the figure. Therefore, with the material properties and physical size of the object constant, the heat transfer coefficient, h, must also be constant. Each variable in τ must be independent of time, so the τ is the "time constant." Figure 6.2 demonstrates how the temperature response varies for different time constants. If the solid object does not change, the only variable is the heat transfer coefficient. As the heat transfer coefficient decreases, the time constant will increase. However, each individual data set shown in Figure 6.2 is defined by a single heat transfer coefficient.

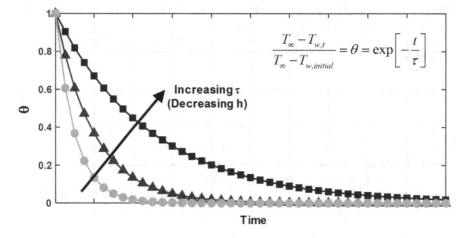

FIGURE 6.2 Temperature response during a lumped capacitance experiment with varying time constants.

The convective heat transfer coefficient must be constant with time, to produce the given temperature response. These figures also demonstrate as the heat transfer coefficient increases, the temperature of the solid more quickly approaches the fluid temperature.

6.2.2 IMPLEMENTATION OF LUMPED CAPACITANCE USING THERMOCOUPLES

Figure 6.3 shows a generic transient lumped capacitance model. This technique is one of the oldest techniques used by researchers to calculate the heat transfer coefficients using a transient method. This technique involves the use of a metal piece with high thermal conductivity and low volume to surface area ratio (thin wall thickness). By using the appropriate material and geometry, the Biot number of the test surface can be manipulated to a magnitude of less than 0.1, such that the lumped capacitance principle may be applied. Typically, copper, which has a very high thermal conductivity (~400 W/mK), is chosen as the material. The test surface is heated to a certain initial temperature with an embedded heater beneath the copper test surface. The temperature of the test surface can be monitored using several thermocouples. The thermocouples placed throughout the volume of the surface should read approximately the same temperature, as the Biot number is less than 0.1. Once the predetermined initial temperature is reached, the experiment begins by allowing the fluid at ambient temperature to come to contact with the test surface. Once the fluid contacts the surface, convective heat transfer will take place and the temperature of the surface will fall exponentially with time as given by Equation (6.3). The transient response of the temperature of the test surface is recorded using a suitable data acquisition system. The fluid temperature must also be recorded. The heat transfer coefficient at any given time can be calculated by Equation (6.4). Equations (6.3) and (6.4) assume that the mainstream fluid temperature does not change with time. If significant change is observed in this temperature, these equations should be corrected using the Duhamel's superposition principle. More details of Duhamel's principle of superposition are given in Chapter 7. The measured initial and transient metal temperatures can be inserted in Equation (6.4) along with the mainstream fluid temperature to calculate the heat transfer coefficient at any given time. Since fluid flow is steady, the measured/calculated heat transfer coefficient at a given surface location is a constant value, which is independent of time.

6.2.3 EXPERIMENT EXAMPLE

Example: Lumped Capacitance with a Heated Surface

We are interested in the regional averaged heat transfer coefficients at different locations on the pressure surface of the turbine airfoil shown in Figure 6.3. The airfoil is instrumented such that its pressure surface is divided into different regions through thin copper plates with embedded T-type thermocouples and insulation strips. Each thin copper plate is 10" long, 0.75" wide, and 0.0625" thick. A custom, flexible rubber heater is attached behind the copper plates to provide a uniform heat flux across the surface. The heater is able to raise the copper temperature to 50°C uniformly under steady-state. When this temperature is reached, the heater is

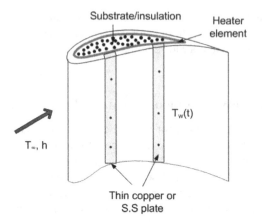

FIGURE 6.3　Typical test model for lumped capacitance measurements.

turned off, and mainstream air at 25°C begins flowing over the airfoil. During this time, the copper temperature (as a function of time) at two locations is recorded:

	Time	
Location	10 s	30 s
30% chord	48.1°C	44.8°C
60% chord	47.8°C	43.9°C

The properties of copper can be taken as:
Density = 8960 kg/m³
Conductivity = 385 kg/m³
Specific heat = 385 J/kg-K

Based on the measured temperatures, calculate the heat transfer coefficient using the lumped capacitance technique at the two locations and verify this method is applicable. Also, estimate the uncertainty for the calculated heat transfer coefficient.
Characteristic length (L) of the thin copper plate:

$$L = \frac{V}{A_s} = \frac{10 \times 0.75 \times 0.0625}{10 \times 0.75} = 0.0625"$$

For lumped capacitance technique,

$$\frac{T_\infty - T_w(t)}{T_\infty - T_i} = \exp\left[-\frac{hA_s}{\rho V C_p} t\right] = \exp\left[-\frac{h}{\rho C_p L} t\right]$$

$$h = -\frac{\rho C_p L}{t} \ln\left(\frac{T_\infty - T_w(t)}{T_\infty - T_i}\right)$$

For 30% chord location and t = 10 s,

$$h = -\frac{\left(8960\,\frac{kg}{m^3}\right)\left(385\,\frac{J}{kgK}\right)(0.0625\,\text{in})\left(\frac{0.0254\,m}{1\,\text{in}}\right)}{10\,s}\ln\left(\frac{(25-48.1)\,°C}{(25-50)\,°C}\right)$$

$$h = 43.3\,\frac{W}{m^2K}$$

For 30% chord location and t = 30 s,

$$h = -\frac{\left(8960\,\frac{kg}{m^3}\right)\left(385\,\frac{J}{kgK}\right)(0.0625\,\text{in})\left(\frac{0.0254\,m}{1\,\text{in}}\right)}{30\,s}\ln\left(\frac{(25-44.8)\,°C}{(25-50)\,°C}\right)$$

$$h = 42.6\,\frac{W}{m^2K}$$

For 60% chord location and t = 10 s,

$$h = -\frac{\left(8960\,\frac{kg}{m^3}\right)\left(385\,\frac{J}{kgK}\right)(0.0625\,\text{in})\left(\frac{0.0254\,m}{1\,\text{in}}\right)}{10\,s}\ln\left(\frac{(25-47.8)\,°C}{(25-50)\,°C}\right)$$

$$h = 50.4\,\frac{W}{m^2K}$$

For 60% chord location and t = 30 s,

$$h = -\frac{\left(8960\,\frac{kg}{m^3}\right)\left(385\,\frac{J}{kgK}\right)(0.0625\,\text{in})\left(\frac{0.0254\,m}{1\,\text{in}}\right)}{30\,s}\ln\left(\frac{(25-43.9)\,°C}{(25-50)\,°C}\right)$$

$$h = 51.1\,\frac{W}{m^2K}$$

We can see the heat transfer coefficient for a given location is roughly a constant. Therefore, based on the calculated heat transfer coefficient, the Biot number can be calculated.

$$Bi = \frac{hL}{k} = \frac{h\cdot(0.0625\,\text{in})\left(\frac{0.0254\,m}{1\,\text{in}}\right)}{\left(385\,\frac{J}{kgK}\right)} \sim 4\times10^{-6}\cdot(h) < 0.1$$

Therefore, use of the lumped capacitance method is validated.

For the uncertainty of the heat transfer coefficient:

$$h = -\frac{\rho V C_p}{A_s t} \ln\left(\frac{T_\infty - T_w(t)}{T_\infty - T_i}\right)$$

$$\frac{w_h}{h} = \left\{ \begin{array}{l} \left(\frac{w_\rho}{\rho}\right)^2 + \left(\frac{w_V}{V}\right)^2 + \left(\frac{w_{C_p}}{C_p}\right)^2 + \left(\frac{w_{A_s}}{A_s}\right)^2 + \left(\frac{w_t}{t}\right)^2 + \left[\frac{\left(T_w(t) - T_i\right)}{\left(T_\infty - T_w(t)\right)\left(T_\infty - T_i\right)} \cdot \frac{w_{T_\infty}}{\ln\left(\frac{T_\infty - T_w(t)}{T_\infty - T_i}\right)}\right]^2 \\[2em] + \left[\frac{1}{\left(T_\infty - T_w(t)\right)} \cdot \frac{w_{T_w}}{\ln\left(\frac{T_\infty - T_w(t)}{T_\infty - T_i}\right)}\right]^2 + \left[\frac{1}{\left(T_\infty - T_i\right)} \cdot \frac{w_{T_i}}{\ln\left(\frac{T_\infty - T_w(t)}{T_\infty - T_i}\right)}\right]^2 \end{array} \right\}$$

For 30% chord location and $t = 10$ s, as an example

$$\left\{ 0.001^2 + 0.001^2 + 0.001^2 + 0.001^2 + 0.02^2 + \left[\frac{(48.1 - 50)}{(25 - 48.1)(25 - 50)} \times \frac{0.2}{\ln\left(\frac{25 - 48.1}{25 - 50}\right)}\right]^2 \right.$$
$$\left. + \left[\frac{1}{(25 - 48.1)} \times \frac{0.2}{\ln\left(\frac{25 - 48.1}{25 - 50}\right)}\right]^2 + \left[\frac{1}{(25 - 50)} \times \frac{0.2}{\ln\left(\frac{25 - 48.1}{25 - 50}\right)}\right]^2 \right\}$$

$$\frac{w_h}{h} = 15.2\%$$

The uncertainty of other cases can be calculated using the same method with different temperatures and times.

Example: Lumped Capacitance with a Heated Fluid

Martin et al. [2] used the lumped capacitance method in a jet impingement experiment. This series of tests investigated the effect of jet temperature on the impingement. Therefore, the experiments were conducted with heated jets impinging on a relatively cool surface. Modeling jet impingement on the internal surface of a gas turbine blade, the target surface was a concave surface.

Compressed air flows through an ASME orifice flow meter, into a pipe heater, and through a three-way valve. After the three-way valve, the air can be diverted either away from test section or into a plenum supplying air to the impingement array. The researchers were able to remove the instrumented target surface from the flow loop, so the hot air could be used to pre-heat the plenum and surrounding plumbing. Therefore, when the test began at $t = 0$, the jet temperature was instantaneously at the desired, elevated temperature.

The stagnation region, target surface was fabricated from a strip of 6061 Aluminum. For a given test, the following experimental data were obtained.

Reynolds Number (based on the Jet Diameter):	$Re_{jet} = 10,000$
Jet Diameter:	$d = 0.484$ inches
Jet Array Dimensions:	$s/d = 4, l/d = 4, D/d = 3.6$
Mass of Aluminum Strip:	$m = 23.0$ grams
Convective Surface Area of Aluminum Strip:	$A_s = 0.00119$ m^2
Initial Temperature of Aluminum:	$T_{w,i} = 22.4°C$
Time of Measurement:	$t = 1.6$ s
Wall Temperature at Time, t:	$T_{w,t} = 22.8°C$
Jet Temperature:	$T_{jet,t} = 55.0°C$
Density of 6061 Aluminum:	$\rho_{Al} = 2700$ kg/m^3
Specific Heat of 6061 Aluminum:	$C_{p,Al} = 896$ J/kgK
Conductivity of 6061 Aluminum:	$k_{Al} = 200$ W/mK
Conductivity of air at T_{jet}:	$k_{air} = 0.02865$ W/mK

Based on the given and measured values, is the lumped capacitance method valid for evaluating the heat transfer coefficients on the impingement target surface? Calculate the heat transfer coefficient and Nusselt number on the target surface.

A well-known correlation, used by gas turbine designers, is available as a design tool before completing impingement experiment. According to Chupp et al. [3], the Nusselt number on the stagnation region of a curved target surface can be approximated by:

$$Nu_{stag} = 0.44Re_{jet}^{0.7}\left(\frac{d}{s}\right)^{0.8}\exp\left[-0.85\left(\frac{l}{d}\right)\left(\frac{d}{s}\right)\left(\frac{d}{D}\right)^{0.4}\right]$$

With the given geometry and flow parameters, the expected Nusselt number can be approximated:

$$Nu_{stag} = 0.44(10000)^{0.7}\left(\frac{1}{4}\right)^{0.8}\exp\left[-0.85(4)\left(\frac{1}{4}\right)\left(\frac{1}{3.6}\right)^{0.4}\right] = 55.0$$

In the Nusselt number equation, the characteristic length is the jet diameter. Therefore, the predicted heat transfer coefficient can be calculated:

$$Nu_{stag} = 55.0 = \frac{hD_{jet}}{k_{air}}$$

$$h = 55.0 \cdot \frac{k_{air}}{D_{jet}} = 55.0 \cdot \frac{0.02865 \frac{W}{mK}}{(0.484 \text{ in})\left(\frac{0.0254 \text{ m}}{1 \text{ in}}\right)} = 128.2 \frac{W}{m^2 K}$$

Using this expected heat transfer coefficient, the Biot number for the aluminum, stagnation region, target surface can be calculated.

$$Bi = \frac{hL_c}{k_{Al}} \quad \text{where} \quad L_c = \frac{V}{A_s}$$

Based on the given information, the volume is not directly known. However, the researchers used a mass balance to determine the mass of the strip. Using the aluminum density and mass, the volume, and characteristic length, can be determined. Note the characteristic length used in the Biot number definition is different from that used in the Nusselt number.

$$L_c = \frac{V}{A_s} = \frac{m}{\rho_{Al}} \cdot \frac{1}{A_s}$$

$$L_c = \frac{(23.0 \text{ g})\left(\frac{1 \text{ kg}}{1000 \text{ g}}\right)}{2700 \frac{\text{kg}}{\text{m}^3}} \cdot \frac{1}{0.00119 \text{ m}^2} = 0.0072 \text{ m}$$

$$Bi = \frac{hL_c}{k_{Al}} = \frac{\left(128.2 \frac{W}{m^2 K}\right)(0.0072 \text{ m})}{200 \frac{W}{mK}} = 0.005$$

Before beginning the experiment, the estimated Biot number is significantly less than 0.1. Therefore, use of the lumped capacitance method is justified for the aluminum strips. Based on the measured temperatures, the actual heat transfer from the experiment can now be calculated. Notice the driving fluid temperature is the jet temperature (measured at the outlet of the plenum) for this internal flow experiment.

$$h = -\frac{\rho_{Al} L_c C_{P,Al}}{t} \ln\left(\frac{T_{jet} - T_{w,t}}{T_{jet} - T_{w,initial}}\right) = -\frac{\left(2700 \frac{\text{kg}}{\text{m}^3}\right)(0.0072 \text{ m})\left(896 \frac{J}{\text{kgK}}\right)}{1.6 \text{s}}$$

$$\ln\left(\frac{(55.0 - 22.8)°C}{(55.0 - 22.4)°C}\right)$$

$$h = 134.4 \frac{W}{m^2 K}$$

The heat transfer coefficient can be non-dimensionalized in the form of the Nusselt number. Again, care should be taken with the selection of the characteristic length and thermal conductivity.

$$Nu_d = \frac{hd}{k_{air}} = \frac{\left(134.4\,\frac{W}{m^2 K}\right)(0.484\,\text{in})\left(\dfrac{0.0254\,\text{m}}{1\,\text{in}}\right)}{0.02865\,\dfrac{W}{mK}}$$

$$Nu_d = 57.7$$

Both the measured heat transfer coefficient and Nusselt number are in good agreement with those predicted by correlation from Chupp et al. Therefore, not only is the use of the lumped capacitance method validated in principle, the method is also validated against published results. The researchers moved forward confidently to other flow conditions or impingement geometries. The validation is also needed due to the relatively short time period for this transient test (1.6 s). The entire experimental rig was preheated, and an electric solenoid value was used to direct the flow to the test section. Therefore, the "ramp-up" time to begin the transient event was minimal. In addition, the time history of the thermocouples was recorded with a data acquisition system capable of measuring at 100 Hz. The small time of 1.6 s could be resolved into a sufficiently large number of measurement points, so the time was well-defined.

6.3 TRANSIENT HEAT FLUX MEASUREMENTS FOR LOCAL HEAT TRANSFER COEFFICIENTS

6.3.1 FUNDAMENTAL PRINCIPLE

Transient response of heat flux on the order of 10^{-8} s can be measured using small heat flux gauges. Due to their fast response, and limitations of the analytical models for calculating the heat flux, this technique is most commonly used in short duration experiments. This technique gives reliable data for heat transfer coefficients with low uncertainty. The transient response, due to unsteady flow, can be investigated using this technique because they quickly respond to changes in the local flow conditions. Another major advantage is that if properly mounted, the thin film heat flux gauges are robust and can be used at high operating temperatures up to 800°C. The primary disadvantage of these transducers is that they provide only point measurements. Thus, the data collected using these gauges can only be used as a design check (local hot or cold spots may not be captured). Another disadvantage of this measurement technique is the necessity of extensive instrumentation. A dexterous experimentalist may be required, as the gauges themselves are tiny and delicate. If the gauges is damaged when it is applied to a surface, the experiments may yield inaccurate measurements. To increase the spatial resolution of the measurements, additional gauges can be attached to a surface, but this will increase the complexity of the instrumentation.

Thin film heat flux gauges have been used extensively over the last five decades to measure high frequency, fluctuating heat flux on a test surface under harsh operating

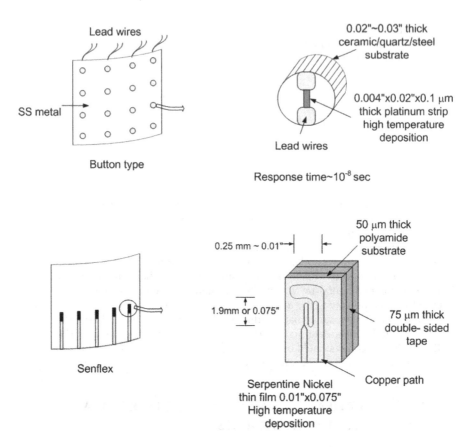

FIGURE 6.4   Sketches of typical button type and continuous strip-type gauges.

conditions. The thin film sensor is manufactured from a highly conductive material such as platinum or copper alloys. The gauges are available as button type and continuous strip-type gauges as shown in Figure 6.4 [4,5]. The thickness of the thin film sensor is typically expressed in fractions of a micrometer, which gives them a frequency response up to 50 kHz. The thin film sensors are manufactured by depositing or painting the material on a substrate in very fine layers and curing them in an autoclave at high temperatures.

Thin-film heat flux gauges have been used by researchers for measuring heat-transfer rates on actual gas turbine blades under engine-simulated conditions [4–7]. Thin-film heat flux gauges provide accurate information for actual turbine rigs, which are difficult to instrument with standard thermocouples. The thin-film heat flux gauges can easily be attached to a heat transfer surface (flush mounted with the surface) and connected to a data acquisition unit. Figure 6.5 shows the typical surface temperature and heat flux outputs from of a thin-film heat flux gauge. Figure 6.6 shows the transient heat flux responses for several gauges on a turbine blade [7]. The movement of the unsteady wake flow across the blade passage is captured by the heat flux responses at different axial locations on the blade surface.

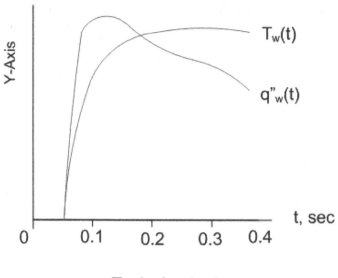

FIGURE 6.5 Typical surface temperature and heat flux outputs out of a thin-film heat flux gauge.

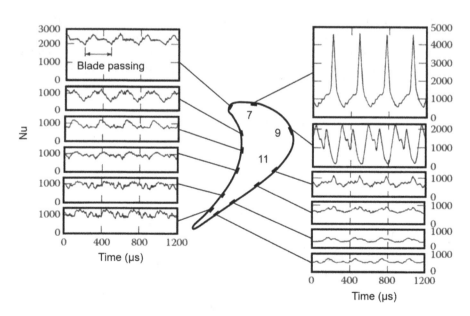

FIGURE 6.6 Heat transfer measured by several heat flux gauges at different locations on a blade.

The following section provides a fundamental background on how to determine time dependent heat transfer coefficients, in unsteady flows, using thin film heat flux gauges. Since the fluid flow is unsteady, the measured/calculated heat flux or heat transfer coefficient at a given surface location is not a constant value. Responding to the unsteady flow conditions, the measured heat transfer coefficients are also time dependent.

6.3.2 THIN FILM HEAT FLUX GAUGES

The determination of the heat transfer coefficient with a heat flux gauges requires the knowledge of three components, as shown in Equation (6.6): the fluid temperature, the wall temperature, and the heat flux at the wall. The fluid temperature can be measured using a thermocouple inserted into the mainstream.

$$h = \frac{q''(t)}{T_w(t) - T_\infty} \tag{6.6}$$

The wall temperature is measured using the same principle as a variable resistance thermometer (RTD). Temperature changes in the material of the thin film sensor will cause its resistance to change according to linear approximation given as $R(t) = a \cdot T(t) + b$, where a and b are constants. The film sensor is connected as one leg of a Wheatstone bridge circuit, operating in the constant current mode (Figure 6.7). A Wheatstone bridge is a divided electrical circuit used to measure the dynamic change in resistance of the thin film sensor. Using V (voltage) = I (current) $\times R$ (resistance), the fluctuating voltage output $V(t)$ from this circuit is proportional to the fluctuations in wall temperature with time $T(t)$ due to the linear approximation between the film resistance, $R(t)$, and its temperature, $T(t)$. A calibration curve between the resistance and the temperature will give the wall temperature. An amplifier can be

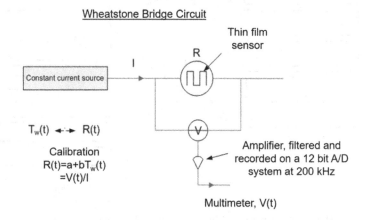

FIGURE 6.7   Sketch of typical thin film sensor operation principle.

connected to boost the output voltage from the bridge circuit. A suitable gain may also be applied. Thin film heat flux gauges can also be combined with a separate resistance-temperature sensing element to measure the surface temperature mounted on the same substrate.

In order to convert the measured, fluctuating, voltage signals to heat flux, appropriate analytical models are used to determine the wall heat flux history. As the wall heat flux changes, the wall temperature will also change rapidly. The substrate for the thin film is designed such that the temperature on the backside of the substrate does not change even if the wall temperature changes, for the duration of the experiment. If the substrate thickness is large and the thermal conductivity is low, the wall heat flux cannot conduct to the backside of the gauges, and the semi-infinite solid model can be assumed for the substrate. The one-dimensional, unsteady, conduction equation is given in Equation (6.7). A solution to the differential equation can be obtained using the wall temperature history and the uniform initial temperature on the backside of the substrate as shown in Figure 6.8 [1]. The semi-infinite model assumption is critical for the accurate determination of the heat flux.

$$\frac{\partial^2 T(x,t)}{\partial x^2} = \frac{1}{\alpha(x)} \frac{\partial T(x,t)}{\partial t} \tag{6.7}$$

where, α is the thermal diffusivity of the substrate. The initial and boundary conditions necessary to solve this equation are

Initial condition: $\quad\quad T(x,0) = T_i \quad\quad (x \geq 0)$
Boundary conditions:

$$T(\infty,t) = T_i \quad\quad (t \geq 0) \tag{6.8}$$

$$T(0,t) = T_w \quad\quad (t \geq 0)$$

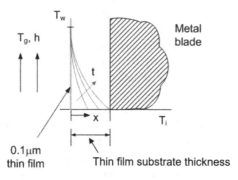

FIGURE 6.8 Sketch of typical thin film sensor one-dimensional unsteady conduction model.

The analytical solution to the above equation can be given as

$$\frac{T-T_w}{T_i-T_w} = erf\left(\frac{x}{2\sqrt{\alpha t}}\right)$$ (6.9)

The resulting surface heat flux obtained from Fourier's law of conduction can be expressed as

$$q_w''(t) = \frac{k_{substrate}\left[T_w(t)-T_i\right]}{\sqrt{\pi \alpha t}}$$ (6.10)

Equation (6.10) gives the wall heat flux on the test surface from the thin film heat flux gauge. There is a criterion to determine the minimum substrate thickness needed to satisfy the semi-infinite model assumption. The criterion states that with a thickness of $x = 3.648\sqrt{\alpha t}$, the wall heat flux ratio, $q(x)/q_w(x=0)$ is less than 1% [5].

To increase the sensitivity of the thin film heat flux gauge, several hundred thermocouple pairs linked in series are mounted on the same substrate. This array of thermocouples is collectively referred to as a thermopile. The total voltage across the thermopile is measured and is the sum of the individual voltages across each thermocouple pair.

The gauges are embedded into the test surface into grooves such that the film is flush with the test surface. Gauges not mounted flush with the surface will alter the near wall flow, and give misleading data. The presence of a transducer should never change the flow field. Before mounting the gauge onto the test surface, they first need to be attached to an insulating substrate, such as quartz or ceramic. Thus, the semi-infinite model can be applied to calculate the heat flux from the thin film sensor. Using insulating materials such as plastics should be avoided, as they cannot withstand high operating temperatures. Quartz and ceramic are selected, as their thermal property variation at high temperatures is both small and well-documented. These materials can be also machined or cast in complicated shapes. The low thermal diffusivity of the substrate will ensure that the change in its temperature is small for the duration of the experiment. This is particularly important when the gauge is mounted on a metal wall. If the test surface is made from an insulating material, the substrate will then become redundant.

In some experiments, such as heat flux investigation on a rotating turbine blade, the application of the gauge may require that the thin film sensor and substrate are exposed to large stresses. In these situations, the thin film sensor may be directly mounted onto a metal surface for increased strength or the substrate size may be reduced to minimize the stress concentration on the test surface (the machining of a groove to flush mount the surface can create a stress concentration) [4]. However, the semi-infinite assumption may not be true in these cases. Appropriate corrections for heat conduction into the metal surface may be used to offset this concern. Several researchers have introduced numerical models to correct the data from the sensor for heat conduction. Duration of the experiment can also be decreased to minimize heat conduction errors.

6.3.3   EXPERIMENT EXAMPLE

Example: Thin Film Heat Flux Gauge

Unsteady heat transfer behavior can be found on a rotating turbine blade because it is constantly passing in and out of the wake created by the upstream ring of non-rotating vanes. Researchers desire to capture this unsteady heat transfer behavior due to the unsteady wake flow. A turbine blade is instrumented with one-layer, thin film heat flux gauges. The gauges are flush mounted on the turbine blade surface. A short duration wind tunnel with hot air at 600°C (measured by a K-type thermocouple) is used to simulate the unsteady wake effect.

An initial temperature of 25°C is measured using a thermocouple mounted to the surface near the heat flux gauge. The test begins when the hot gas is directed through the rig and around the blade. During the short duration test, the following temperature history is measured with the thin film heat flux gauge.

		Time	
Location	0.3 s	0.35 s	0.4 s
1	530.1°C	515.3°C	531.8°C

The following properties are available for the gauge substrate:

	Material
Property	polyimide
Conductivity (W/mK)	0.12
Density (kg/m³)	1420
Specific heat (J/kgK)	1090

Calculate the corresponding time dependent heat transfer coefficient at different times and include the uncertainty analysis.

Using the thin film gauge, from the one-dimensional, unsteady conduction model, we have the analytical solution of the heat flux,

$$q_w''(t) = \frac{k_{substrate}\left(T_w(t) - T_i\right)}{\sqrt{\pi \alpha t}} \qquad \alpha = \frac{k}{\rho C_p}$$

The heat transfer coefficient can be obtained from the definition:

$$h(t) = \frac{q_w''(t)}{T_\infty - T_w(t)}$$

First, the time dependent heat flux is calculated:

$$q_w''(0.3\,s) = \frac{\left(0.12\dfrac{W}{mK}\right)\cdot(530.1-25)\,°C}{\sqrt{\pi\cdot\dfrac{\left(0.12\dfrac{W}{mK}\right)}{\left(1420\dfrac{kg}{m^3}\right)\cdot\left(1090\dfrac{J}{kgK}\right)}\cdot 0.3s}} = 224{,}000\dfrac{W}{m^2}$$

$$q_w''(0.35\,s) = \frac{\left(0.12\dfrac{W}{mK}\right)\cdot(515.3-25)\,°C}{\sqrt{\pi\cdot\dfrac{\left(0.12\dfrac{W}{mK}\right)}{\left(1420\dfrac{kg}{m^3}\right)\cdot\left(1090\dfrac{J}{kgK}\right)}\cdot 0.35\,s}} = 202{,}000\dfrac{W}{m^2}$$

$$q_w''(0.40\,s) = \frac{\left(0.12\dfrac{W}{mK}\right)\cdot(531.8-25)\,°C}{\sqrt{\pi\cdot\dfrac{\left(0.12\dfrac{W}{mK}\right)}{\left(1420\dfrac{kg}{m^3}\right)\cdot\left(1090\dfrac{J}{kgK}\right)}\cdot 0.40\,s}} = 195{,}000\dfrac{W}{m^2}$$

Next, the heat transfer coefficients can be calculated:

$$h(0.3\,s) = \frac{224{,}000\dfrac{W}{m^2}}{(600-530.1)\,°C} = 3200\dfrac{W}{m^2K}$$

$$h(0.35\,s) = \frac{202{,}000\dfrac{W}{m^2}}{(600-515.3)\,°C} = 2380\dfrac{W}{m^2K}$$

$$h(0.40\,s) = \frac{195{,}000\dfrac{W}{m^2}}{(600-531.8)\,°C} = 2860\dfrac{W}{m^2K}$$

For the uncertainty (here, we do not consider heat loss for thin film gauge), Taking $t = 0.3$ s as an example:

$$h = \frac{q_w''(t)}{T_\infty - T_w(t)}$$

$$\frac{W_h}{h} = \sqrt{\left(\frac{W_{q_w''(t)}}{q_w''(t)}\right)^2 + \left(\frac{W_{T_\infty}}{T_\infty - T_w(t)}\right)^2 + \left(\frac{W_{T_w(t)}}{T_\infty - T_w(t)}\right)^2}$$

And

$$\frac{W_{q_w''(t)}}{q_w''(t)} = \sqrt{\left(\frac{W_k}{k}\right)^2 + \left(\frac{W_{T_w(t)}}{T_w(t) - T_i}\right)^2 + \left(\frac{W_{T_i}}{T_w(t) - T_i}\right)^2 + \left(\frac{1}{2}\frac{W_\alpha}{\alpha}\right)^2 + \left(\frac{1}{2}\frac{W_t}{t}\right)^2}$$

However, the uncertainty of wall temperature comes from the Wheatstone bridge circuit, where the measured temperature is proportional to the sensor resistance,

$$R(t) = \alpha + bT_w(t) = \frac{V(t)}{I}$$

So

$$\frac{W_{T_w(t)}}{T_w(t)} = \sqrt{\left(\frac{W_{V(t)}}{V}\right)^2 + \left(\frac{W_I}{I}\right)^2} = \sqrt{(0.01)^2 + (0.01)^2} = 1.41\%$$

$$\frac{W_{q_w''(t)}}{q_w''(t)} = \sqrt{(0.001)^2 + \left(\frac{530.1 \cdot 0.0141}{530.1 - 25}\right)^2 + \left(\frac{0.2}{530.1 - 25}\right)^2 + \left(\frac{1}{2} \cdot 0.001\right)^2 + \left(\frac{1}{2} \cdot 0.02\right)^2}$$

$$\frac{W_{q_w''(t)}}{q_w''(t)} = 1.79\%$$

$$\frac{W_h}{h} = \sqrt{(0.0179)^2 + \left(\frac{0.2}{600 - 530.1}\right)^2 + \left(\frac{530.1 \cdot 0.0141}{600 - 530.1}\right)^2}$$

$$\frac{W_h}{h} = 10.85\%$$

The uncertainty at different times can be calculated accordingly with different temperature inputs.

6.4 TRANSIENT TECHNIQUES FOR MEASUREMENT OF DETAILED HEAT TRANSFER COEFFICIENTS

6.4.1 FUNDAMENTAL PRINCIPLE

For convective heat transfer to occur between a fluid and a test surface, a temperature differential must exist between them. Either the fluid or the test surface can be at the higher temperature. If the fluid is heated, the test surface, which is initially at ambient conditions, will get heated as heat is convected to it. On the other hand, if the test surface is initially hot, the ambient fluid will cool the test surface. It should be noted that when the test surface is being heated, its entire volume should be uniformly heated to its initial temperature before the start of the fluid flow. After the fluid flow is started, the surface heater should be switched off to let the surface cool down as the ambient fluid comes into contact with it. If a heat-generating source is used on the test surface during the experiment, the energy equation used to derive the governing equations will not be applicable, and thus cannot be used.

The test surface is typically made of low conductivity materials such as Plexiglas, polycarbonate, or other plastics. This is necessary so the wall can be modelled as a one-dimensional, semi-infinite solid. Figure 6.9 shows a schematic of a semi-infinite solid exposed to a convection boundary. The semi-infinite wall assumption indicates the penetration depth of the thermal pulse is small compared to the thickness of the wall. The local surface temperature rises for an impulsively started heat transfer experiment, which can be related to the time, thermo-physical properties of the wall, and the convective heat transfer coefficient. Equations (6.11)–(6.14) show the governing differential equation, the assumed boundary and initial conditions, and the solution to the differential equation at $x = 0$ (the surface exposed to the fluid) [1]. T_w is the wall temperature at any instant in time at every pixel location, T_i is the initial temperature of the test surface before the transient event occurs, T_∞ is the oncoming air temperature, t is the time at which the temperature is measured, and (ρ, C_p, k, and α) are the thermophysical properties of the test surface. The heat-transfer coefficient at every pixel location can be determined if all other values in the equation are known. Equation (6.14) cannot be solved explicitly for the heat transfer coefficient, but numerical methods are available to solve this equation.

$$k\frac{\partial^2 T}{\partial x^2} = \rho c_p \frac{\partial T}{\partial t} \tag{6.11}$$

$$\text{at } t = 0, \ T = T_i \tag{6.12}$$

$$\text{at } x = 0, \ -k\frac{\partial T}{\partial X} = h(T_w - T_m); \ \text{as } x \to \infty, \ T = T_i \tag{6.13}$$

The solution of the above equation at the convective boundary surface ($x = 0$) is the following:

$$\frac{T_w - T_i}{T_m - T_i} = 1 - \exp\left(\frac{h^2 \alpha t}{k^2}\right) erfc\left(\frac{h\sqrt{\alpha t}}{k}\right) \tag{6.14}$$

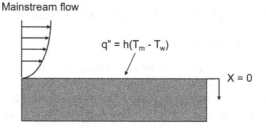

FIGURE 6.9 One-dimensional semi-infinite solid material with surface convection flow.

If the mainstream temperature changes with time, the varying temperature can be represented as a series of step change by using Duhamel's superposition theorem, and this is discussed in more detail in Chapter 7 [8]. Since the flow is steady, the local heat transfer coefficient at a selected surface location is a constant value, which is independent of time.

6.4.2   ONE-DIMENSIONAL TRANSIENT HEAT CONDUCTION
WITH SURFACE CONVECTION

The experiment is performed by first recording the initial temperature distribution on the test wall at time $t = 0$. While the surface temperature is measured, the initial condition shown in Equation (6.12) requires a uniform, initial temperature through the entire volume of the test surface. Fluid flow over the test surface commences immediately after recording the initial temperature. The temperature change of the surface is recorded as a function of time over the duration of the experiment. The wall temperature changes exponentially with time (Equation (6.14)), so the largest temperature gradient occurs during the first few seconds of the experiment. The local (and detailed) wall temperature distribution can be measured using liquid crystal paint, temperature sensitive paint, or an infrared camera. Upon recording the wall temperature as a function of time, the detailed heat transfer coefficient distribution can be calculated. These advanced methods for measuring wall temperatures will be discussed in later chapters.

Regions with high heat transfer coefficients will show a faster time response to temperature. With increasing time, the test surface temperature will approach the fluid temperature. The experiment can be concluded when the difference between these temperatures is small. The duration of the experiment depends on the heat transfer coefficient magnitude and the temperature of the heated medium. As the heat transfer coefficient is the quantity to be determined, the duration can be controlled by adjusting the initial temperature differential between the fluid and the test surface. A large difference will result in a sharper change in the surface temperature, and the surface will approach the fluid temperature quickly. However, a smaller temperature difference will tend to increase the duration of the experiment. The length of the experiment should be controlled to minimize the uncertainties associated with time and heat conduction errors in the test surface. If the test duration is very short (<30 s), a slight error in the measurement of time will result in relatively large errors causing higher uncertainties. On the other hand, if the duration of the experiment is long (3 to 4 minutes), lateral heat conduction may occur in the insulating substrate used on the underside of the test surface. As heat conducts through the insulation, with increasing time, the adjacent measurement points will be affected. The local measured temperature will be biased resulting in large errors. Experiments with a time duration of around 1 to 2 minutes yield the most reliable results.

One of the primary advantages of this method is that the test duration is very short, generally on the order of a few minutes. If performed carefully, this technique can give uncertainty values less than 8%. Accurate measurement of time is necessary for low uncertainty. A requirement of this technique is that an automated temperature and time data acquisition system is needed to capture the transient behavior of the test surface. This results in high initial costs in setting up the experiment.

As mentioned earlier, a time lag may exist from the time the flow is started until it attains the desired pressure and velocity conditions. This time lag should be short compared to the duration of the transient experiment to reduce uncertainty. A long time lag can cause the surface temperature to change considerably due to the developing flow. Final temperatures and the time are recorded at the end of the experiment. Either the final surface temperature can be specified and the time needed to achieve this end condition can be recorded or the duration of the experiment can be fixed and the final surface temperatures can be measured. A time lag may also exist in the thermocouple response. The thermocouples response time can be estimated as discussed in the Chapter 3. It is around 0.5 to 1.0 s for a bead diameter around 0.5 mm.

6.4.3 EXPERIMENT EXAMPLE

Example: One-Dimensional, Semi-Infinite Solid

Researchers are interested to obtain the local heat transfer distribution on the suction side of a turbine vane through the one-dimensional, transient conduction method. The vane is to be heated to a constant temperature through an embedded heating element within the structure of the vane. The vane test section is made of polycarbonate and installed in a wind tunnel. After the turbine vane temperature reaches steady-state at 50°C, the heating element is turned off, and room temperature air (25°C) is directed to the wind tunnel, cooling down the heated vane under steady flow conditions. Assume instrumentation is available to capture the temperature history of vane surface (more details will come in later chapters) and the mainstream temperature is kept constant.

The measured temperature history is shown below:

		Time	
Location	2 s	10 s	20 s
Leading edge	30.6°C	27.7°C	26.9°C
20% chord	31.6°C	28.2°C	27.3°C
80% chord	31.0°C	27.9°C	27.0°C

Thermophysical Properties of Polycarbonate:

Density:	$\rho = 1210 \text{ kg/m}^3$
Specific heat:	$C_p = 1250 \text{ J/kgK}$
Conductivity:	$k = 0.205 \text{ W/mK}$

Calculate the heat transfer coefficients at different times and locations and investigate if the heat transfer coefficient is a time dependent quantity or not. Also include the uncertainty estimation of the calculation.

From

$$\frac{T_w - T_i}{T_m - T_i} = 1 - \exp\left(\frac{h^2 \alpha t}{k^2}\right) \text{erfc}\left(\frac{h\sqrt{\alpha t}}{k}\right)$$

The time, temperatures, and thermophysical properties can be used to determine the heat transfer coefficient.

For the leading edge at $t = 2$ s:

$$\frac{(30.6-50)\,^\circ\text{C}}{(25-50)\,^\circ\text{C}} = 1-\exp\left(h_2^2 \cdot \left[\frac{\left(0.205\,\frac{\text{W}}{\text{mK}}\right)}{\left(1210\,\frac{\text{kg}}{\text{m}^3}\right)\left(1250\,\frac{\text{J}}{\text{kgK}}\right)} \cdot 2\,\text{s}\right]}{\left(0.205\,\frac{\text{W}}{\text{mK}}\right)^2}\right)$$

$$\cdot\,erfc\left(h_2 \cdot \frac{\sqrt{\dfrac{\left(0.205\,\frac{\text{W}}{\text{mK}}\right)}{\left(1210\,\frac{\text{kg}}{\text{m}^3}\right)\left(1250\,\frac{\text{J}}{\text{kgK}}\right)} \cdot 2\,\text{s}}}{\left(0.205\,\frac{\text{W}}{\text{mK}}\right)}\right)$$

$$h = 915\,\frac{\text{W}}{\text{m}^2\text{k}}$$

In similar way:

Location	Heat Transfer Coefficient (W/m²K)		
	2 s	10 s	20 s
Leading edge	915	903	913
20% chord	756	754	750
80% chord	846	833	860

For a given location, each instant in time gives approximately the same heat transfer coefficients; therefore, this shows under steady conditions, the heat transfer coefficient is independent of time.

For the uncertainty, take leading edge at $t = 2$ s data as an example,

$$\frac{w_x}{x} = \frac{1-\exp\left(x^2\right)erfc\left(x\right)}{\left[\dfrac{2x}{\sqrt{\pi}} - 2x^2\exp\left(x^2\right)erfc\left(x\right)\right]}\frac{w_\theta}{\theta} = \Gamma\frac{w_\theta}{\theta}$$

We can plug in the corresponding x value, for region on interest at $t = 2$ s as example,

$$x = \frac{h\sqrt{\alpha t}}{k} = \frac{915\frac{W}{m^2K} \cdot \sqrt{\dfrac{\left(0.205\frac{W}{mK}\right)}{\left(1210\frac{kg}{m^3}\right)\left(1250\frac{J}{kgK}\right)} \cdot 2\,s}}{\left(0.205\frac{W}{mK}\right)} = 2.32 \rightarrow \Gamma = 4.0 \rightarrow \frac{w_x}{x} = 4.0\frac{w_\theta}{\theta}$$

Therefore,

$$\frac{w_\theta}{\theta} = \sqrt{\left[\frac{w_{T_w}}{(T_w - T_i)}\right]^2 + \left[\frac{w_{T_i}}{(T_w - T_i)}\right]^2 + \left[\frac{w_{T_i}}{(T_m - T_i)}\right]^2 + \left[\frac{w_{T_m}}{(T_m - T_i)}\right]^2}$$

$$\frac{w_\theta}{\theta} = \sqrt{\left[\frac{30.6 \times 0.02}{(30.6 - 50)}\right]^2 + \left[\frac{0.2}{(30.6 - 50)}\right]^2 + \left[\frac{0.2}{(25 - 50)}\right]^2 + \left[\frac{0.2}{(25 - 50)}\right]^2} = 3.5\%$$

$$\frac{w_h}{h} = \sqrt{\left(\frac{w_x}{x}\right)^2 + \left(\frac{w_k}{k}\right)^2 + \left(\frac{1}{2}\frac{w_\alpha}{\alpha}\right)^2 + \left(\frac{1}{2}\frac{w_t}{t}\right)^2}$$

$$\frac{w_h}{h} = \sqrt{(4.0 \times 0.035)^2 + (0.001)^2 + \left(\frac{1}{2} \times 0.001\right)^2 + \left(\frac{1}{2} \times 0.02\right)^2} = 14.04\%$$

The uncertainty at different times and locations can be calculated accordingly with different temperature and time inputs.

PROBLEMS

1. In a wind tunnel, the air is heated above the room temperature. The air temperature is measured with a T-type thermocouple. The bead of the thermocouple can be approximated as a 0.5 mm sphere. Use the following information and apply the lumped capacitance technique to determine the time required for this thermocouple to reach 85% of the initial temperature difference.

Information:

a. The properties of the material in T-type thermocouple are:

Property	Material	
	Copper	Constantan
Conductivity (W/mK)	385	19.5
Density (kg/m³)	8960	8900
Specific heat (KJ/kgK)	0.385	0.39

Use the average value between two materials as properties for calculation.

b. The convective heat transfer coefficient between the air and the bead: 788 W/m²K.

2. Regionally averaged heat transfer coefficients at different locations on the pressure surface of a turbine airfoil are to be measured. The pressure side surface of the airfoil is divided into different regions with thin copper plates embedded with T-type thermocouples and insulation bars. Each thin copper plate is 10" long, 0.75" wide, and 0.0625" thick. A custom, flexible rubber heater is attached to inside of the airfoil to provide a uniform heat flux to all the copper plates, as shown in the figure below. The heater is able to raise the surface temperature to 50°C uniformly under steady-state conditions. After the vane is at a uniform temperature, the heater is turned off, and 25°C, room temperature air flows over the airfoil. As the airfoil cools, the temperature at two locations on the pressure surface is recorded as a function of time.

	Time	
Location	**10.5 s**	**30.5 s**
30% chord	48.1°C	44.8°C
60% chord	47.8°C	43.9°C

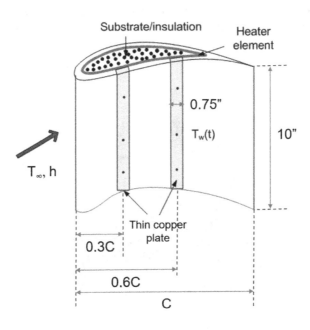

Information:
The material properties

	Material
Property	Copper
Conductivity (W/mK)	385
Density (kg/m³)	8960
Specific heat (J/kgK)	385

Calculate the surface heat transfer coefficient using the lumped capacitance technique at the two locations and verify that it is indeed applicable. Also, include the uncertainty estimation for your calculations.

3. Researchers want to study the heat transfer characteristics of flow impinging on a flat surface. The target surface is divided into concentric copper plate regions (0.125" thick). The copper plates are thermally isolated from one another, and a thermocouple is embedded in each plate. A circular, flexible, rubber heater is attached to the backside of the target plate to provide a uniform heat flux to all the copper plates, as shown in the figure below. The heater power is sufficient to raise the target surface temperature to 50°C uniformly at steady-state conditions. The transient test begins by turning off the heater and directing 25°C air through a 0.25" diameter hole (1.9375" above the target plate) toward the target surface, so the air impinges on plate. The jet Reynolds number is 25,000 (based on the jet diameter). The temperature history of two locations, the jet center and five diameters away from the center, are recorded.

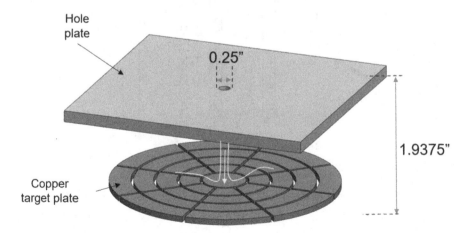

	Time	
Location	**10 s**	**20 s**
Center	42.4°C	37.1°C
5 diameters away from center	45.1°C	41.1°C

Calculate the heat transfer coefficient at these two locations and verify that lumped capacitance technique is applicable. Also, include an uncertainty estimation for your calculated heat transfer coefficients.

4. Unsteady heat transfer behavior can be found on a rotating turbine blade because it is constantly passing in and out of the wake created by the upstream ring of non-rotating vanes. Researchers want to capture this unsteady heat transfer behavior. A turbine blade is instrumented with one-layer, thin film heat flux gauges. The gauges are flush mounted on the turbine blade surface. A short duration wind tunnel with hot air at 600°C (measured by a K-type thermocouple) is used to simulate the unsteady wake effect. An initial temperature of 25°C is measured using a thermocouple mounted to the surface near the heat flux gauge. The test begins when the hot gas is directed through the rig and around the blade. During the short duration test, the following temperature history is measured with the thin film heat flux gauge.

	Time		
Location	**0.4 s**	**0.45 s**	**0.5 s**
1	530.1°C	515.3°C	531.8°C

Information:
Thin film gauge material properties

	Material
Property	**polyimide**
Conductivity (W/mK)	0.12
Density (kg/m³)	1420
Specific heat (J/kgK)	1090

Calculate the corresponding time dependent heat transfer coefficient at the specified location and include an uncertainty analysis.

5. There is a similar experimental setup as described in Problem 4, but the frequency of the upstream, unsteady wake device has changed to simulate the blade passing effect at a higher RPM. A different temperature history, compared to Problem 4, was measured at a different location on the blade surface.

	Time		
Location	0.1 s	0.2 s	0.3 s
2	380.9	397.2	406.6

Calculate the corresponding time dependent heat transfer coefficient and include an uncertainty analysis.

6. The von Karman vortex street is a periodic, unsteady, flow separation behavior due to flow around a blunt body. There is interest to study the heat transfer characteristics of this vortex street, so a 3" diameter cylinder is placed in a wind tunnel. Half of the cylinder surface (trailing side) is instrumented with one-layer heat flux gauges (the same gauge described in Problem 4), as the figure shows. Initially, the cylinder is at room temperature (25°C), and is surrounded by room temperature air in the tunnel. At $t = 0$, hot, 100°C, air is directed through the tunnel to the cylinder. Both the initial temperature of the cylinder and the mainstream air temperature are measured using thermocouples. The temperature response at one point on the cylinder (on the backside approximately $\theta = 160°$ from the stagnation point) is measured. The table shows the measured temperature from the heat flux gauge.

	Time			
Location	0.25 s	0.5 s	0.75 s	1.0 s
1	35.8°C	36.3°C	51.5°C	45.9°C

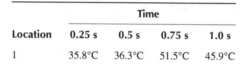

Cylinder

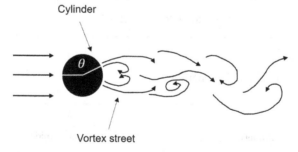

Vortex street

Calculate the heat transfer coefficient as a function of time and estimate the uncertainty level of calculated value.

7. Researchers are interested to measure the local heat transfer distribution for flow inside a square channel with rib turbulators using the 1D, transient conduction method. The channel configuration is the same as the example in Chapter 5. Both the ribs and the test channel are made of Plexiglas. The test channel is kept at room temperature (25°C) and the mainstream air is first preheated to 45°C through a bypass loop (both are recorded by thermocouples). After the air temperature reaches steady-state, the bulk flow is diverted to the ribbed channel. Assume instrumentation is available to capture the surface temperature history of the channel area between the

ribs and on top of the rib (more details will be provided in later chapters) and the mainstream temperature remains constant (45°C). The measured temperature history at different locations is shown below.

| | Time | |
Location	20 s	40 s
Top of the rib	39.7°C	41.1°C
Between two ribs	37.8°C	39.5°C

Information:
Test section material properties

| | Material |
Property	Plexiglas
Conductivity (W/mK)	0.2
Density (kg/m³)	1180
Specific heat (J/kgK)	1450

Calculate the heat transfer coefficient in locations at different times and determine if the heat transfer coefficient is independent of time. Include the uncertainty estimation for the calculations.

8. It is desired to measure the local heat transfer distribution for flow impinging on a curved surface using the 1D, transient conduction method. The test channel with a curved surface is made of polycarbonate. The test channel is kept at room temperature (25°C) while the mainstream air is first preheated to 55°C through a bypass loop (both are recorded by thermocouples). After the air temperature reaches steady-state, it is immediately directed to the test section, so it impinges on the curved target surface. Assume instrumentation is available to capture the surface temperature history of the curved surface (more details will be provided in later chapters) and the mainstream temperature is kept constant. The measured temperature history at the center of the curved surface is as below.

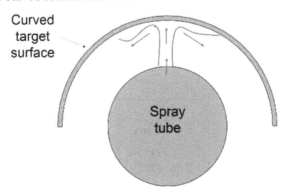

	Time		
Location	5 s	50 s	100 s
Center of curved surface	31.5°C	39.6°C	42.5°C

Information:
Test section material properties

	Material
Property	Polycarbonate
Conductivity (W/mK)	0.205
Density (kg/m³)	1210
Specific heat (J/kgK)	1250

Calculate the heat transfer coefficient at different times and comment if the heat transfer coefficient is a time dependent quantity or not. Also, include an uncertainty estimation of the calculation.

9. Researchers would like to measure the heat transfer coefficient variation for laminar flow over the flat plate by using the 1D, transient conduction method. The flat plate is made of Plexiglas with a heater mat attached behind the surface and thermocouples are embedded in the plate to monitor its temperature. The length of the plate is 1 m. The plate is wide enough so we can assume it is a 2D problem. This flat plate test section is put inside a wind tunnel with a Reynolds number, based on plate length, of 100,000. When the plate temperature reaches steady-state at 127°C, measured with thermocouples, the heater power is turned off and suddenly steady room temperature air (25°C) flows into the wind tunnel to cool the plate. Assume instrumentation is available to capture the surface temperature history of the flat plate (more details will be provided in later chapters) and the mainstream temperature remains constant. The measured temperature history is as below.

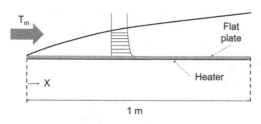

	Time	
Location	30 s	50 s
X = 0.1 m	117.1°C	114.5°C
X = 0.8 m	123.3°C	122.3°C

Calculate the heat transfer coefficient at different times and locations and determine if the heat transfer coefficient is a time dependent quantity or not. Also, include an uncertainty estimation of the calculated heat transfer coefficients.

10. The local heat transfer distribution on the suction surface of a turbine vane will be investigated using the 1D, transient conduction method. The vane is heated to a constant temperature using an embedded heating element inside the vane. The vane test section is made of polycarbonate and installed in a wind tunnel. After the turbine vane temperature reaches steady-state at 50°C, the heating element is turned off and room temperature air (25°C) is directed through the tunnel to cool the vane. Assume that instrumentation is available to capture the surface temperature history of the vane (more details will be provided in later chapters) and the mainstream temperature remains constant. The measured temperature history is as below.

	Time		
Location	**2.5 s**	**10.5 s**	**20.5 s**
Leading edge	30.6°C	27.7°C	26.9°C
20% chord	31.6°C	28.2°C	27.3°C
80% chord	31.0°C	27.9°C	27.0°C

Calculate the heat transfer coefficient at different times and locations and confirm if the heat transfer coefficient is a time dependent quantity or not. Also, include the uncertainty estimation of the calculations.

REFERENCES

1. Incropera, F. and Dewitt, D., *Fundamentals of Heat and Mass Transfer*, 5th ed., John Wiley & Sons, New York, 2002.
2. Martin, E.L., Wright, L.M., and Crites, D.C., "Impingement Heat Transfer Enhancement on a Cylindrical, Leading Edge Model with Varying Jet Temperatures," *ASME Journal of Turbomachinery*, Vol. 135, No. 031021, 2013, 8 pages.
3. Chupp, R.E., Helms, D.E., McFadden, P.W., and Brown, T.R., "Evaluation of Internal Heat Transfer Coefficients for Impingement Cooled Turbine Airfoils," *AIAA Journal*, Vol. 6, No. 3, 1969, pp. 203–208.
4. Dunn, M.G., "Turbine Heat Flux Measurements: Influence of Slot Injection on Vane Trailing Edge Heat Transfer and Influence of Rotor on Vane Heat Transfer." ASME Paper 84-GT-175, 1984.
5. Iliopoulou, V., Denos, R., Billiard, N., and Arts, T., "Time-Averaged and Time-Resolved Heat Flux Measurements on a Turbine Stator Blade using Two-Layered Thin-Film Gauges," *ASME Journal of Turbomachinery*, Vol. 126, October, 2004, p. 570.

6. Doorly, D.J., and Oldfield, M.L.G., "Simulation of the Effects of Shock-Waves Passing on a Turbine Rotor Blade," *ASME Journal of Engineering for Gas Turbines and Power*, Vol. 107, 1985, pp. 998–1006.

7. Guenette, G.R., Epstein, A.H., Giles, M.B., Hanes, R., and Norton, R.J.G., "Fully Scaled Transonic Turbine Rotor Heat Transfer Measurements," *ASME Journal of Turbomachinery*, Vol. 111, 1989, pp. 1–7.

8. Ekkad, S.V., and Han, J.C., "A Transient Liquid Crystal Thermography Technique for Gas Turbine Heat Transfer Measurements," *Measurement Science and Technology*, Vol. 11, July, 2000, pp. 957–968.

7 Liquid Crystal Thermography Techniques

7.1 MEASUREMENT THEORY

7.1.1 LIQUID CRYSTAL THERMOGRAPHY OVERVIEW

Liquid crystal thermography (LCT) has been a popular method for measuring surface temperatures for the past several years. Liquid crystals are thermochromic, i.e., they change color with temperature. Thus, a surface coated with thermochromic liquid crystals (TLC) will show a color change when exposed to certain temperatures. Due to this characteristic of a tangible color response, TLC's are used in a wide range of applications. Gas turbine blades, electronic component cooling, and boiling heat transfer, are some areas of research performed using TLC's. Even the medical field has utilized their response to measure body temperature.

The liquid crystal technique requires the use of three major components: the liquid crystal coated test surface, a white light source(s) to illuminate the surface, and a camera to capture the local distribution of color on the test surface. The use of a camera can sometimes be avoided by tracking the color change manually using several predetermined grid points. More details of this particular method are discussed in the section on the "yellow band tracking method." However, for detailed local measurements on the test surface, a camera may be necessary.

Liquid crystals are composed of a chiral, or twisted, molecular structure which becomes optically active when they change phase. Below the event temperature (temperature at which the first color change starts), the liquid crystals are in a solid phase and appear transparent. The liquid crystals change phase from solid to liquid state over a range of temperatures. Based on the relative solid-liquid content on the phase change curve, the two-phase solution refracts light and appears to change color. When the temperature of the liquid crystal coated surface reaches its event temperature, melting of the liquid crystals on the surface is initiated and they will start reflecting visible light at a certain wavelength when illuminated with white light. At this temperature, the liquid crystal will change color from clear to red. As the temperature increases further, more quantities of the unmelted liquid crystals enter the liquid phase and start reflecting color at wavelengths proportional to the mass of the melted liquid crystals. Thus, the reflected wavelength or color changes with increasing temperature until all the crystals change to liquid. At this stage, the

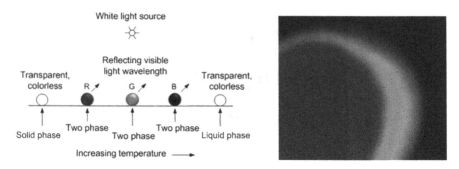

FIGURE 7.1   Conceptual representation of TLC color change with temperature and the reflected color from a coated surface.

TLC becomes transparent again with no appearance of color with further increases in temperature. Figure 7.1 demonstrates the color change behavior of the liquid crystal coating.

The color change is reversible, and during cooling processes, the sequence is reversed (colorless, blue, green, red, colorless). Upon re-heating, the phase and color changes occur at the same temperature, so the coating offers a repeatable measurement tool. This fact is important as the experiments performed using this technique give reproducible results. Generally, liquid crystals are viewed against a non-reflective black background; this minimizes the interference of the light transmission from the illumination source and the reflected light from the liquid crystals. There are two categories in which TLC's can be classified; cholesteric and chiral-nematic. Due to their inherently oily form, the true liquid crystals are microencapsulated in tiny protective capsules and are suspended in an aqueous binder material. This makes the liquid crystals easier to work with and chemical contamination is avoided.

The range of temperature when the TLC reflects light from red until the beginning of blue is called the bandwidth. TLC's are grouped by two color/temperature designations to describe their relationship between color and temperature. A typical example would be "R35C5W" which is a commonly used TLC formulation. In this case, "R35C" means that the start of the red color or the event temperature is at 35°C while "5W" indicates that the start of the blue color is 5°C above the event temperature. Thus, the bandwidth for this type of liquid crystal is 5°C. TLC's are available either as narrow band with a bandwidth between 0.5°C to 2°C or as wide band with bandwidth ranging from 5°C to 30°C. Depending on the application, a TLC with an appropriate event temperature and bandwidth can be selected.

The selection of a narrow band or a wide band TLC depends on the type of application being investigated. Very accurate temperature measurements can be obtained using a narrow band TLC as the color change is very sharp at a specific temperature. Also, image processing is easier as a single color can be tracked for different input heat fluxes. More details about using narrow band liquid crystals are given in the section on "yellow band tracking method." The chief advantage of using a wide band liquid crystal is that the entire isotherm pattern of the surface can be mapped in a single image. A single image can also capture large variations in surface temperature at high spatial resolution. Careful calibration of the color change to temperature response is needed to obtain high accuracy.

The color reflected from the TLC is dependent on the light source used to illuminate the test surface. A sufficiently bright and stable white light source is needed. When the test surface is curved, more than one light source might be needed to ensure uniform light intensity over the test surface. The light sources selected should not have infrared or ultraviolet light in their output spectrum. The presence of infrared light in the light source can cause unwanted heating of the TLC coated test surface through radiation. This will give inaccurate measurements of the surface temperature. Ultraviolet radiation on the other hand can cause rapid deterioration of the liquid crystal response resulting in an inconsistent response of color change to temperature. Typically, the use of fluorescent lights should also be avoided as they emit light periodically but at high frequencies. This periodicity is not observable by the human eye but it is magnified when viewed using a camera with a relatively small exposure time. This periodic nature results in an inconsistent intensity distribution on the test surface over time.

7.1.2 LIQUID CRYSTAL THERMOGRAPHY CALIBRATION

Calibration of the TLC can either be performed in situ or in a separate calibration facility. In situ calibration is recommended as the color perceived from a true TLC is dependent on the lighting and viewing arrangement at a given temperature. Thus, due to the difficulty of having two identical setups for measurement and calibration, in situ calibration is preferred. It should be also noted that the color play of the TLC is dependent on the viewing angle. If the viewing angle is very large, the observed color may be different than the color seen from a normal viewing angle, even if the surface is at the same temperature. However, the error induced when the viewing angle is less than ±30° from the normal angle is small, and hence, can be neglected. Additional calibrations should be performed for large viewing angles to account for any discrepancy when the camera is not oriented normal to the heat transfer surface.

The calibration is performed by applying the TLC to an isothermal surface whose temperature can be controlled. An example of such a surface would be a copper plate with an embedded thermocouple and a heater (attached on the opposite side of the copper from the TLC). The high conductivity of copper ensures that the surface temperature on the TLC coated surface is uniform. By controlling the power supplied to the heater, successive increments in temperature can be obtained and the color response can be recorded. This process is repeated from the event temperature through the clearing point temperature. The hue of the color when plotted against temperature is typically linear from the start of red until the beginning of blue with a sharp, varying slope through the various shades of blue.

The color image of the surface is typically recorded with an RGB, CCD camera. If detailed measurements of the entire surface are to be performed, the camera should have a sufficiently high spatial resolution. The required camera resolution depends on the size of the test surface and the smallest surface feature that needs to be captured in detail. The pixel size, on the test surface, should be at least less than half the size of the smallest feature on the test surface. More pixels will give an image with better resolution. However, high resolution cameras are expensive and frequently a compromise must be reached between cost and the camera resolution. For example, an image resolution of 0.5 mm/pixel will be able to resolve a hole of diameter 2 mm on the test surface in sufficient detail in most cases. However, if the pixel resolution is 2 mm/pixel, only one pixel will be available to cover the entire hole, which will yield what appears to be a blurred image from the camera. Also, the camera chosen should be able to take several images of the test surface in a relatively short time, so that the color or hue magnitudes at each pixel can be averaged to get the true surface color distribution. Several hundred images may need to be averaged to reduce noise from the camera CCD.

TLC's are available as a sprayable liquid or as a coated (printed) sheet. Sprayable liquid crystal is typically used when the surface is not flat. A coating of a non-reflective, black paint on the test surface is necessary before spraying the TLC for good visualization and to accurately interpret the color response. Sufficient care should be taken while spraying the paint in order to obtain a uniform layer thickness of the black paint and liquid crystal. The liquid crystal and black paint are aqueous so that they can easily be removed by water or alcohol. A coated sheet consists of a thin film of liquid crystal sandwiched between a transparent substrate (sheet), and a black background. They are usually made by printing an ink containing microencapsulated TLC onto the opposite side of the substrate. A black ink is then applied on top of the dry TLC coating and color change effects are viewed from the uncoated side. A coated sheet has a much longer life as compared to a surface sprayed with TLC, as the transparent coating prevents the direct contact of the TLC with hot air, and it protects the paint from debris.

Due to the property of changing color with temperature, TLC's have been successfully used to measure heat transfer coefficients on a surface. Steady-state as well as transient techniques have been utilized. The following section discusses the steady-state methods available for measuring heat transfer along with some applications in research.

7.2 STEADY-STATE LOCAL HEAT TRANSFER COEFFICIENT MEASUREMENT

7.2.1 Fundamental Principal

In this method, a foil heater is coated with liquid crystal paint and the heat applied to the surface is adjusted through a series of step increments to obtain a predetermined color (temperature) on the test surface. The temperature at the predetermined color is obtained via a calibration curve. At a specific heat input, the region with heat transfer coefficients corresponding to the preset wall temperature will change color to form an iso-color, or iso-temperature, line on the test surface. Regions which have a high heat transfer coefficient will require a larger heat input to attain the same constant, predetermined color. By measuring the heat input for a particular iso-temperature line, the heat transfer coefficient can be calculated on this line. The position of this line is recorded and the heat input is adjusted to another value. The iso-temperature line will shift corresponding to the heat transfer coefficient distribution on the test surface. Thus, iso-temperature lines can be tracked over the test surface. After traversing through all the heat input increments, the local heat transfer can be obtained over the entire test surface.

A major drawback of this method is that the heat input has to be adjusted, and allowed to reach steady-state, through several steps to determine all the iso-temperature lines on the test surface. This can be cumbersome when the heat transfer coefficient distribution is highly non-uniform with large gradients thus requiring a large number of steps in the heat input.

7.2.2 YELLOW-BAND TRACKING TECHNIQUE

This is one of the earliest techniques employed by researchers to obtain detailed local heat transfer coefficients using liquid crystals. The liquid crystal paint is first calibrated to obtain the temperature which corresponds to the "yellow" color. The color yellow can be defined in terms of an RGB (Red-Green-Blue) scale with $R = 255$, $G = 255$ and $B = 0$. The color yellow is chosen as the transition between orange to yellow and yellow to green. This is a well-defined, sharp (and intense) color at a narrow temperature band. Thus, accurate temperature readings can be obtained. Figure 7.2 shows the yellow band tracking technique.

In this method, a uniform heat flux heater, such as from a thin foil heater as described in the section on thin foil heaters with thermocouples (Chapter 5), can be used. As liquid crystals will be employed for measuring temperature, the use of thermocouples can be avoided. A narrow band TLC is typically used for greater accuracy. The heater is then covered with liquid crystals. If a sprayable liquid crystal is used, the heater should be covered with a non-reflective coating of black paint. During the experiment, at a certain heat flux, the wall temperature will increase, and it will cross the event temperature of the TLC. When the color turns yellow, the position of the color band can be recorded. A predetermined grid mapped on the test surface can be used to indicate the location of the yellow band. This position can be recorded either with a camera or by eye. The local heat flux is then manipulated to change the location of this yellow band so that all the grid points are covered.

The main advantage of this technique is that it is inexpensive as it does not require a sophisticated image acquisition system. The uncertainty for this technique is typically around $\pm 8\%$ which can be further improved by accurately estimating heat

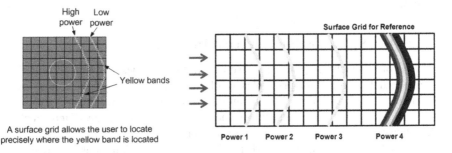

FIGURE 7.2 Conceptual schematic of the steady state, yellow band tracking method.

losses and by careful calibration. However, this technique is time-consuming, as several experimental runs are needed to cover all the grid points on the test surface and the experimentalist has to wait for steady-state for each case.

7.2.3 EXPERIMENT EXAMPLE

One of the earliest studies that used this technique was performed by Hippensteele et al. [1], on a large-scale model of a turbine blade airfoil. The researchers also discussed the effect of viewing angle on the perceived liquid crystal color. They found that if the angle between the camera-view and the strobe aiming was less than 30°, there was no significant change in the yellow color, regardless of the angle to the liquid crystal sheet. However, for angles greater than 30°, the color was shifted from yellow to blue; thus, erroneously indicating a higher temperature. For this reason, the included angles were restricted to less than 30° during data acquisition.

7.3 STEADY-STATE DETAILED HEAT TRANSFER MEASUREMENT TECHNIQUES

7.3.1 FUNDAMENTAL PRINCIPAL

Detailed, local measurements of the convective heat transfer coefficients can be obtained on the entire test surface in one experimental run, as opposed to the several runs needed for the yellow band tracking method. A foil heater, similar to the one discussed previously (Chapter 5), can be used to heat the surface. Either a TLC sheet or sprayable TLC can be applied to the surface of the heater. If sprayable TLC's are used, the test surface is first prepared by coating the foil heater with a layer of non-reflective black paint. The TLC can then be uniformly sprayed onto the surface (usually with an airbrush). Depending on the test surface geometry, large variations in the local heat transfer coefficients may occur over the test surface which will consequently result in large variations in the local wall temperature. A narrow band liquid crystal may not cover the entire range of wall temperatures on the test surface and may go out of range. In other words, if the temperature is above or below the range of the TLC, the liquid crystals will be transparent, and only the black background will be seen. Without color being reflected by the paint, it is impossible to know the its surface temperature in these regions. Hence, wideband TLC's are used in these experiments.

7.3.2 HUE SATURATION INTENSITY (HSI) TECHNIQUE

The use of this technique in convective heat transfer studies was first presented by Camci et al. [2]. From basic principles in colorimetric theory, any color can be specified in terms of three independent quantities, which the healthy, human eye can distinguish. These quantities, identified as the colorimetric, coordinate systems include the intensities of the three primary colors: red, blue, and green (RGB) or equivalently as hue, saturation and intensity (HSI). Hue refers to the dominant wavelength of color in the visible spectrum which can be separated into groups such as red, yellow, green, etc. Saturation may be defined as a color's purity or the amount of white contained in that color under a set light intensity. Here, intensity is the relative brightness of the color as observed by the sensor (human eye or camera).

Figure 7.3 represents the color of a TLC from the initial appearance of red through all the colors into the blue shades leading to violet. While the data shown is for a narrow band paint (R25C1), the behavior of any bandwidth will be similar, but for wide band with paints, the change will happen over a wider temperature range. Images have been recorded over the full range of the paint. Black paint and liquid crystal paint were applied to a copper plate, and the plate was heated from the backside. Multiple thermocouples were embedded in the copper, to provide the temperature corresponding to each image. A region of interest was selected on the copper surface, and the color quantities were averaged over this region for each image/temperature. In the top three plots, of Figure 7.3, the individual red, green, and blue components have been extracted from each image. In other words, for a given image, the composite color is determined based on a fraction of red, a fraction of green, and a fraction of blue. These individual contributions have been separated in these figures (the range of 0–255 is used). Comparing all three figures, it is first noticeable that each color reaches its peak value at a different temperature. The maximum red occurs around 25.5°C, the maximum green is observed at approximately 25.75°C, and the maximum blue at 27.5°C. The one-degree width of the paint is based on the beginning of red to the beginning of blue; therefore, the actual, visible spectrum is greater than 1°C. Secondly, the magnitude of the maximum green is greater than that for the red or the blue. This is why the "yellow" to "green" transition is often used as the color of choice, as it is the most distinct color. Finally, the width of the curves varies for each color. The red and green colors show very narrow distributions with a well-defined peak. Blue is a much flatter curve; with blue showing up over a wide range of temperatures, it is often difficult to distinguish between various shades. Therefore, researchers are cautioned to minimize the high temperature, blue colors shown on a test surface.

The bottom two plots of Figure 7.3 show the hue and intensity components from the same images used for the RGB components. From the hue distribution, it is shown at low temperatures (red → green), large changes in hue represent small changes in temperature. However, around 26.5°C, the curve flattens (the paint begins to appear blue), and small changes in hue yield relatively large changes in temperature. To improve the accuracy of the liquid crystal measurements, it is desirable to avoid this region for the liquid crystal coating. The final figure shows the measured intensity of the paint over its active range. The three peaks corresponding to red, green, and blue are also shown in this intensity distribution. The green color provides the most intense signature, followed by red, and blue. This figure again demonstrates why the "yellow" to "green" transition is usually the color of choice for TLC experiments. Regardless of the measurement method, RGB or HSI, this yellow to green color is the most intense and the most distinct. Therefore, this color can be identified with the most certainty.

Camci et al. [2] indicated that the use of this HSI coordinate system provides all the color information for a certain stimulus and is convenient and appropriate for an image processing system involved in measuring convective heat transfer coefficients

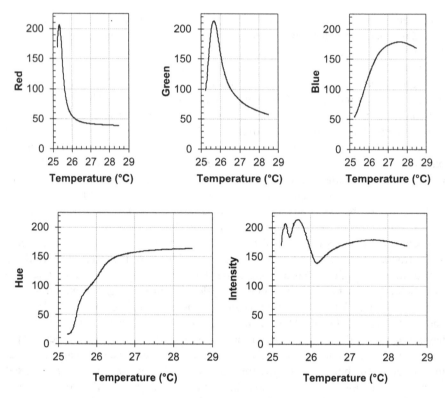

FIGURE 7.3 Liquid crystal color descriptions (R25C1)—Red, green, and blue versus hue and intensity.

using liquid crystals. This is primarily the result of the local hue exhibited by the liquid crystal being directly related to the local temperature. The relation between the RGB and HSI coordinate systems is depicted in Figure 7.4 as a triangular model. Pritchard [3] has expressed the conversion of RGB values to HSI magnitudes using the triangular model in terms of simple trigonometric expressions given in the following equations.

$$I = \frac{R+G+B}{3}$$

$$S = 1 - \frac{Min(R,G,B)}{I}$$

$$H = \frac{\pi}{2} - \arcsin\left[\frac{\left(0.5 \cdot \left((R-G)+(R-B)\right)\right)}{\sqrt{\left((R-G)^2 +(R-G)\cdot(G-B)\right)}}\right]$$

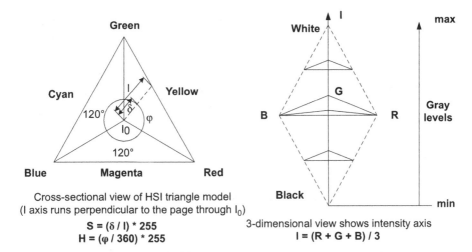

FIGURE 7.4   Triangular model representation of HSI implementation system.

$$H = H \cdot \frac{180}{\pi} \quad \text{if} \quad H > 180 \quad \text{then} \quad H = 360 - H$$

$$H = \frac{H}{360} \cdot 255$$

By using the hue as the measured quantity from the liquid crystal, several colors such as red, green, blue, etc., can be used to measure the temperature of a surface as opposed to the single color, yellow band tracking, method discussed in the previous section. Typically, a wide band liquid crystal is used with this method resulting in a high spatial resolution. The test surface does not need to be mapped into a grid, as the wide band liquid crystal can capture large differences in temperature on the test surface. This reduces the time required for data acquisition, making it a more convenient, alternative for acquiring local temperature distributions, as compared to the yellow band tracking technique. Before the experiment, the bandwidth of the liquid crystal should be chosen. For a steady-state experiment, the surface under investigation should have a temperature approximately 20°C higher than the fluid temperature, to avoid excessive heat losses and to minimize uncertainty. Based on this temperature difference and the expected variation in heat transfer coefficient distribution on the surface, the liquid crystal bandwidth should be selected.

Calibration for hue against the temperature for the liquid crystal needs to be performed in situ with the experiment. This is essential as the perceived color by the camera sensor can change if the saturation or intensity is different between the test surface and the calibration test piece. For example, a brighter light source used during calibration will make the color appear whiter. Similarly, adding saturation will also make the color appear whiter. Thus, differences in the saturation and intensity

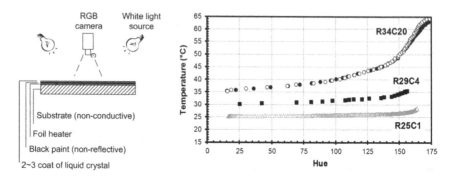

FIGURE 7.5   Calibration setup and sample hue-temperature calibration curves.

during calibration and experiment can result in erroneous results. In situ calibration will ensure that the calibration light conditions and camera angles are same as those during the experiment. Once the calibration is performed, this setup should not be disturbed until all the necessary experiments are complete.

Figure 7.5 shows the typical setup for liquid crystal calibration and the calibration curves for narrow band and wide band liquid crystals. Both wideband and narrowband paints are shown for comparison. The data shown in these curves are specific to given setup and lighting conditions; these are not universal curves that can be used for any application of these paints. Regardless of what bandwidth paint is being used, when the hue exceeds 150, the slope of the curve quickly changes. This corresponds to the onset of shades of blue, and small changes in color yield large changes in temperature. Therefore, it is desirable, as much as possible, to limit measurements to hue values less than 150. The repeatability of the R34C20 calibration is also shown with two separate sets of data shown for the same experimental setup. In the range of red to green, the data is very repeatable, but within the shades of blue (hue > 150), there is separation between the sets of calibration data. Although the full calibration of the narrowband paint, R25C1, is shown, generally when a narrowband paint is used, only the temperature corresponding to the yellow color is needed (a one-point calibration is sufficient compared to the full spectrum calibration).

The test surface is instrumented with a uniform heat flux heater and then covered with liquid crystals of the required bandwidth. The setup is same as that for the yellow band tracking technique. Once the experiment is started, the heat flux is controlled such that the resulting temperature distribution will capture an agreeable color distribution with minimum black spots. Once steady-state is obtained, images can be obtained on the test surface using an RGB-CCD camera mounted normal to the surface. The acquired images are then stored and processed to obtain the local hue level for each pixel. Using the calibration curve, these hue magnitudes can be converted to temperatures thus giving a temperature distribution for all the image pixels. By estimating the heat losses, the heat transfer coefficients can then be obtained for all the pixels.

7.4 TRANSIENT DETAILED HEAT TRANSFER MEASUREMENT TECHNIQUE

Transient methods combined with liquid crystal thermography have been popular for measuring heat transfer. A few techniques frequently used by researchers [4,5] are presented in the following sections.

7.4.1 TRANSIENT HSI TECHNIQUE

The scheme used for distinguishing colors based on their hue, saturation, and intensity is same as that described in the section using the steady-state HSI technique. These coordinates can be obtained from the red, green, and blue color primaries by using the triangular model. In this technique, the time from the start of the mainstream flow ($t = 0$) until the time to capture the hue distribution is constant for every pixel. In other words, a single image is captured, at time, t, and the hue is recorded at this time. Hue magnitudes are calculated for each pixel in the image and using a calibration curve, the surface temperature at every pixel can be determined. The surface temperature can thus be mapped for the entire surface at one instant. The calibration procedure for this technique is similar to that described in the previous section on the steady-state HSI technique. A wide band liquid crystal may be used as it can cover a broad range of temperatures. Use of a wide band TLC though gives low temperature resolution as small temperature changes will not be captured. Figure 7.6 shows a surface temperature distribution from the hue-temperature calibration [6].

7.4.2 TRANSIENT SINGLE-COLOR CAPTURING TECHNIQUE

This technique is based on the semi-infinite solid model and involves the time measurement for the appearance of a particular color at every pixel location on the test surface. A liquid crystal coated surface, when heated, will change color from colorless before the event temperature to red, to green and then to blue. Upon further heating, it will revert back to its colorless form. Thus, the imaging system can be programmed to look for a certain color composed of specific red, green and blue color intensities. This specified color will correspond to a temperature which can be determined from the calibration of that TLC. Once this color is obtained during the experiment, the time required from the start of the experiment to this instant is recorded. Thus, by substituting the time and surface temperature for each pixel in the solution for the semi-infinite solid, the heat transfer coefficient can be calculated. During the experiment, the time required for each pixel to reach the specified color will be different and is proportional to the heat transfer coefficient at that pixel. Thus, many images are required during the experiment. This can be done by recording a movie file from the camera. This movie file can be imported to a computer and digitally separated into individual frames (images). The color information for each pixel from each frame is obtained and converted to temperature by utilizing the calibration data. In the solution to the semi-infinite solid model, the wall temperature will be a

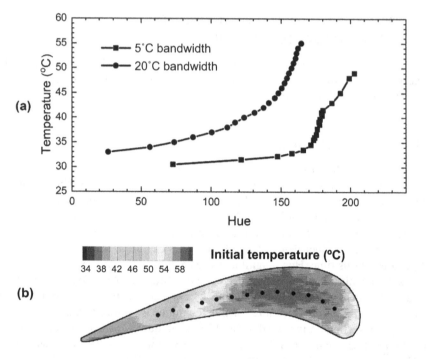

FIGURE 7.6   (a) Hue-temperature calibration, (b) Blade tip surface temperature distribution.

constant value; therefore, only a single point calibration is required. However, the time to reach this set temperature will vary from pixel to pixel. It is often beneficial to plot the time distribution before solving for the detailed heat transfer coefficient distributions. Areas of high heat transfer coefficients will change color quickly, and regions with relatively low heat transfer coefficients will require more time to reach their event temperature.

The test surface, typically made from a plastic material, is coated with liquid crystal and the initial temperature is measured. In most experiments, a TLC with a different bandwidth is necessary to measure the initial temperature. For example, a wide band liquid crystal may be used to map the surface temperature whereas a narrow band liquid crystal may be used for transient measurements for greater temperature resolution. Wide band TLC can be chosen mainly as the temperature distribution on the entire surface can be mapped in a single image. A narrow band liquid crystal will not cover this distribution if the initial temperature variation is larger than the band width of the TLC. As the time required to reach a specific temperature is measured during the test, a narrow band TLC may be used during the transient test. The initial temperature can be measured in a separate test and can be assumed to be unchanging for each transient test. If the test surface is heated, careful control of the surface temperature is needed. Different initial temperatures can cause large uncertainties in the calculated heat transfer coefficients. If the mainstream air is heated then it should be ensured that the test surface attains the ambient temperature before each test run. It should be noted that the camera position, surface illumination, etc., should not change for both cases (initial temperature and flow tests).

Another option is to mix the wide band and narrow band TLC's in an appropriate proportion and to spray this mixture on the test surface. Thus, the initial temperature of the test surface can be recorded for each different test run. This method requires taking additional images to measure the initial temperature. Thus, mixing the two TLC's can be cumbersome due to the image processing involved but at the same time, greater flexibility and control over the transient experiment can be obtained.

7.4.3 EXPERIMENT EXAMPLE

7.4.3.1 Background Information and Heat Transfer Theory [4,5]

The test surface is initially at a uniform temperature (T_i, initial temperature, usually room temperature) and then is suddenly exposed to the desired flow at a different temperature (T_m, mainstream temperature, usually higher). The local wall temperature (T_w) starts to change color over time. The wall temperature changes faster when it has a higher, local heat transfer coefficient (h). At a certain time, the local wall temperature can be obtained via the color change of the TLC coated surface. Under the condition that the material of the test surface is made of low thermal conductivity material (such as Plexiglas), the local heat transfer coefficient on the test surface can be obtained by using the 1-D semi-infinite assumption (similar to the method shown in Chapter 6):

$$k \frac{\partial^2 T}{\partial x^2} = \rho C_P \frac{\partial T}{\partial t} \tag{7.1}$$

And the boundary conditions are:

$$\text{at } t = 0, \ T = T_i$$

$$\text{at } x = 0, \ -k \frac{\partial T}{\partial x} = h\left(T_w - T_m\right) \tag{7.2}$$

$$\text{at } x \to \infty, \ T = T_i$$

Solving the above equation with the prescribed initial and boundary conditions, one derives the non-dimensional temperature on the surface exposed to the convective boundary ($x = 0$):

$$\frac{T_w - T_i}{T_m - T_i} = 1 - \exp\left(\frac{h^2 \alpha t}{k^2}\right) erfc\left(\frac{h\sqrt{\alpha t}}{k}\right) \tag{7.3}$$

where $\alpha = k/(\rho C_P)$ is the thermal diffusivity of the material, and k is the thermal conductivity of the material.

7.4.3.2 Heat Transfer Measurement with Time-Dependent Mainstream Temperature

In a true experiment, there is no perfect step change of the mainstream temperature (T_m is a function of time). Thus, the variations of the mainstream temperature should be considered. Figure 7.7 shows the time history of the mainstream temperature.

The time history of the mainstream temperature can be simulated by a series of time steps, which are incorporated into Equation (7.3) using Duhamel's superposition theorem. The solution for the local heat transfer coefficient can then be expressed as:

$$T_w - T_i(t) = \sum_{n=1}^{N} \left(1 - \exp[\beta] \cdot erfc\left(\sqrt{\beta}\right)\right) \cdot \left(T_{m,n} - T_{m,n-1}\right) \qquad (7.4)$$

$$\beta = \frac{h^2 \alpha \cdot (t - t(n))}{k^2} \qquad (7.5)$$

7.4.3.3 Film Cooling Effectiveness

The film cooling effectiveness is a dimensionless parameter used to quantify the effectiveness of film cooling. It is defined as:

$$\eta = \frac{T_m - T_f}{T_m - T_c} \qquad (7.6)$$

$$\text{or } T_f = \eta(T_c - T_m) + T_m = \eta T_c + (1 - \eta)T_m$$

The value of η falls between 0 and 1. $\eta = 0$ represents no film coverage, the film temperature, T_f, equals the mainstream temperature, T_m. $\eta = 1$ represents highest effectiveness, the film temperature, T_f, equals the coolant temperature, T_c. When film cooling takes place, the driving temperature for the convective heat

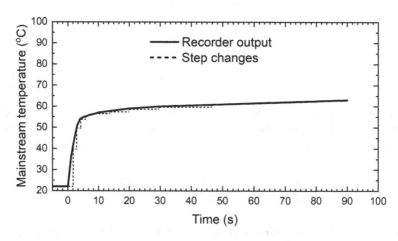

FIGURE 7.7   Typical time history of the mainstream temperature.

transfer becomes the film temperature, instead of the mainstream temperature. Thus, T_m in Equation (7.3) is replaced with the film temperature, T_f. The governing equation becomes:

$$T_w - T_i = \left[1 - \exp\left(\frac{h^2 \alpha t}{k^2}\right) erfc\left(\frac{h\sqrt{\alpha t}}{k}\right)\right] \cdot \left[\eta T_c + (1-\eta)T_m - T_i\right] \qquad (7.7)$$

The above equation has two unknowns: η and h. In order to solve the two unknowns, two separate tests, under different conditions, are needed. In the first test, the mainstream is heated to a desired temperature and the coolant and surface are at room temperature. In the second test, the mainstream and coolant are both heated to a desired temperature and the surface is at room temperature. The heated mainstream temperature (T_m) varies as a function of time in the first test. Both the heated mainstream and coolant temperatures (T_c) vary with time in the second test. Duhamel's superposition method is again used to account for the temperature variations and the governing equations for the two tests can be written as:

$$T_w - T_i(t1) = \sum_{n=1}^{N} \left(1 - \exp[\beta 1] \cdot erfc\left(\sqrt{\beta 1}\right)\right) \cdot \left(\eta T_{c1} + (1-\eta)\Delta T_{m1} - T_{i1}\right) \qquad (7.8)$$

$$\beta 1 = \frac{h^2 \alpha \cdot (t1 - t(n))}{k^2} \qquad (7.9)$$

$$T_w - T_i(t2) = \sum_{n=1}^{N} \left(1 - \exp[\beta 2] \cdot erfc\left(\sqrt{\beta 2}\right)\right) \cdot \left(\eta \Delta T_{c2} + (1-\eta)\Delta T_{m2} - T_{i2}\right) \qquad (7.10)$$

$$\beta 2 = \frac{h^2 \alpha \cdot (t2 - t(n))}{k^2} \qquad (7.11)$$

where ΔT_m and ΔT_c are step changes in the mainstream temperature and coolant temperature, respectively. Equations (7.8) and (7.10) are iteratively solved using a numerical solver. The h obtained is used to calculate η in Equation (7.10).

Example 7.1: Jet Impingement on Channel Surface with Spent Air Exiting from Two Different Directions [7]— Transient, Yellow Band Tracking Method

The experimental setup is shown in Figure 7.8. The data acquisition system includes a CCD camera, a computer to acquire the color image of the test surface, and a stable, white light to illuminate the test surface. The test section consists of a liquid crystal coated surface, an impingement jet plate, and a plenum (thermally insulated). The air flow comes from an upstream compressor. The mass flow rate can be measured by an orifice flowmeter. An inline heater is used to preheat the air. A valve is attached downstream of the heater to direct the heated flow to the bypass loop or the test loop [7].

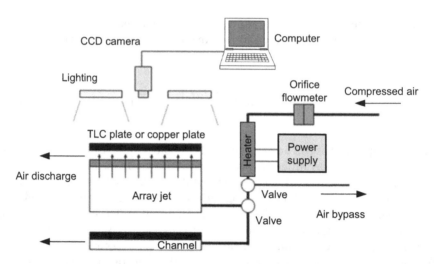

FIGURE 7.8   Experimental setup of impingement cooling with a liquid crystal imaging system.

The following steps describe the experimental procedure:

1. Switch the valve to direct the flow to the test section.
2. Use the flow regulator to control the flow rate to a desired value (the desired jet Reynolds number).
3. Switch the valve to divert the flow to the bypass loop. Remember the pressure difference of the orifice flowmeter in case fluctuation of the flow occurs (note that this pressure difference will not be the same as the pressure difference when the flow was directed to the test section). If the flow rate is changing significantly, you can always adjust it back to the desired value.
4. Turn on the heater, and preheat the air gradually until it reaches about 45°C to 65°C (depending on the Reynolds number, liquid crystal range, and the time required for color change).
5. Turn off the room lights and turn on the white lights surrounding the test section.
6. Start temperature acquisition (jet temperature).
7. Switch the valve to route the preheated air to the test section. Start image recording at the same time ($t = 0$) at a rate of at least 1–5 frames/second. To ensure the validity of the one-dimensional, semi-infinite solid, heat transfer assumption, the entire transient, heat transfer test will last no more than 90 s.

After the experiment, all the acquired images will be analyzed by a computer program. The time required for each pixel (location) to reach a specific temperature of 32.7°C (green color) will be recorded. With the information of time, wall temperature, impinging jet temperature, one can solve the local heat transfer coefficient implicitly (Equation (7.4)) using a numerical solver.

Figure 7.9 shows the concept of crossflow effect on impingement cooling within jet array. The measured Nusselt number distribution is shown in Figure 7.10. Heat transfer coefficients are the highest under impinging jets and then gradually decreased along the spent air direction from left to right.

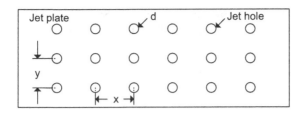

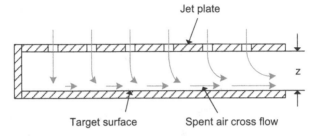

FIGURE 7.9   Concept of crossflow on impingement cooling within a jet array.

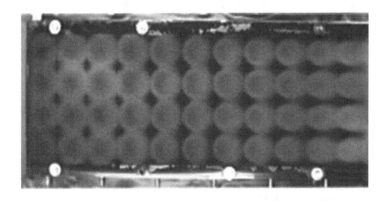

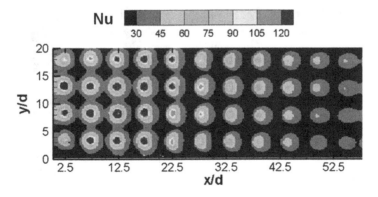

FIGURE 7.10   Temperature (upper) and Nusselt number (lower) distributions under 4 by 12 impinging jet array using transient liquid crystal image technique.

Example 7.2: Two-Pass Square Channel with Smooth and Rib-Roughened Walls [8]—Transient, Yellow Band Tracking Method

A two-pass square channel with a 180° turn is used to simulate the internal flow channels of a gas turbine blade. A smooth channel and two different rib configurations were tested, as shown in Figure 7.11. The study also included the effect of local coolant extraction for film cooling. Detailed Nusselt number ratio (Nu/Nu_0) distributions for all configurations (including the smooth case) were presented. The flow Reynolds number was 30000. The results show detailed flow/heat

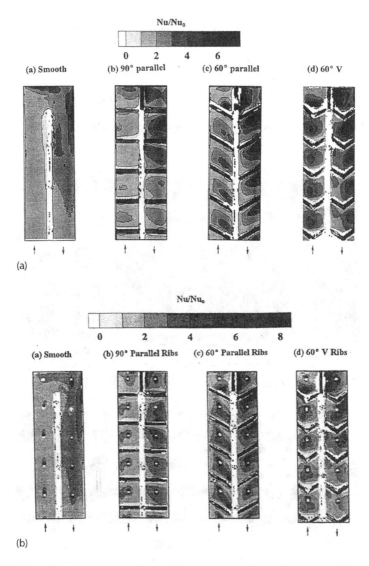

FIGURE 7.11 Heat transfer enhancement distribution in two-pass smooth and ribbed channels with and without film coolant extraction holes: (a) Without bleed holes and (b) With bleed holes.

transfer phenomena such as turn effect, rib induced secondary flow, flow separation, and reattachment. These details cannot be seen with a conventional non-optical method.

The experimental setup and procedures are similar to Example 7.1. Readers may refer to the reference for more details [8].

Example 7.3: Flat Surface Film Cooling [9]—Transient, Yellow Band Tracking Method

Film cooling has become one of the primary measures to lower the wall temperature of turbine components. Film cooling on a flat surface is the most fundamental way to simulate the film cooling on a turbine blade or other components. Figure 7.12 shows the schematic of the experimental setup, top view of the flat surface, film hole configuration, and a cross-sectional view of the test plate [9].

The experimental procedures are similar to Example 7.1 except that two tests are required in order to solve Equations (7.6) and (7.7,) and the coolant and bypass loop valves are switched simultaneously. The results are shown in the Figure 7.13. At a fixed blowing ratio ($M = 1$), both the heat transfer coefficient and the film cooling effectiveness are enhanced at a larger compound angle, β.

Additional References: The aforementioned transient, TLC measurements have been applied to determine the local heat transfer coefficient and film cooling effectiveness distributions when air, or carbon dioxide, is injected through the film-cooling holes, by measuring the time required to reach a specific hue

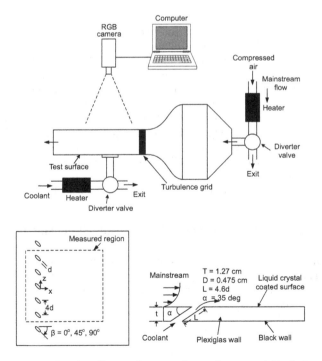

FIGURE 7.12 Typical flat plate film cooling experimental setup and film hole configuration.

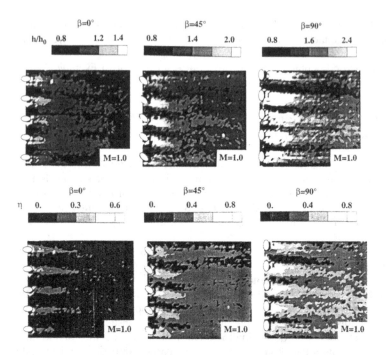

FIGURE 7.13 Heat transfer coefficient and film cooling effectiveness distributions for various compound angle holes.

(surface temperature) value. Several selected papers have been published using this methodology for cylindrical, leading-edge film cooling by Ekkad et al. [10], unsteady flow effect on turbine blade film cooling by Du et al. [11], turbine blade film cooling with shaped holes by Teng et al. [12], turbine blade tip film cooling by Kwak and Han [6], and turbine blade trailing edge film cooling by Choi et al. [13]. This transient TLC measurements have also been applied to determine the local heat transfer coefficient distributions for cylindrical leading-edge region with internal normal and tangential impinging jets without film cooling holes by Wang et al. [14] as well as with film cooling holes by Zhang et al. [15], and a realistic turbine blade internal cooling system by Shiau et al. [16]. Interested readers can read these papers for their detailed flow loop design, test geometry, experimental procedure, and data analysis and results presentation.

Example 7.4: Flat Surface Film Cooling—Steady-State, HSI Technique

Detailed film cooling effectiveness distributions will be obtained on a flat plate using the HSI steady-state liquid crystal technique. The current steady-state liquid crystal method utilizes a true heat transfer experiment to determine adiabatic film cooling effectiveness. Using the traditional definition, the film cooling effectiveness, η, is defined as

$$\eta = \frac{T_{aw} - T_{\infty}}{T_c - T_{\infty}}$$

With the steady-state, liquid crystal method, the coolant (T_c) and mainstream temperatures (T_∞) are measured using thermocouples placed in these respective fluid streams. A single measurement is taken for these temperatures after the entire experimental setup has achieved steady-state. The adiabatic wall temperature (T_{aw}) is obtained from the color distribution of the liquid crystal. Therefore, this steady-state technique is capable of providing detailed film cooling effectiveness distributions.

Based on the decades of studies, groups have shown a simplified representation of a film cooling jet: the film cooling jet is essentially a jet in crossflow that can be approximated as mainstream flow around a cylinder. The flow in the region of the jet is highly three-dimensional, with vortices developing both in the jet and along the cooled surface. In an effort to weaken the effect of the vortices forming on the film cooled surface and within the jet, a second row of film cooling holes are introduced upstream of the primary row of holes. This upstream row of holes should protect the primary, inclined holes. The following figures show the geometry which will be considered in this example (Figure 7.14).

Figure 7.15 shows a schematic of the small scale, low speed wind tunnel used for this liquid crystal experiment. The tunnel is used to experimentally investigate the proposed, double hole geometry. The tunnel has a cross section of 6" × 4", and the mainstream velocity through the tunnel is approximately 7.2 m/s (as monitored with a pitot-static tube). As measured using a hot wire anemometer, the freestream turbulence intensity in the tunnel is approximately 1.2%.

The following raw data were obtained for this flat plate, film cooling example:

1. Coolant-to-mainstream mass flux ratio: $M = \dfrac{\rho_c V_c}{\rho_\infty V_\infty} = 0.5$
2. $T_c = 52.27°C$
3. $T_\infty = 21.47°C$
4. Raw bitmap image (red, green, blue) of the plate at steady-state conditions could be obtained
5. Calibration curve for the R25C15 liquid crystal paint: Higher hue = Higher temperature (Figure 7.16)

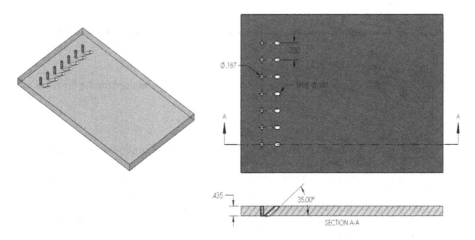

FIGURE 7.14 Anti-vortex hole geometry for steady state, liquid crystal experiment (units shown in inches).

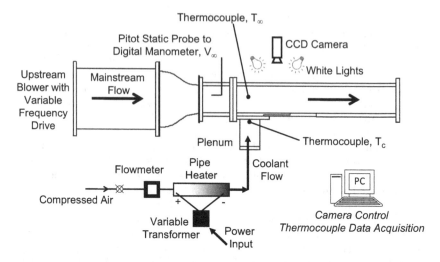

FIGURE 7.15 Overview of the experimental setup for the steady state liquid crystal experiment.

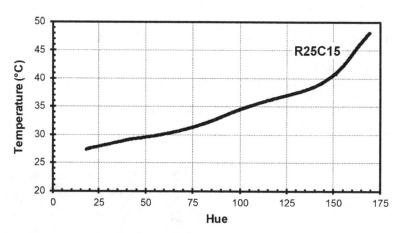

FIGURE 7.16 Hue—Temperature calibration for the R25C15 paint used for the film cooling experiment.

From the given information, the following steps can be completed to ultimately calculate the detailed, film cooling effectiveness distribution on the surface.

The raw bitmap image can be analyzed, so the hue at every pixel is evaluated: Higher hue values around and downstream of film holes (Figure 7.17).

Combining the hue distribution, with the calibration data, the surface temperature at every pixel is calculated: Higher surface temperatures around and downstream of film holes (Figure 7.18).

Finally, the surface temperature can be combined with the thermocouple measurements of the coolant and mainstream temperatures to yield the detailed film

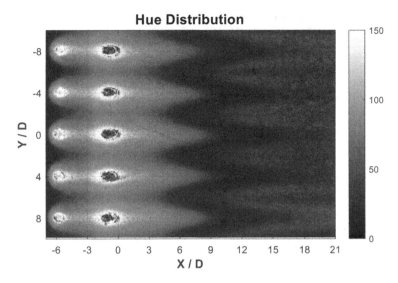

FIGURE 7.17   Steady state, hue distribution on the film cooled surface.

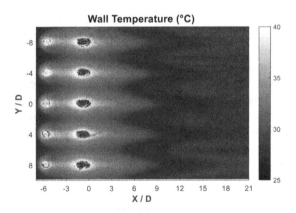

FIGURE 7.18   Steady state, temperature distribution on the film cooled surface.

cooling effectiveness on the surface of the flat plate: Higher film effectiveness around and downstream of film holes (Figure 7.19).

These results highlight one key disadvantage of this type of steady-state experiment. With the hot air flowing into the plenum and through the film cooling holes, the entire volume of the test plate is heating up. Therefore, conduction through the plate is occurring in all directions. The data presented has not been corrected for conduction. The elevated film cooling effectiveness upstream of the film cooling holes is an artifact of the method. Finite element methods can be used to remove the conduction through the plate.

Also, given the way this experiment was conducted, only the film cooling effectiveness was measured. In order to obtain a distribution of the heat transfer coefficients on the surface, a heater and power source would be needed to heat the surface. The power supplied to the heater could be measured using a digital multimeter.

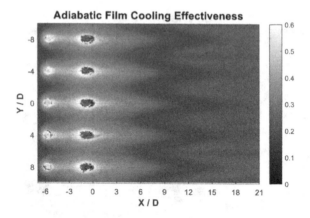

FIGURE 7.19   Film cooling effectiveness distribution on the flat plate.

PROBLEMS

1. You are asked to prepare an experimental setup to evaluate the proposed cylindrical, leading-edge region, film cooling geometry for gas turbine blade applications. The required compressed air (25°C), heaters, piping, flow meters, materials, thermocouples, CCD camera, and equipment for calibrations and measurements are available for your use. Please use the transient TLC method to map the surface temperature, heat transfer coefficient, and film cooling effectiveness distributions on the leading-edge surface as shown in below image. Based on the wind tunnel flow and cooling/heating flow capability, one steady-state blowing ratio (cooling to mainstream mass flux ratio) will be tested: $M = 1.0$.

   a. Show an overall sketch of your experimental test loop and the associated equipment and instrumentation required to conduct the proposed test, i.e., sketch your overall experimental setup along with a suitable flow measuring device and detail your test section design with proper materials using the transient TLC imaging technique.
   b. Describe the transient TLC measurement principle, TLC calibration method, and detail the TLC measurement procedures using the transient technique with heated air.
   c. Sketch and discuss the expected target surface temperature distributions (contours), heat transfer coefficient and film cooling/heating effectiveness distributions (contours) for one tested blowing ratio $M = 1.0$ and estimate the possible uncertainty.
2. Repeat Problem 1 by using the steady-state TLC method to map the surface temperature, heat transfer coefficient, and film cooling/heating effectiveness distributions (contours) on the leading-edge surface as shown on the figure from Problem 1. Compare the heat transfer coefficient and film cooling effectiveness contours between Problems (1) and (2) at the same blowing ratio. Are they the same or difference? Why?
3. You are asked to prepare an experimental setup to evaluate the proposed circular heating disk through five impinging jets for industrial applications. The required compressed air (25°C), heaters, piping, flow meters, materials, thermocouples, CCD camera, and equipment for calibrations and measurements are available for your use. Use the Liquid Crystal Thermography (LCT) method to map the surface temperature and heat transfer coefficient distributions on the top target wall in the region of interest as shown in below sketch.

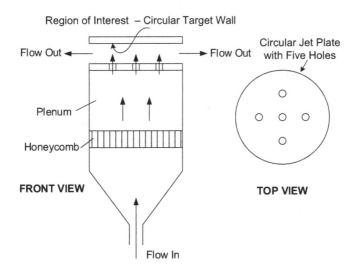

   a. Show an overall sketch of your experimental test loop and the associated equipment and instrumentation required to conduct the proposed test, i.e., sketch your overall experimental setup along with a suitable flow measuring device and detail your test section design with proper materials using the transient liquid crystal imaging technique. Based on the jet-hole diameter (1 cm), calculate the required mass flow rates for two steady flow Reynolds numbers at $Re = 10,000$ and $30,000$, respectively.

b. Describe the LCT measurement principle, LCT calibration method, and detail the LCT measurement procedures by using the transient technique with heated air.

c. Sketch and discuss the expected target surface temperature distributions (contours) as well as heat transfer coefficient distributions (contours) for two tested Reynolds numbers at time, $t = 50$ seconds and 100 s and estimate the possible uncertainty.

4. Repeat Problem 3 by using the steady-state, Liquid Crystal Thermography (LCT) method to map the surface temperature and heat transfer coefficient distributions on the top target wall in the region of interest as shown in sketch provided with Problem 3.

5. Prepare experimental methods to evaluate the surface temperature and heat transfer coefficient distributions for steady, incompressible turbulent air flow through a two-passage, rectangular cooling channel with a 180-deg turn for heat exchanger applications, as sketched below. You should pay close attention to the region of interest shown in the sketch.

a. Use the Liquid Crystal Thermography (LCT) method to map the surface temperature distributions on the bottom wall as shown in the region of interest.

b. Sketch the primary components of the experimental setup; describe the measurement principle using the steady-state heating method from a constant heat flux heater.

c. Sketch and discuss the expected experimental results (surface heat transfer coefficient distributions), and perform an uncertainty analysis.

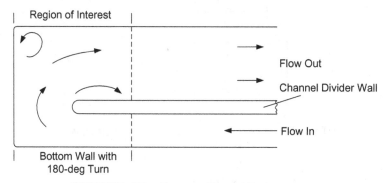

TOP VIEW of Two-Passage Channel Design

6. Repeat Problem 5 by using the transient Liquid Crystal Thermography (LCT) method to map the surface temperature and heat transfer coefficient distributions in the turn region of the cooling channel, as shown in the sketch with Problem 5.

7. You are asked to prepare experimental methods to evaluate the surface temperature and heat transfer coefficient distributions for steady, incompressible, turbulent air flow through a two-dimensional channel with a sudden expansion outlet for heat exchanger applications, as sketched below. The required compressed air (25°C), heaters, piping, flow meters, materials, thermocouples, CCD camera, and equipment for calibrations and measurements are available for your use. Use the steady-state, TLC method to map the surface temperature and heat transfer coefficient distributions on the bottom wall in region of interest, as shown in below sketch.

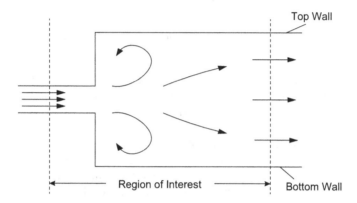

a. Show an overall sketch of your experimental test loop and the associated equipment and instrumentation required to conduct the proposed test, i.e., sketch your overall experimental setup along with a suitable flow measuring device and detail your test section design with proper materials using the steady-state TLC camera technique.

b. Based on channel hydraulic diameter, two steady flow Reynolds numbers will be tested: Re = 10,000, and 30,000. Describe the TLC measurement principle, TLC calibration method, and detail the TLC measurement procedures by using the steady-state, surface heating technique.

c. Sketch and discuss the expected temperature distributions (contours) as well as heat transfer coefficient distributions (contours) downstream of the expansion, for the two tested Reynolds numbers at steady-state and estimate the possible uncertainty.

8. Repeat Problem 7 using the transient Liquid Crystal Thermography (LCT) method to map the surface temperature and heat transfer coefficient distributions through the expansion, as shown with the sketch for Problem 7.

9. Prepare experimental methods to evaluate the temperature and heat transfer coefficients for steady, incompressible, turbulent air flow through a two-dimensional channel with protrusions on one wall, as shown in the following sketch. Focus on the region interest shown in the sketch. Use the Liquid Crystal Thermography (LCT) method to map the surface temperature distributions on the bottom wall as shown in the region of interest.

a. Sketch the experimental setup including the primary pieces of instrumentation, and describe the measurement principle using the steady-state, uniform heat flux, heating technique.

b. Sketch and discuss the expected experimental results (surface heat transfer coefficient distributions) and perform uncertainty analysis.

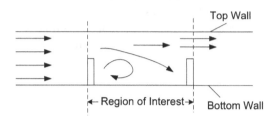

10. Repeat Problem 9 using the transient Liquid Crystal Thermography (LCT) method to map the surface temperature and heat transfer coefficient distributions on the rib roughened wall, in the region of interest, as shown in sketch with Problem 9.

11. Prepare an experimental method to the evaluate surface temperature and heat transfer coefficient distributions for steady, incompressible, turbulent air flow through a square channel with nine circular tubes running between the top and bottom walls, as sketched below. Pay attention to the region of interest shown in the sketch. Use the Liquid Crystal Thermography (LCT) method to map the surface temperature distributions on the bottom wall.

a. Sketch the experimental setup with all necessary instrumentation and hardware; describe the measurement principle using the transient technique with heated air.

b. Sketch and discuss the expected experimental results (surface heat transfer coefficient distributions) and perform an uncertainty analysis.

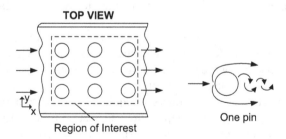

12. Repeat Problem 11 by using the steady-state, Liquid Crystal Thermography (LCT) method to map the surface temperature and heat transfer coefficient distributions on the bottom wall of the channel, as indicated by the region of interest in the sketch with Problem 11.

REFERENCES

1. Hippensteele, S.A., Russell, L.M., and Torres, F.J., "Local Heat-Transfer Measurements on a Large Scale-Model Turbine Blade Airfoil Using a Composite of a Heater Element and Liquid Crystals," *ASME, the 30th International Gas Turbine Conference*, Houston, Texas, March 18–21, ASME Paper No. 85-GT-59, 1985

2. Camci, C., Kim, K., and Hippensteele, S.A., "A New Hue Capturing Technique for the Quantitative Interpretation of Liquid Crystal Images Used in Convective Heat Transfer Studies," *ASME Journal of Turbomachinery*, Vol. 114, 1992, pp. 765–775.

3. Pritchard, D.H., "US Color Television Fundamentals—A Review," *IEEE Transactions on Consumer Electronics*, V. CE-23, 1977, pp. 467–478.

4. Ekkad, S.V. and Han, J.C., "A Transient Liquid Crystal Thermography Technique for Gas Turbine Heat Transfer Measurements," *Measurement Science and Technology*, Vol. 11, 2000, pp. 957–968.

5. Han, J.C., Dutta, S., and Ekkad, S.V., Chapter 6-Experimental Methods, in *Gas Turbine Heat Transfer and Cooling Technology*, Taylor & Francis, Inc., New York, December, 2000.

6. Kwak, J.S. and Han, J.C., "Heat Transfer Coefficient and Film Cooling Effectiveness on the Squealer Tip of a Gas Turbine Blade," *ASME Journal of Turbomachinery*, Vol. 125, 2003, pp. 648–657.

7. Huang, Y., Ekkad, S.V., and Han, J.C., "Detailed Heat Transfer Distributions under an Array of Orthogonal Impinging Jets," *AIAA Journal of Thermophysics and Heat Transfer*, Vol. 12, No. 1, 1998, pp. 73–79.

8. Ekkad, S.V., Huang, Y., and Han, J.C., "Detailed Heat Transfer Distributions in Two-Pass Smooth and Turbulated Square Channels with Bleed Holes," *The International Journal of Heat Mass Transfer*, Vol. 41, No. 23, 1998, pp. 3781–3791.

9. Ekkad, S.V., Zapata, D., and Han, J.C., "Heat Transfer Coefficients over a Flat Surface with Air and CO_2 Injection through Compound Angle Holes Using a Transient Liquid Crystal Image Method," *ASME Journal of Turbomachinery*, Vol. 119, No. 3, 1997, pp. 580–586.

10. Ekkad, S.V., Han, J.C., and Du, H., "Detailed Film Cooling Measurements on a Cylindrical Leading Edge Model: Effect of Free-Stream Turbulence and Coolant Density," *ASME Journal of Turbomachinery*, Vol. 120, No. 4, 1998, pp. 799–807.

11. Du, H., Han, J.C., and Ekkad, S.V., "Effect of Unsteady Wake on Detailed Heat Transfer Coefficient and Film Effectiveness Distributions for a Turbine Blade," *ASME Journal of Turbomachinery*, Vol. 120, No. 4, 1998, pp. 808–817.

12. Teng, S., Han, J.C., and Poinsatte, P., 2001, "Effect of Film-Hole Shape on Turbine Blade Film Cooling Performance," *AIAA Journal of Thermophysics and Heat Transfer*, Vol. 15, No. 3, 1998, pp. 266–274.

13. Choi, J., Mhetras, S.P., Han, J.C., Lau, S.C., and Rudolph, R.J., "Film Cooling and Heat Transfer on Two Cut-Back Trailing Edge Models with lateral Perforated Blockages," *ASME Journal of Heat Transfer*, Vol. 130, No.1, 2008, Article No. 012201, 13 pages.

14. Wang, N., Chen, A.F., Zhang, M., and Han, J.C., "Turbine Blade Leading Edge Cooling with One Row of Normal or Tangential Impinging Jets," *ASME Journal of Heat Transfer*, Vol. 140, No. 6, 2018, Article No. 062201, 10 pages.

15. Zhang, M., Wang, N., and Han, J.C., "Internal Heat Transfer of film-cooled Leading Edge Model with Normal and Tangential Impinging Jets," *International Journal of Heat and Mass Transfer*, Vol. 139, 2019, pp. 193–204.

16. Shiau, C.C., Chen, A., Han, J.C., and Krewinkel, R., "Detailed Heat Transfer Coefficient Measurements on a Scaled Realistic Turbine Blade Internal Cooling System," *ASME Journal of Thermal Science and Engineering Applications*, Vol. 12, 2020, Article No. 031015, 9 pages.

8 Optical Thermography Techniques

8.1 INFRARED THERMOGRAPHY (IR) TECHNIQUE

Infrared energy was first discovered in 1800 by an astronomer named Sir William Herschel. He split light into its spectrum of colors and measured the temperature for each color. He found that the temperature increased from violet to red with the hottest temperature beyond red. This energy beyond the visible color spectrum is called infrared energy. Thus, infrared radiation stretches from a wavelength of approximately 1 micron (near-infrared) to 200 microns (far-infrared). Figure 8.1 shows the spectral distribution for different classes of radiation [1,2].

All objects in the universe absorb and emit thermal radiation and the amount an object will absorb or emit depends on the temperature of that object. The radiated infrared energy will be zero when the object is at absolute zero temperature and increases with increasing temperature. The heat radiated by a "black body" per unit time is given by the Stefan-Boltzmann law of thermal radiation. The amount of radiation emitted by a surface is directly related to the temperature of the body as shown in Figure 8.2 and represented by Equation (8.1) [1,2].

$$\dot{Q} = \sigma A T^4 \tag{8.1}$$

where $\dot{Q}$ is the emitted radiation per unit time, σ is the Stefan-Boltzmann constant, A is the surface area of the black body (m^2), and T is its absolute temperature, in Kelvin, of the object's surface. The net heat flow between two black bodies at different temperatures of T_1 and T_2 is given by

$$\dot{Q} = \sigma A \left(T_1^4 - T_2^4 \right) \tag{8.2}$$

The net heat flow between two bodies that emit less radiation than a black body can be expressed by introducing a parameter known as the emissivity. The emissivity of an object is a measure of the objects efficiency in emitting radiation as compared to a black body.

$$\dot{Q} = \varepsilon \sigma A \left(T_1^4 - T_2^4 \right) \tag{8.3}$$

Cameras which measure infrared radiation utilize the above equation to measure the surface temperature of an object. If its emissivity is known, the infrared sensor in the camera, composed of quantum detectors, measures the net heat flux from the object, thus calculating its surface temperature. The quantum detectors convert the photon energy of the particular range of wavelengths into an electrical signal by releasing electrons. The electrical output generated by these detectors is very small and can be overshadowed by the background noise generated by the device. Hence, to improve the signal to

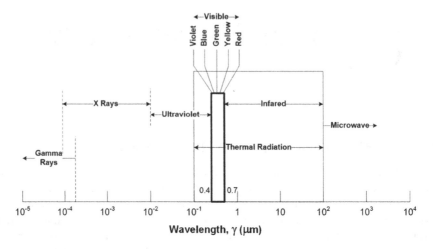

FIGURE 8.1   Spectrum of electromagnetic radiation.

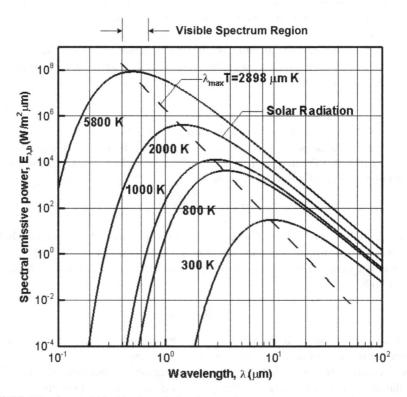

FIGURE 8.2   Spectral blackbody emissive power.

noise ratio from these detectors, they must be cooled to cryogenic temperatures. Using an array of sensors, the IR camera can resolve the surface into individual pixels and produce a monochromatic image of the temperature distribution on the surface. To get a better visual perception of the temperature distribution on the surface, false color is often assigned to different intensities of the monochrome image to convert it to color.

8.2 FUNDAMENTAL PRINCIPLE

The first IR cameras were developed for military use and became available in the commercial arena in 1960s. Several advancements in IR measurement technology have been made in the last few decades in terms of improving the resolution and accuracy of the measured temperatures and also making them more affordable. The infrared thermography technique is based on the measurement of surface radiation using an infrared camera. Figure 8.3 shows the simplified concept of the IR camera principle. When the camera is facing the target surface (known ε), the heat flux (~V voltage) from emitted photon is captured by IR sensor, T_1 is thus calculated from the above-mentioned radiation exchange equation (for a known target surface emissivity and a known sensor temperature T_2). In reality, one can calibrate the relationship between T_1 and voltage V where T_1 is first measured by the thermocouple and the voltage V is measured by IR sensor. The IR sensor material is very similar to those of photovoltaic cells.

The temperature of the test surface can be recorded using an the IR camera and the distributions converted to color images. These color images can then be subsequently processed to give an accurate temperature map of the surface by converting the color magnitudes into temperature using the available calibration relationship. Most advanced IR cameras come pre-calibrated from the factory for temperature and color relationships. If a calibration is not available, they can be calibrated using a simple setup composed of a heated surface with thermocouples for temperature measurement. The known temperature from the thermocouples can be then calibrated against the color depicted by the IR camera. Accurate knowledge of the surface emissivity is critical for reading the correct temperature. For this reason, the surface may be coated with a non-reflective black paint for high emissivity. Figure 8.4 shows an IR calibration setup and calibration curve between the IR camera and thermocouples.

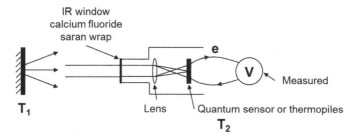

FIGURE 8.3 Sketch of simplified concept of IR camera principle.

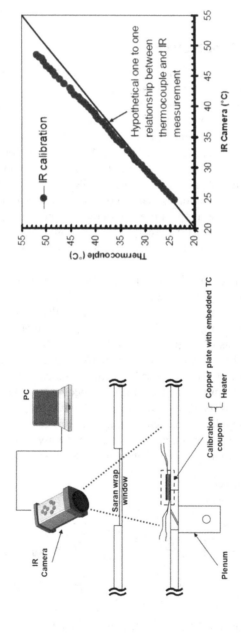

FIGURE 8.4 IR calibration setup and calibration curve between IR camera and thermocouples.

The infrared thermography technique is a powerful method, which can give detailed distributions of the local heat transfer coefficients on the surface. It has several inherent advantages over other optical techniques such as liquid crystal thermography [3]:

1. The IR method is not bound by an operating temperature range, as in the case of liquid crystals.
2. Typically, only black paint needs to be sprayed on the test surface for IR thermography. In case of liquid crystal thermography, the liquid crystal also needs to be painted on the test surface. Thus, the IR technique is free from errors due to non-uniform paint distribution, thickness of coating, etc.
3. For transient tests, the IR method can give the initial temperature of the test surface which is difficult to obtain using liquid crystals.
4. IR cameras do not require a sophisticated illumination system as in the case of LCT.
5. Most IR cameras are pre-calibrated for temperature, which reduces the time and cost for a detailed calibration for hue against temperature required for wide band liquid crystals.

A major disadvantage of the IR thermography technique is the high cost and sophistication of the IR camera itself. In addition, windows made from special materials such as zinc selenide, transparent to infrared radiation, must be fabricated in order to capture surface temperatures. Blair and Lander [4] used this technique for film effectiveness measurements in gas turbine applications. Other researchers such as Gritsch et al. [5] and Kohli and Bogard [6] have since used IRT for adiabatic wall temperature measurements. Since 2000, the following selected research groups have been continuously using the IR camera technique for turbine airfoil surface heat transfer coefficient, film cooling effectiveness, and internal cooling channel heat transfer measurements: Ekkad et al. [7], Lynch and Thole [8], Wright et al. [9], and Gao et al. [10]. Interested readers can read these papers for their detailed flow loop design, test geometry, experimental procedure, data analysis, and results presentation.

8.3 STEADY-STATE, INFRARED THERMOGRAPHY TECHNIQUE

In this method, the heat transfer coefficients are obtained on a heated surface. A thin foil heater is generally used as the heat source. By applying a suitable voltage across the heater, the desired temperature distribution can be obtained. A non-reflective black paint with a known emissivity is applied to the test surface on top of the heater. The black paint will result in a high emissivity uniformly across the surface. The test surface is placed in the test section or wind tunnel and the mainstream flow is turned on. Heat is applied to the heater such that a temperature difference, between the surface and the fluid, of approximately 20°C is maintained. Once temperature equilibrium conditions are obtained on the test surface, an image of the surface is captured by the IR camera through a zinc selenide (or other material transparent to IR) window. The power input to the heater, the mainstream flow rate, and mainstream temperature are measured. With the knowledge of heat loss from the heater to ambient surroundings, the net heat input can be calculated. The heat transfer coefficients for each pixel can then be calculated using Equation (8.4).

$$h = \frac{q''}{T_{w,IR} - T_\infty} = \frac{q''_{gen} - q''_{loss}}{T_{w,IR} - T_\infty} \qquad (8.4)$$

Additional procedural details for the steady-state method can be found in Chapter 5.

8.4　TRANSIENT, INFRARED THERMOGRAPHY TECHNIQUE

The general transient techniques described in Chapter 6 can be used with infrared cameras. The IR camera can be used to measure detailed surface temperature distributions at known times and from the relationship between surface temperature and time, the heat transfer coefficients can be calculated.

Ekkad et al. [7] demonstrated the use of an IR camera in transient tests to simultaneously measure detailed heat transfer coefficient and film cooling effectiveness distributions. The technique was conducted in a manner similar to the transient HSI technique with liquid crystals. At a given instant in time, the surface temperature distribution is measured. In the case of Ekkad et al. [7], two unknowns are present, so two data sets are needed. With the IR camera, these can easily be obtained from a single test but with two different times. This method is more challenging with the wide band liquid crystal, as the temperature may exceed the range of the paint. IR cameras are advantageous over liquid crystal paints, as IR cameras are not limited to relatively small temperature ranges. The IR camera is essentially taking the place of an infinitely, wideband liquid crystal. In addition to measuring the wall temperature, as with any transient technique, the researchers must have an accurate way of measuring time and a working knowledge of the material properties of the test surface.

This transient method is similar to the IR method developed by Ekkad et al. [7], and it is based on the transient liquid crystal method described by Vedula and Metzger [11]. With the film temperature replacing the mainstream temperature in the solution to the transient, one-dimension, semi-infinite solid equation, the following equations can relate the surface heat transfer coefficient and film cooling effectiveness.

$$\frac{T_w(t) - T_i}{T_f - T_i} = 1 - \exp\left(\frac{h^2 \alpha t}{k^2}\right) erfc\left(\frac{h\sqrt{\alpha t}}{k}\right)$$

$$\eta = \frac{T_f - T_m}{T_c - T_m}$$

$$\frac{T_w(t) - T_i}{\eta T_c + (1-\eta)T_m - T_i} = 1 - \exp\left(\frac{h^2 \alpha t}{k^2}\right) erfc\left(\frac{h\sqrt{\alpha t}}{k}\right)$$

Both the film cooling effectiveness, η, and the heat transfer coefficient, h, are unknown in the above equation. Therefore, two equations are required to solve for the two unknowns. The two data sets can be obtained by using the IR camera to measure the wall temperature at two separate times. With both the heat transfer coefficient and the film cooling effectiveness being independent of time, the following equations represent the system of two equations and two unknowns that can be solved simultaneously.

$$\frac{T_w(t_1)-T_i}{\eta T_c+(1-\eta)T_m-T_i}=1-\exp\left(\frac{h^2\alpha t_1}{k^2}\right)erfc\left(\frac{h\sqrt{\alpha t_1}}{k}\right)$$

$$\frac{T_w(t_2)-T_i}{\eta T_c+(1-\eta)T_m-T_i}=1-\exp\left(\frac{h^2\alpha t_2}{k^2}\right)erfc\left(\frac{h\sqrt{\alpha t_2}}{k}\right)$$

8.5 EXPERIMENT EXAMPLE

Example 8.1: Steady-State IR Measurement of the Film Cooling Effectiveness on a Flat Plate [9]

In this experiment, surface temperature distributions were measured on a film cooled, flat plate using an IR camera. Based on the surface temperature measurements, the detailed film cooling effectiveness was calculated. The coolant flow was heated (T_c) while the mainstream was unheated (T_m). From the mixture of the hot coolant and cold mainstream, the film effectiveness on the plate surface can be calculated using the following equation at every pixel.

$$\eta=\frac{T_f-T_m}{T_c-T_m}\cong\frac{T_{aw}-T_m}{T_c-T_m}$$

A single row of seven cylindrical holes design (Figure 8.5) were used for this study. The holes ($D = 4$ mm) are equally spaced 12 mm apart, and the hole length-to-diameter ratio is 9.92. The axial angle (θ on x–z plane) of the holes is 30°, and the compound angle (β on x–y plane) is 45°. The effect of the coolant blowing ratio ($M = 0.4$, 0.6, 1.2, and 1.8) and mainstream turbulence intensity are considered by measuring effectiveness measurements.
　　Experiential Procedure

1. A low speed, suction-type, wind tunnel facility (Figure 8.6) with a mainstream velocity of 25 m/s is used for this study. The test channel cross section is 30.48 × 15.24 cm where the film cooling holes are located 21.75 cm ($x/D = 54.4$) from the turbulence grid. A turbulence grid is set upstream of the test surface to generate a turbulence intensity of 6% near the film cooling test plate; without that grid the freestream turbulence is approximately 0.5% near the film cooling holes. The test plate material is Plexiglas.
2. The coolant air, supplied from a compressor, passes through a flow control valve and orifice flow meter. The coolant then passes through a 5-kW pipe heater and bypass valve before it enters the air plenum, located underneath the film cooling plate.

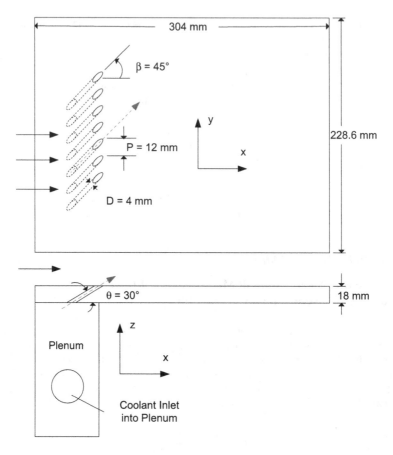

FIGURE 8.5   Details of the compound angle, film cooling holes.

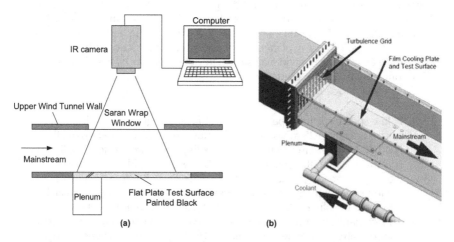

FIGURE 8.6   Schematic of the overall test section. a) Infrared set up b) 3D overview of wind tunnel.

3. A T-type thermocouple is used to measure the inlet mainstream temperature. For the coolant temperature, two T-type thermocouples are attached on the bottom of the film cooling plate—one at each entrance of the two outside holes. The thermocouple readings are measured by either a Fluke 2285 B Data Logger or with National Instruments' LabVIEW software.

4. A central air-conditioning system maintained the mainstream temperature (T_m) at 22°C. The coolant temperature (T_c) is heated to 43.3°C.

5. A Mikron Thermo Tracer 6T62 is used to measure the surface temperatures. The IR-system consists of an optical scanner which directs the incoming infrared radiation line by line onto the detector working in a wavelength bandwidth of 8–13 microns. The IR camera views the test surface through a sheet of Vinylidene Chloride-Vinyl Chloride copolymer (Saran Wrap food wrap), which serves as an infrared window in the top wall of the wind tunnel.

6. In the calibration procedure, a thermocouple was attached to the surface of the copper block, and the copper block was painted black to increase the emissivity. The IR camera views the test surface through a sheet of Vinylidene Chloride-Vinyl Chloride copolymer (Saran Wrap food wrap), which serves as an infrared window in the top wall of the wind tunnel. Temperatures recorded by the thermocouple were compared to temperatures recorded by the IR camera. The relationship between the thermocouple and IR measurements is shown in Figure 8.4. With this calibration curve, the data obtained from the IR camera in the actual film cooling experiment was corrected.

7. After the experiment, all the temperatures are recorded and used to calculate the effectiveness pixel by pixel. A sample calculation is demonstrated for selected pixels in Table 8.1 below. The final results are shown in Figure 8.7.

In Chapter 9, these results will be compared to another steady-state heat transfer method (steady-state, TSP) and the methods relying on mass transfer, rather than heat transfer (PSP). With the steady-state heat transfer method, the plate is heated by the hot air passing through the plenum until the test surface reaches thermal equilibrium. To achieve steady-state, the plate may be heated by the hot air passing through the cooling holes for approximately one hour. During this time, conduction is occurring through the plate (as also shown with the steady-state

TABLE 8.1

Sample Film Cooling Effectiveness Distributions from Measured Temperatures (IR and Thermocouples)

Location	T_m	T_c	T_w	η
A	22	43.3	30.31	0.39
B	22	43.3	25.62	0.17
C	22	43.3	23.92	0.09

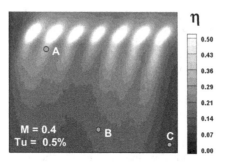

FIGURE 8.7   Sample film cooling effectiveness distribution.

liquid method in Chapter 6). Upstream of the film cooling holes, the effectiveness should be zero; however, the steady-state IR method clearly provides a value much greater than zero. Without numerical correction of the results, this steady-state method is affected by conduction through the plate. The area far downstream of the holes is expected to be less effected by conduction.

Example 8.2: Steady-State IR Measurements with a Foil Heater on a Roughened Flat Plate [12]

For this experiment, the convective heat transfer enhancement of roughness elements is measured in a turbulent flow. The conductivity of the elements is varied to investigate the effect of roughness material on the heat transfer enhancement. The measurements are performed using a low-speed, subsonic wind tunnel. A sheet of Mylar film with a very-thin layer of gold deposited on the surface is used to create a constant heat flux boundary condition on a flat Plexiglas plate. Because the roughness elements will be attached to the plate, the Mylar is glued to the Plexiglas such that the gold deposition layer is in contact with the glue and Plexiglas.

Figure 8.8 demonstrates the general test plate construction. The heater has been painted with black paint to maximize the surface emissivity.

Staggered arrays of both hemispherical and conical roughness elements are considered. In addition, the elements are constructed of both ABS plastic and aluminum. Using an IR camera placed above the wind tunnel, detailed temperature distributions on and around the elements will be measured under steady conditions. The flat plate has a knife edge, leading edge, and the roughness elements are placed on the downstream half of the plate. The mainstream velocity is monitored with a Pitot-static probe, and a thermocouple is used to monitor the mainstream temperature. Figure 8.9, shows the placement of the flat plate in the wind tunnel.

During the tests, the following raw data are measured and required for full data analysis:

ΔP: Pitot-static probe for calculation of mainstream velocity
P_{atm}: atmospheric pressure in the laboratory
ϕ: relative humidity in the laboratory
V: voltage supplied to the gold-Mylar foil heater

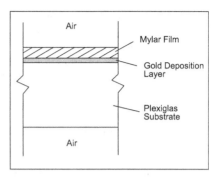

FIGURE 8.8   Basic construction of the flat plate for steady state IR measurements.

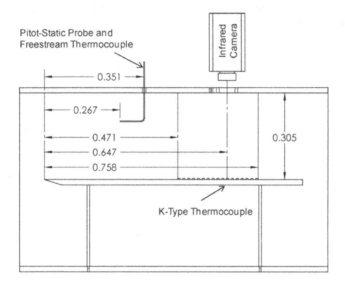

FIGURE 8.9   Placement of the flat plate in the low speed wind tunnel.

I: current supplied to the gold-Mylar foil heater
T_∞: mainstream air temperature
T_p: bottom Plexiglas temperature
T_m: temperature of Mylar surface during test (distribution from IR camera)

The bottom of the Plexiglas is measured during the tests for a real-time calculation of the heat loss through the plate. As shown in Figure 8.8, the foil is attached to Plexiglas and adjacent to the air. However, some fraction of the heat input will conduct through the plate, and this must heat transfer must be taken into account when calculating the heat transfer on the surface.

From these measured quantities, the thermal performance of the roughness elements can be evaluated and compared to flow over a smooth surface.

Reflecting on the previous example and envisioning how this type of experiment should be fabricated, instrumented, and conducted, the following questions can be considered:

1. What is the freestream velocity determined from the Pitot-static pressure measurement?
2. What is the Reynolds number of the flow based on the distance from the leading edge of the plate to the center of the infrared image?
3. Neglecting radiation from the surface and neglecting conduction through the plate, what are the average heat flux from the metered Mylar area and the average heat transfer coefficient?
4. Assuming a value of 0.97 for the emissivity of the painted surface and assuming all radiation interactions are with the walls of the room at the free stream temperature, what is the rate of radiation from the heated surface? What is the corrected convection coefficient?
5. Accounting for the conduction through the Plexiglas yields a data reduction equation of the form

$$h_M = \left(\frac{\dfrac{\dot{W}_E}{A}\dfrac{t_P}{k_P} - (T_M - T_P)}{(T_M - T_\infty)} \right) \left(\frac{t_M}{k_M} + \frac{t_P}{k_P} \right)^{-1} - \frac{\varepsilon\sigma\left(T_M^4 - T_w^4\right)}{(T_M - T_\infty)}$$

Correcting for both the conduction through the Plexiglas and the radiation from the surface of the plate, what is the average convection coefficient?

6. The test plate has an unheated starting length of 0.75 in. The correlation for the constant flux convection coefficient for a flat plate turbulent flow is

$$Nu_x = 0.0296 Re_x^{4/5} Pr^{1/3}$$

with the unheated starting length correction of

$$Nu_x = \frac{Nu_{x,uncor}}{\left[1 - \left(\dfrac{\xi}{x}\right)^{9/10}\right]^{1/9}}$$

Based on the smooth surface convection coefficients from these two correlations, what is the convective enhancement of the distribution using either h_{rough}/h_{smooth} or Nu_{rough}/Nu_{smooth}?

Figure 8.10 shows sample temperature distributions on the roughness elements along with the Stanton number distribution along the length of the plate. The steady-state IR method is capable of detecting heat transfer enhancement due to different types of roughness elements. When modeling a variety of surface features (from heat exchangers to surface icing), it is necessary to include the effects materials.

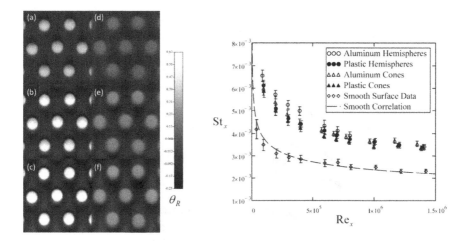

FIGURE 8.10 Sample results for heat transfer enhancement on a roughened surface.

Example 8.3: Transient IR Measurement of the Film Cooling Effectiveness on a Circular Cylinder

In this experiment, cylindrical, leading edge, surface temperature distributions were measured using an IR camera at discrete times.

The film cooled cylinder is shown in Figure 8.11. The polycarbonate cylinder has a diameter of 7.62 cm and is 25.4 cm tall. It is hollow (0.64 cm thick), and a copper cylinder is placed inside the polycarbonate cylinder. Six cartridge heaters are imbedded in the copper cylinder to uniformly heat the cylinder. The copper interior is insulated by a wood cylinder. The film cooling holes are drilled through the wood, copper, and polycarbonate cylinders. Coolant from the pipe heater enters the cylinder at the bottom of the cylinder, and the air is discharged through the film cooling holes. The film holes are ±15° from the stagnation line of the cylinder, and each hole has a diameter of 0.475 cm. The holes are inclined 30° in spanwise direction and 90° in streamwise direction.

Experiential Procedure

1. A low speed suction-type wind tunnel facility (Figure 8.12) with a mainstream Reynolds number of 109,000 is used for this study. The center of the cylinder is located 79.4 cm downstream of the nozzle. A turbulence grid is set upstream of the test surface to generate a turbulence intensity of 7% near the cylinder.
2. The coolant air, supplied from a compressor, passes through a flow control valve and orifice flow meter. The coolant then passes through a 5-kW pipe heater and bypass valve before it enters the cylinder.
3. A T-type thermocouple is used to measure the mainstream temperature. For the coolant temperature, a T-type thermocouple is placed within the internal cavity at the cylinder near its base. The thermocouple readings are measured with National Instruments data acquisition hardware and monitored with the LabVIEW software.

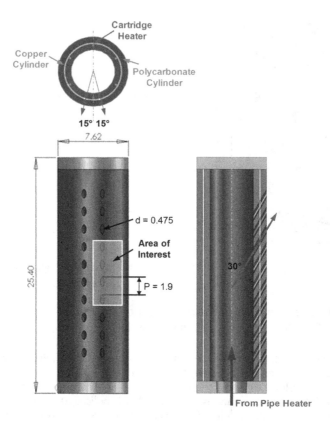

FIGURE 8.11 Details of the film cooled cylinder.

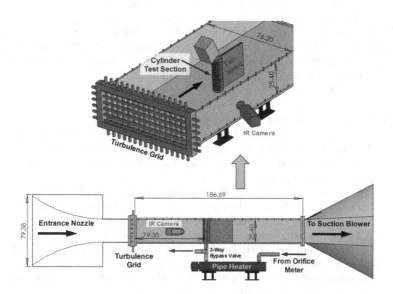

FIGURE 8.12 Overview of the experimental setup.

4. A central air-conditioning system maintained the mainstream temperature (T_m) at 22°C. The surface of the cylinder is initially heated with the embedded cartridge heaters to a surface temperature of approximately 43.3°C (monitored with thermocouples placed on the outer surface of the cylinder).

5. While the cylinder is preheating, the coolant air is heated using the pipe heater while being diverted to the room, away from the cylinder. The coolant is heated to approximately 43.3°C to match the surface temperature of the cylinder.

6. When both the cylinder surface and coolant flow reach the desired temperatures, the transient test begins by simultaneously starting the wind tunnel blower, diverting the cooling flow to the test cylinder, and shutting off the cartridge heaters used to heat the surface of the cylinder. The coolant and mainstream temperatures are monitored with LabView, and the surface temperature, at discrete times, is measured with an IR camera. The first image recorded by the IR camera serves as the initial temperature of the cylinder (heated above the ambient room).

7. A Mikron Thermo Tracer 6T62 is used to measure the surface temperatures. The IR-system consists of an optical scanner which directs the incoming infrared radiation line by line onto the detector working in a wavelength bandwidth of 8–13 microns. The IR camera views the test surface through a sheet of Vinylidene Chloride-Vinyl Chloride copolymer (Saran Wrap food wrap), which serves as an infrared window on the side of the wind tunnel.

8. In the calibration procedure, a thermocouple was attached to the surface of the copper block, and the copper block was painted black to increase the emissivity). The IR camera views the test surface through a sheet of Vinylidene Chloride-Vinyl Chloride copolymer (Saran Wrap food wrap), which serves as an infrared window in the top wall of the wind tunnel. Temperatures recorded by the thermocouple were compared to temperatures recorded by the IR camera. The relationship between the thermocouple and IR measurements is shown in Figure 8.4. With this calibration curve, the data obtained from the IR camera in the actual film cooling experiment was corrected.

9. Using at least two combinations of surface temperature distributions, both the film cooling effectiveness and heat transfer coefficient distributions can be obtained at every pixel in the region of interest.

Figure 8.13 shows the wall temperature distributions measured (and corrected from the calibration) during a single transient test. As shown by the scales, the cylinder is initially hot, and cools during the transient experiment.

From these temperature distributions, and the material properties of polycarbonate (d,k), detailed heat transfer coefficient and film cooling effectiveness distributions can be obtained by solving the system of transient equations. To follow the examples shown in other chapters, only the resulting film cooling effectiveness distributions are shown here. Figure 8.14 shows the calculated film cooling effectiveness at each pixel using different combinations of wall temperatures.

Comparing the three distributions shown in Figure 8.14, the magnitudes and the trends for the film cooling effectiveness are similar across all three time combinations. The 30s/60s combination gives the distribution that is most expected. Near the stagnation region on the cylinder, the effectiveness

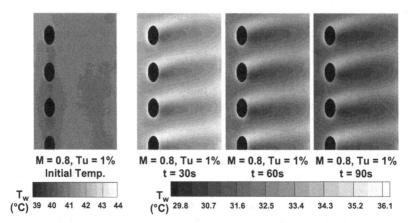

FIGURE 8.13   Temperature distributions obtained on the surface of the cylinder during the transient IR experiment.

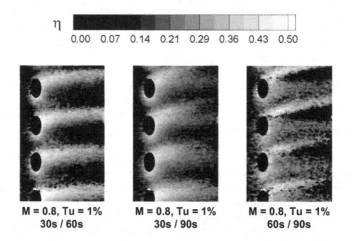

FIGURE 8.14   Sample film cooling effectiveness distributions from the transient IR method.

approaches zero, and moving around the cylinder, between the jets, the effectiveness also approaches zero. At longer periods of time, the effectiveness in these areas where coolant is not expected is gradually rising (similar to the behavior observed in steady-state tests). Therefore, minimizing the time for the test is beneficial.

Comparing the effectiveness plots to the temperature plots shown in Figure 8.13, the spatial resolution appears to have declined. The calculated effectiveness in Figure 8.14 is the combination of three, raw IR images, the initial image, and two separate images from the test. Any pixel-to-pixel variation on the surface can mean a misalignment of pixels between images, and thus appear as noise in the final result.

Using an IR camera capable of recording at a higher frame rate could improve the results for a transient IR test. By comparing many combinations of images, it is possible to "over-define" the problem and thus, capture the true effectiveness and heat transfer coefficient for each pixel. Also, increasing the frame rate could allow several images to be averaged at a given time to reduce noise in the process (this can only be done if the surface temperature does not change significantly over the time when the images are averaged).

PROBLEMS

1. You are asked to prepare an experimental setup to evaluate the proposed turbine blade film cooling for gas turbine applications. The required compressed air (25°C), heaters, piping, flow meters, materials, thermocouples, IR camera, and equipment for calibrations and measurements are available for your use. Use the transient, Infrared Thermography (IRT) method to map the surface temperature and heat transfer coefficient distributions on the pressure side surface, as shown in below sketch. Based on the wind tunnel flow and cooling/heating flow capability, one steady-state blowing ratio (cooling to mainstream mass flux ratio) will be tested: $M = 1.0$.

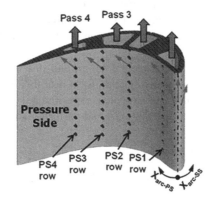

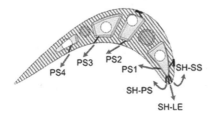

   a. Show an overall sketch of your experimental test loop and the associated equipment and instrumentation required to conduct the proposed test, i.e., sketch your overall experimental setup along with a suitable flow measuring device and detail your test section design with proper materials for the transient, infrared thermography image technique.

    b. Describe the transient IRT measurement principle, IRT calibration method, and detail the IRT measurement procedures by using the transient air heating technique.

    c. Sketch and discuss the expected target surface temperature distributions (contours), heat transfer coefficient, and film cooling/heating effectiveness distributions (contours) for one tested blowing ratio $M = 1.0$ and estimate the possible uncertainty.

2. Repeat Problem 1 by using the steady-state, Infrared Thermography (IRT) method to map the surface temperature, heat transfer coefficient and film cooling/heating effectiveness distributions (contours) on the pressure side surface as shown in the sketch with problem 1. Can you compare the film cooling effectiveness contours between (1) and (2) at the same blowing ratio? Are they the same or difference? Why?

3. Prepare an experiment to evaluate surface temperature and heat transfer coefficient distributions for steady, incompressible turbulent air flow through a two-passage, square, angled ribbed, cooling channel with a 180-deg turn, for heat exchanger applications, as sketched below. Focus on the region of interest shown in the sketch.

    a. Use the steady-state, Infrared Thermography (IRT) method to map the surface temperature distributions on the bottom wall as shown in the region of interest.

    b. Sketch the major experimental setup, and describe the measurement principle using the steady-state uniform heat flux heating technique.

    c. Sketch and discuss the expected experimental results (surface heat transfer coefficient distributions) and perform uncertainty analysis.

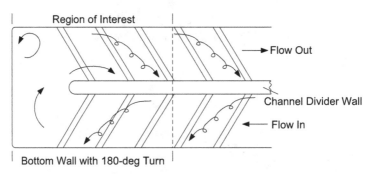

TOP VIEW of Two-Passage Channel Design with Angled Rib

4. Repeat Problem 3 by using the transient Infrared Thermography (IRT) method to map the surface temperature and heat transfer coefficient distributions on the top target wall in region of interest as shown in the sketch with Problem 3.

5. You are asked to prepare experimental methods to evaluate the surface temperature and heat transfer coefficient distributions for steady, incompressible, turbulent air flow through a two-dimensional channel with a

sudden expansion outlet for heat exchanger applications, as shown in the following sketch. The required compressed air (25°C), heaters, piping, flow meters, materials, thermocouples, IR camera, and equipment for calibrations and measurements are available for your use. Use the steady-state, IRT method to map the surface temperature and heat transfer coefficient distributions on the bottom wall in region of interest, as shown in the following sketch.

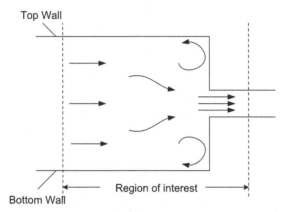

a. Show an overall sketch of your experimental test loop and the associated equipment and instrumentation required to conduct the proposed test, i.e., sketch your overall experimental setup along with a suitable flow measuring device and detail your test section design with proper materials using the steady-state, IRT camera technique.

b. Based on channel hydraulic diameter, two steady flow Reynolds numbers will be tested: $Re = 10,000$ and $30,000$. Describe the IRT measurement principle, IRT calibration method, and detail the IRT measurement procedures by using the steady-state and surface heating technique.

c. Sketch and discuss the expected temperature distributions (contours) as well as heat transfer coefficient distributions (contours) downstream of the expansion, for the two tested Reynolds numbers at steady-state and estimate the possible uncertainty.

6. Repeat Problem 5 using the transient Infrared Thermography (IRT) method to map the surface temperature and heat transfer coefficient distributions through the expansion, as shown with the sketch for Problem 5.

7. Prepare experimental methods to evaluate the temperature and heat transfer coefficients for steady, incompressible, turbulent air flow through a two-dimensional channel with protrusions on one wall, as shown in the following sketch. Focus on the region interest shown in the sketch. You particularly pay attention to the region of interest shown in the sketch. Use the Infrared Thermography (IRT) method to map the surface temperature distributions on the bottom wall as shown in the region of interest.

   a. Sketch the experimental setup including the primary pieces of instrumentation, and describe the measurement principle using the steady-state, uniform heat flux, and heating technique.

   b. Sketch and discuss the expected experimental results (surface heat transfer coefficient distributions) and perform uncertainty analysis.

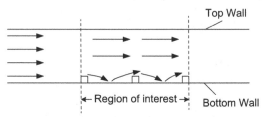

8. Repeat the Problem 7 using the transient Infrared Thermography (IRT) method to map the surface temperature and heat transfer coefficient distributions on the rib roughened wall, in the region of interest, as shown in sketch with Problem 7.

9. You are asked to prepare experimental methods to evaluate surface temperature and heat transfer coefficient distributions for steady, incompressible, turbulent air flow through a square channel with short pins on the bottom wall, as sketched below. Focus on the region of interest shown in the sketch. Use the transient Infrared Thermography (IRT) method to map the surface temperature distributions on the bottom wall.

   a. Sketch the major experimental setup, and describe the measurement principle using the transient heating technique.

   b. Sketch and discuss the expected experimental results (surface heat transfer coefficient distributions) and perform uncertainty analysis.

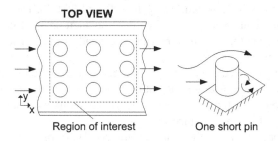

10. Repeat Problem 9 using a steady-state, Infrared Thermography (IRT) method to map the surface temperature and heat transfer coefficient distributions on the bottom target wall in the region of interest, as shown in the sketch with Problem 9.

REFERENCES

1. Incropera, F.P., DeWitt, D.P., Bergman, T.L., and Lavine, A.S., *"Fundamentals of Heat and Mass Transfer,"* 7th ed., John Wiley & Sons, Hoboken, NJ, 2011.
2. Han, J.C., *Analytical Heat Transfer*, CRC Press, Taylor & Francis Group, Boca Raton, FL, 2012.
3. Han, J.C., Dutta, S., and Ekkad, S.V., "Chapter 6-Experimental Methods" in *Gas Turbine Heat Transfer and Cooling Technology*, Taylor & Francis, Inc., New York, 2000.
4. Blair, M.F. and Lander, R.D., "New Techniques for Measuring Film Cooling Effectiveness," *ASME Journal of Heat Transfer*, Vol. 97, 1975, pp. 539–543.
5. Gritsch, M., Schulz, A., and Wittig, S., "Adiabatic Wall Effectiveness Measurements of Film Cooling Holes with Expanded Exits," *ASME Journal of Turbomachinery*, Vol. 120, 1998, pp. 549–556.
6. Kohli, A. and Bogard, D.G., "Fluctuating Thermal Field in the Near-Hole Region for Film Cooling Flows," *ASME Journal of Turbomachinery*, Vol. 120, 1998, pp. 86–91.
7. Ekkad, S.V., Ou, S., and Rivir, R.B., "A Transient Infrared Thermography Method for Simultaneous Film Effectiveness and Heat Transfer Coefficient Measurements from a Single Test," *ASME Journal of Turbomachinery*, Vol. 126, 2004, pp. 597–603.
8. Lynch, S. and Thole, K., "The Effect of Combustor-Turbine Interface Gap Leakage on the Endwall Heat Transfer for a Nozzle Guide Vane," ASME Paper No. GT2007-27867, 2007.
9. Wright, L.M., Gao, Z., Varvel, T.A., and Han, J.C., "Assessment of Steady State PSP, TSP, and IR Measurement Techniques for Flat Plate Film Cooling," ASME Paper No. HT2005-72362, 2005.
10. Gao, Z., Wright, L.M, and Han, J.C., "Assessment of Steady State PSP and Transient IR Measurement Techniques for Leading Edge Film Cooling," ASME Paper No. IMECE 2005-80146, 2005.
11. Vedula, R.P. and Metzger, D.E., "A Method for the Simultaneous Determination of Local Effectiveness and Heat Transfer Distributions in a Three Temperature Convective Situation," ASME Paper No. 91-GT-345, 1991.
12. Mart, S.R., McClain, S.T., and Wright, L.M., "Turbulent Convection from Deterministic Roughness Distributions with Varying Thermal Conductivities," *ASME Journal of Turbomachinery*, Vol. 134, 2012, Article No. 051030, 11 pages.

9 Pressure-Sensitive Paint (PSP) and Temperature-Sensitive Paint (TSP) Techniques

9.1 PRESSURE-SENSITIVE PAINT (PSP) TECHNIQUE

9.1.1 FUNDAMENTAL PRINCIPAL

Pressure-sensitive paint (PSP) is a photo-luminescent material that emits light with an intensity proportional to the surrounding partial pressure of oxygen. Any pressure variation on the PSP coated surface causes the emitting light intensity to change due to an oxygen quenching process. A calibration is needed to relate a known pressure to the emission intensity of the paint. This technique was originally developed for measuring static pressure distributions on surfaces, and it has been widely used by aerodynamicists to measure flow characteristics on a surface. This technique was further developed by Zhang et al. [1] to measure the film cooling effectiveness on a flat plate.

A major advantage of this technique is that detailed distributions of the local static pressure and film cooling effectiveness can be obtained. Due to absence of a temperature difference between the test surface and mainstream or coolant, this measurement technique provides for robust measurements near sharp edges such as film cooling hole break outs. In traditional heat transfer experiments, the data in these regions are adversely affected by heat conduction through the surface. Because the PSP method relies on mass transfer, rather than heat transfer, there are no conduction errors to consider. The pressure-sensitive paints currently available give excellent pressure sensitivity for large static pressure gradients. Thus, PSP has been successfully employed for testing under high-speed flow conditions. However, at low static pressure differences, which typically occur under low speed conditions, PSP is not capable of providing resolved results due to its relatively low sensitivity. Further technological developments of PSP resolution in the near future, may make this technique suitable for low speed pressure measurements. Another major drawback of this technique is that the pressure measured from this paint is sensitive to the surface temperature of the test piece. The degree of sensitivity depends on the type of PSP used. Thus, it should be ensured that isothermal conditions are obtained while taking data. If the temperature cannot be maintained constant, the PSP should be calibrated at different temperatures to account for this change.

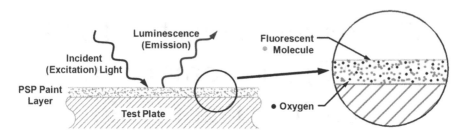

FIGURE 9.1   Fluorescence model for PSP.

9.1.2 PSP FOR PRESSURE MEASUREMENT

Figure 9.1 shows the PSP measurement principle. This is an optical technique and requires a CCD camera to capture images of the surface in order to determine the light intensity emitted from the paint. The painted surface should be illuminated with a stable light source. The maximum emission from the PSP occurs when the PSP is excited at a wavelength of 400–450 nm. A filter can be applied to a white light source, only allowing blue to green light to reach the surface. In recent years, researchers have relied on LED light sources to excite the paint. The LED lights are stable over long periods of time, and they provide a uniform, high-intensity light over the entire surface. This incident light excites the PSP paint, which then emits light with a longer wavelength. This emitted light is generally red (>590 nm). A long pass filter attached to the CCD camera will ensure that the camera sees only this emitted light and not the source light. Special care should be taken in choosing the wavelength range of the filters for the illumination source and the camera to avoid any overlap of the ranges. The emitted light intensity is proportional to the incident light intensity on the surface. Thus, a strong and uniform source light is needed so that the CCD camera can capture a sufficiently bright image.

A black and white image is recorded under grayscale settings, which contains this intensity information. A high performance, cooled, digital, 12-bit CCD camera system with advanced CCD and electronics technology can be used. The system features thermo-electrical cooling of the image sensor (down to −12°C), extremely low noise (down to 4e-rms) and an outstanding quantum efficiency, which achieves a high spectral sensitivity. Exposure time modes (software selectable) range from 500 ns (fast shutter) to 1000 seconds (long exposure). In double shutter mode, two images with the very short interframing time of 500 ns can be recorded. A high-speed serial data link connects the system to the PC (fiber optic link available). This PIV camera system is perfectly suited for many sensitive and low-noise, imaging applications. Figure 9.2 shows a typical arrangement for an experiment using the PSP technique.

There are several vendors in the industry that manufacture pressure-sensitive paint. PSP is relatively expensive, compared to other paints such as liquid crystal. Estimating the amount of paint that will be necessary while planning the experiment will help to reduce costs. Several kinds of pressure-sensitive paints are available depending on the type of application and the range of pressure measurement. These paints are better suited for applications with a large surface pressure variation due to

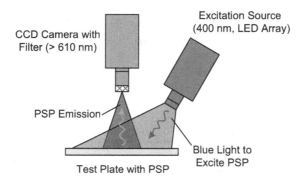

FIGURE 9.2   General optical arrangement for the PSP technique.

their low sensitivity for small pressure changes. Recent advances in this technology have made it possible to investigate even small pressure changes using this paint.

The intensity of light emitted by PSP is inversely proportional to the partial pressure of oxygen. A higher pressure will result in low intensity and vice versa. The intensity emitted by PSP is also dependent on the temperature to a smaller extent. Hence, accurate measurement of pressure requires knowledge of the surface temperature distribution. Performing all experiments under the same thermal conditions also helps in reducing uncertainty due to temperature variation. This becomes critical when small pressure changes need to be measured.

The test surface is first cleaned and painted with the Pressure-sensitive paint. Several coats of the paint may need to be applied in order to get a sufficient response in intensity for small pressure changes. About seven to eight coats may be needed depending on the type of paint used. Spraying the surface with more than eight coats can cause flaking of the paint and hence should be avoided. An air brush should be used for painting as it ensures a uniform distribution of the paint over the surface.

To obtain an intensity ratio from the PSP, three types of images are required. A reference image (with illumination, no mainstream flow, surrounding pressure uniform at 1 atm), an air image (with illumination and mainstream flow), and a black image (no illumination and no mainstream flow) to remove noise effects due to the camera.

The oxygen partial pressure information is obtained from the intensity ratio and calibration curve. This oxygen partial pressure information can be directly converted into a static pressure distribution for the case of air flow, as shown in Equation (9.1)

$$\frac{I_{\text{ref}} - I_{\text{blk}}}{I_{\text{air}} - I_{\text{blk}}} = \text{func}\left(\left(P_{O_2}\right)_{\text{air}}\right) \text{ or func}(P) \tag{9.1}$$

where I denotes the intensity obtained for each pixel at the reference (ref), black (blk), or air (air) conditions, and func(P) is the relation between intensity ratio and pressure ratio obtained after calibrating the PSP. $(P_{O_2})_{\text{air}}$ is the partial pressure of oxygen on the test surface for the air images.

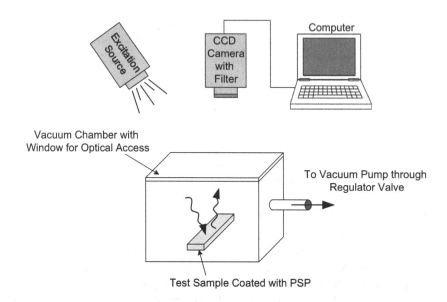

FIGURE 9.3   Schematic of the PSP calibration setup.

The PSP must be calibrated for the range of pressure measurement in the test section. Calibration of the PSP can be performed in a pressure chamber or *in-situ*. When performed in a pressure chamber, such as a vacuum chamber, a test piece painted with PSP is constructed and placed in the chamber. Figure 9.3 depicts a schematic of such a calibration setup. One wall of this chamber should be made from a transparent material such as plexiglass in order to provide optical access to the PSP coated test piece. This chamber should be connected to a pump and a pressure gauge for adjusting and measuring the pressure inside the chamber, respectively. While performing the calibration it should be ensured that the pressure inside the chamber is uniform (i.e., no flow should be occurring internally). The effect of temperature on the PSP response can also be measured by attaching a heater to the test piece. A thermocouple placed on the test piece surface can be used to indicate the surface temperature.

In-situ calibration of PSP can be performed by subjecting the test surface to different pressures and recording the intensities at several pressures using a camera and also measuring the static pressure using pressure taps distributed at several locations along the surface. Thus, a relationship between the static pressure and the emitted intensity can be directly established. An advantage of using an *in-situ* calibration is that both the calibration and the tests can be performed under the same thermal conditions, thus minimizing the errors due to temperature variation.

There is a temperature dependency of PSP. However, if the intensity is normalized by that of the reference image (at 1 atm), the calibration curves, at different temperatures collapse to one curve. Figure 9.4 shows a typical calibration curve of intensity ratio vs. pressure ratio. During testing, it must be ascertained that temperatures of the mainstream air, coolant, and test section are the same while recording the reference and air images to minimize uncertainty. For the temperatures shown

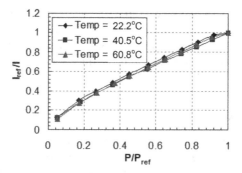

FIGURE 9.4 Typical calibration curve for PSP.

in the curve, both the pressure images and reference images were recorded at the stated temperature (three different references at three different temperatures were used in the calibration). The behavior of the PSP is also shown with the calibration curve. As the pressure decreases from 1 atm ($P/P_{ref} < 1$), the intensity ratio decreases. Therefore, decreasing the pressure increases the intensity emitted by the PSP. Therefore, the light decreases as the pressure increases; this is the oxygen quenching behavior of paint. In the presence of more oxygen, the emission intensity decreases.

Figure 9.5 shows the calibration data points under different viewing angles to the surface. The angles range from 20° to 90°. The result indicates that there is no viewing angle effect on PSP measurements. As a result, PSP is an ideal tool for film cooling effectiveness measurements on curved surfaces. This is a very different result from the sensitivity of liquid crystal measurements with camera viewing angle. With the measurement of "intensity" instead of "color," the PSP technique is insensitive to camera viewing angles.

An image processing tool should be used for analyzing the images recorded by the camera. Images obtained from the camera can be saved as high-resolution TIFF

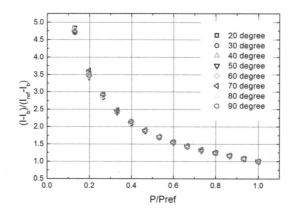

FIGURE 9.5 Measured intensity ratio at viewing angles from 20° to 90°.

images. Several images should be captured for each test case and the average pixel intensity should be calculated from these images. These intensity magnitudes can then be converted to the partial pressure of oxygen using an appropriate calibration curve and then to surface pressure. Results obtained for each pixel can be presented as detailed pressure distributions.

9.1.3 PSP for Film Cooling Measurement

Figure 9.6 shows the measurement of the film cooling effectiveness using the heat (left) and mass (right) transfer analogies [2]. For film cooling effectiveness measurements by the PSP mass transfer technique, four kinds of images are required. A reference image (with illumination, no mainstream flow, surrounding pressure uniform at 1 atm), an air image (with illumination and mainstream flow, air used as coolant), an air/nitrogen image (with illumination and mainstream flow, nitrogen gas used as coolant), and a black image (no illumination and no mainstream and coolant flow) to remove noise effects due to the camera. The intensity ratio for the air and air/nitrogen mixture is calculated using Equations (9.2) and (9.3), respectively.

$$\frac{I_{ref} - I_{blk}}{I_{air} - I_{blk}} = \text{func}\left(\left(P_{O_2}\right)_{air}\right) \text{ or func}(P) \tag{9.2}$$

$$\frac{I_{ref} - I_{blk}}{I_{mix} - I_{blk}} = \text{func}\left(\left(P_{O_2}\right)_{mix}\right) \tag{9.3}$$

where I denotes the intensity measured for each pixel for the reference (ref), black (blk), air (air), and air/nitrogen (mix) images and func(P) is the relation between intensity ratio and pressure ratio obtained after calibrating the PSP. $(P_{O_2})_{air}$ and $(P_{O_2})_{mix}$ are the partial pressures of oxygen on the test surface for air and air/nitrogen mixture images, respectively.

The film-cooling effectiveness can be expressed as a ratio of oxygen concentrations, or partial pressures, measured by PSP and is calculated using the following equation.

$$\eta = \frac{Co_{air} - Co_{mix}}{Co_{air}} = \frac{\left(P_{O_2}\right)_{air} - \left(P_{O_2}\right)_{mix}}{\left(P_{O_2}\right)_{air}} \tag{9.4}$$

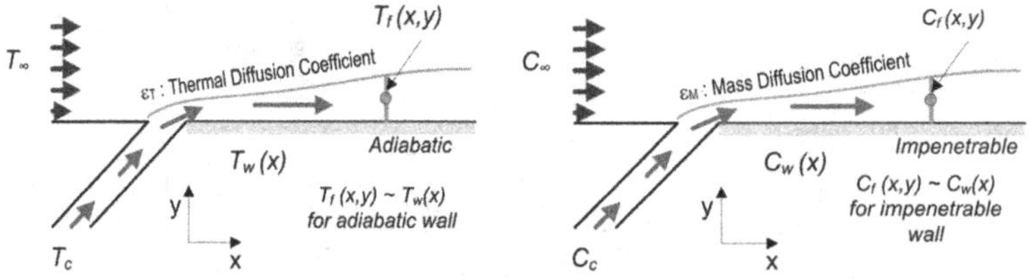

FIGURE 9.6 Measurement of the film cooling effectiveness using heat transfer (left) and mass transfer (right) experiments.

where Co_{air} and Co_{mix} are the oxygen concentrations of the mainstream air and the air/nitrogen mixture on the test surface, respectively. By assuming the molecular weights of air and nitrogen are the same, the effectiveness can be expressed as a ratio of the partial pressures of oxygen due to proportionality between concentration and partial pressure.

When the molecular weight of the foreign gas, used as the coolant, is different from that of air, the equation needs to be modified in order to account for the variation of the effective molecular weight of the mixed gas in the cooling film. The film cooling effectiveness can be obtained by the following equation:

$$\eta = 1 - \cfrac{1}{\left[1 + \left(\cfrac{(P_{O_2})_{air} \big/ (P_{O_2})_{ref}}{(P_{O_2})_{fg} \big/ (P_{O_2})_{ref}} - 1\right) \cdot \cfrac{W_{fg}}{W_{air}}\right]} \qquad (9.5)$$

where W_{air} is the molecular weight of ambient air and W_{fg} is the molecular weight of the foreign gas being used as the coolant. The coolant to mainstream density ratio (DR) effect can be studied by using foreign gases with $DR = 1$ (N_2), 1.5 (CO_2), and 2 (15% SF_6 + 85% Ar).

The uncertainty of the measurement primarily comes from the uncertainty of the pressure measurement during calibration and the uncertainty of the PSP emission intensity. Generally, the uncertainty of the intensity acquired by the CCD camera is lower at higher effectiveness values, where the emission intensity of the paint is higher. On the contrary, when the emission intensity of the paint decreases, the uncertainty increases. Following the Kline and McClintock approach, when the effectiveness is $\eta = 0.3$, the estimated overall uncertainty in effectiveness is around 8% ($\eta = 0.3 \pm 0.024$).

9.1.4 Experiment Example

Film cooling effectiveness is a temperature ratio (dimensionless) that serves as a performance indicator, especially in the gas turbine community, for assessing the efficiency of an external cooling design. Ranging from 0 to 1, when the coolant is perfectly protecting the surface, the surface takes on the temperature of the coolant, and the effectiveness approaches unity. In contrast, if the surface approaches the mainstream temperature, the coolant is not protecting the surface, and the effectiveness approaches zero. As shown in previous chapters, the film cooling effectiveness is defined as:

$$\eta = \frac{T_m - T_f}{T_m - T_c}$$

If we further confine the scenario where the film cooling effectiveness is measured on an adiabatic wall (pure convection), the film (or mixing) temperature can be replaced with the adiabatic wall temperature. The adiabatic film cooling effectiveness can be rewritten as:

$$\eta = \frac{T_m - T_{aw}}{T_m - T_c}$$

The film temperature has been represented by the adiabatic wall temperature by researchers for decades. In modern, gas turbine engines, the temperature difference between the coolant and mainstream can be more than 1000°C. This temperature difference yields flows where the coolant is more than two times more dense than the mainstream. Creating the engine-like temperature difference in a laboratory setting is very challenging. Therefore, researchers developed methods to simulate the temperature difference by re-creating the density difference using "heavy" coolants. By introducing foreign gases, with molecular weights greater than that of air, they were able to develop mass transfer experiments that captured the density ratio effects within the actual engines. Based on the highly turbulent flows encountered in the engines, it was shown the heat transfer can properly be evaluated using a mass transfer analogy if the turbulent Lewis number is approximately one (the ratio of the eddy thermal diffusivity to the eddy mass diffusivity approaches one). If Le ≈ 1, the temperature measured on an adiabatic wall can be approximated by the mass concentration of a particular gas on an impermeable wall. Heat cannot diffuse through an adiabatic wall, and mass cannot diffuse through a solid, impermeable wall. Therefore, the film cooling effectiveness can be rewritten in terms of the gas concentration:

$$\eta = \frac{T_m - T_{aw}}{T_m - T_c} \sim \frac{C_m - C_w}{C_m - C_c}$$

PSP is capable of sensing the surface pressure variation by detecting the change in oxygen partial pressure on the PSP coated surface. This property can be used to obtain the film cooling effectiveness, under an isothermal condition, if the coolant supplied to the test surface is oxygen free, as shown in Figure 9.7. When the coolant is oxygen free ($C_c = 0$), the adiabatic film cooling effectiveness is,

$$\eta \sim \frac{C_m - C_w}{C_m - C_c} = \frac{C_m - C_w}{C_m} = \frac{\left(P_{O_2}\right)_{air} - \left(P_{O_2}\right)_{mix}}{\left(P_{O_2}\right)_{air}}$$

As shown above, the PSP is calibrated to relate a known pressure ratio to a measured intensity ratio (both ratios relative to the atmospheric, reference condition). The film cooing effectiveness can be expressed in terms of the measured intensities:

$$\eta = 1 - \frac{\left(P_{O_2}\right)_{mix}}{\left(P_{O_2}\right)_{air}} = 1 - \frac{\left(P_{O_2}\right)_{mix} / P_{O2,ref}}{\left(P_{O_2}\right)_{air} / P_{O2,ref}} = 1 - \frac{f\left(\dfrac{I_{ref} - I_b}{I_{mix} - I_b}\right)}{f\left(\dfrac{I_{ref} - I_b}{I_{air} - I_b}\right)}$$

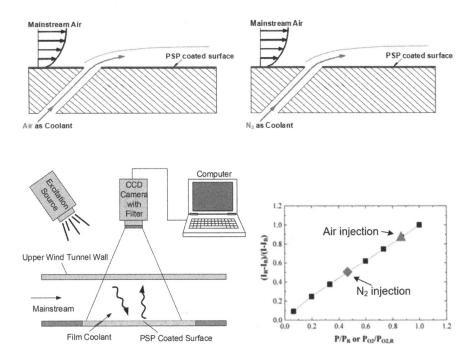

FIGURE 9.7 Measurement of the film cooling effectiveness using the PSP technique.

In summary, each term in the aforementioned equation represents an individual data set in order to comprise the complete information needed for the detailed, film cooling effectiveness calculations from a mass transfer experiment.

Example 9.1: Film Cooling on a Flat Plate [3]—PSP Technique

The following steps detail the preparation and execution of a basic experiment utilizing PSP for detailed film cooling effectiveness distributions:

1. First, the test model, with the area of interest, is painted with PSP (pink color). In this example, the area of interest is a flat plate with compound angle, cylindrical film cooling holes. All the geometry information of the test model is shown in Figure 9.8.
2. The test model is installed within a wind tunnel. The mainstream flow conditions are set to the desired velocity and monitored with a pitot-static probe. A secondary flow loop is also constructed for delivery of the various coolant gases (air and nitrogen). In this example, it is a suction type wind tunnel with an entrance contraction and turbulence grid installed upstream of the film cooling holes, as shown in Figure 9.9.
3. The excitation light source and CCD camera are installed above the wind tunnel, toward the area of interest for PSP activation and to capture the

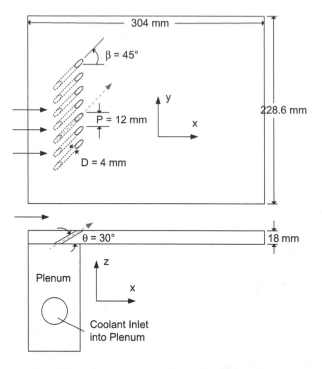

FIGURE 9.8 Details of the compound angle, film cooling holes.

emission intensity from PSP, respectively. In this example, the excitation light source is a strobe light system with a 520 nm wavelength; the camera is equipped with a long pass filter of 610 nm wavelength. To reduce the stray reflection from room light and prevent uncontrolled paint excitation or false signals due to white light's wide wavelength content, the experiment is conducted in a room with the room lights turned off for all images.

4. After determining the location of the light and camera relative to the test surface, a calibration is performed under a similar optical arrangement to generate a correlation between the PSP emission intensity and the surface pressure. A test piece coated with PSP is placed inside a pressure-controlled chamber. With the light source, camera, and test piece replicating the real experiment, the PSP emission intensities at different chamber pressures are recorded (Table 9.1) and a calibration curve is generated, as shown in Figure 9.10.

5. After the setup and calibration have been completed, actual testing can begin. As noted previously, four sets of images are required to calculate the film cooling effectiveness at a given flow condition:

I_{mix}: Excitation light on, mainstream air on, oxygen-free (N_2) coolant injection on

I_{air}: Excitation light on, mainstream air on, air coolant injection on

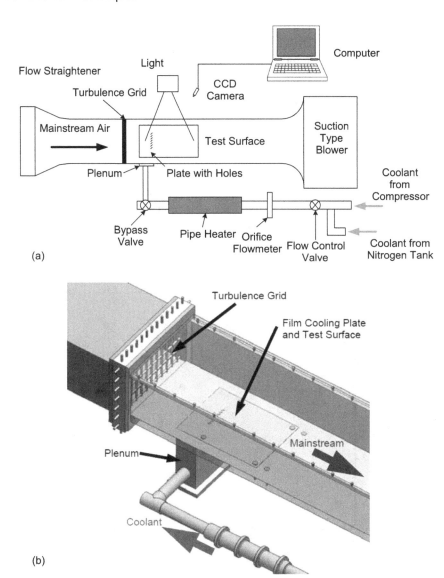

FIGURE 9.9 Schematic of the overall test section.

I_{ref}: Excitation light on, mainstream off, coolant injection off
I_b: Excitation light off, mainstream off, coolant injection off
In practice, the emission intensity of PSP captured by CCD camera will have a slight fluctuation due to nature of the electrical signal that the camera sensor captures (even under the same flow conditions). Therefore, 200 images are recorded for each set to average the emission at each pixel over the 200 images.

TABLE 9.1
Selected PSP Calibration Data

P/P_{ref}	I_{ref}/I
0.06	0.11
0.2	0.22
0.33	0.36
0.47	0.46
0.6	0.6
0.73	0.72
0.87	0.87
1	1

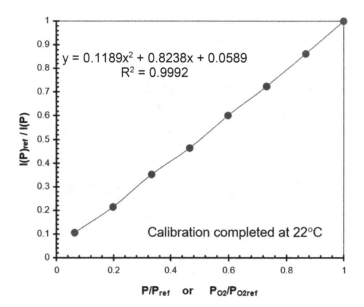

FIGURE 9.10   PSP calibration curve.

6. Finally, the adiabatic film cooling effectiveness in the area of interest (flat plate) can be calculated pixel by pixel to construct an effectiveness distribution, as shown in Figure 9.11. The intensities of four selected locations, from the individual image sets mentioned in Step 5, along with the calculated effectiveness, are listed in Table 9.2.

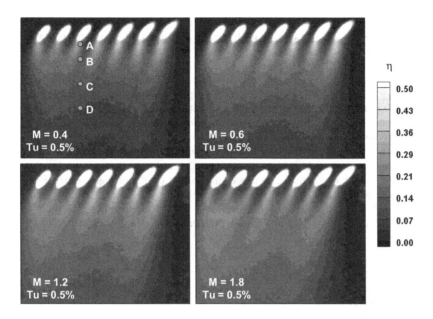

FIGURE 9.11   Sample film cooling effectiveness distributions measured using PSP.

TABLE 9.2
Measured Image Intensities and
Film Effectiveness at Select Points
on the Flat Plate

Location	I_b	I_{ref}	I_{air}	I_{mix}	η
A	54	901	784	1217	0.357
B	51	896	812	987	0.176
C	55	889	822	956	0.139
D	52	893	801	895	0.103

Example 9.2: Film Cooling on a Cylinder Modeling the Leading Edge of a Turbine Airfoil [4]—PSP Technique

Similar to Example 9.1, the following steps detail the preparation and execution of a fundamental experiment utilizing PSP for detailed film cooling effectiveness distributions:

1. The test model is prepared, and the area of interest is painted with PSP (pink color). In this example, the area of interest is the middle portion of a cylinder, to simulate the leading edge of airfoil. The film holes are ±15° from the stagnation line of the cylinder. The details of the geometry are shown in Figure 9.12.

2. The test model is installed inside a wind tunnel to have the cylinder exposed to the desired mainstream flow conditions. A secondary flow loop is also setup control and delivery of the coolant flows (air and nitrogen). In this example, it is a suction type wind tunnel with an entrance contraction and turbulence grid installed upstream of the cylinder, as shown in Figure 9.13. While a pipe heater is referenced in both Figures 9.12 and 9.13, the heater is not required for the PSP measurements. This hardware was used for other measurement techniques presented in this work by Gao et al. [4].

3. The excitation light source and CCD camera are mounted relative to the region of interest for PSP activation and to capture the emission intensity from PSP, respectively. In this example, the excitation light source is a strobe light system with 520 nm wavelength; the camera is equipped with a long pass filter of 610 nm wavelength. To reduce the stray reflection from room light and prevent uncontrolled paint excitation or false

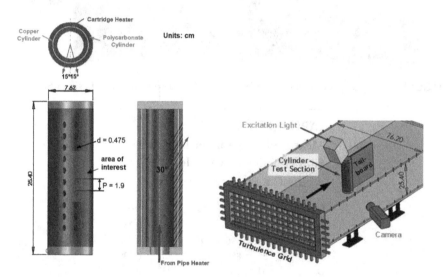

FIGURE 9.12 Cylindrical test section with leading edge film cooling.

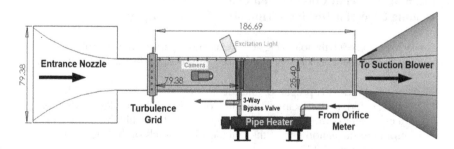

FIGURE 9.13 Overview of the low speed wind tunnel with the cylindrical, leading edge model.

signals due to white light's wide wavelength content, the experiment is conducted in a room with the room lights turned off for all images.

4. After determining the location of the light and camera relative to the test surface, a calibration is performed under a similar optical arrangement to generate a correlation between the PSP emission intensity and the surface pressure. A test piece coated with PSP is place inside a pressure-controlled chamber. With the light source, camera, and test piece replicating the real experiment, the PSP emission intensities at different chamber pressures are recorded (Table 9.3) and a calibration curve is generated, as shown in Figure 9.14.

TABLE 9.3
Selected PSP Calibration Data

P/P_{ref}	I_{ref}/I
0.06	0.11
0.2	0.22
0.33	0.36
0.47	0.46
0.6	0.6
0.73	0.72
0.87	0.87
1	1

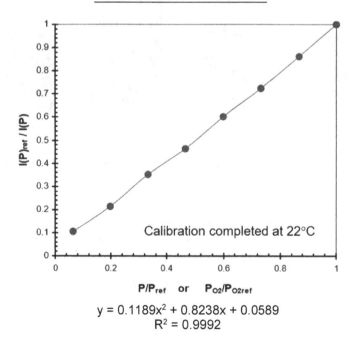

$$y = 0.1189x^2 + 0.8238x + 0.0589$$
$$R^2 = 0.9992$$

FIGURE 9.14 PSP calibration curve.

5. After the setup and calibration have been completed, actual testing can begin. As noted previously, four sets of images are required to calculate the film cooling effectiveness at a given flow condition:

I_{mix}: Excitation light on, mainstream air on, oxygen-free (N_2) coolant injection on

I_{air}: Excitation light on, mainstream air on, air coolant injection on

I_{ref}: Excitation light on, mainstream off, coolant injection off

I_b: Excitation light off, mainstream off, coolant injection off

In practice, the emission intensity of PSP captured by CCD camera will have a slight fluctuation due to nature of the electrical signal that the camera sensor captures (even under the same flow conditions). Therefore, 200 images are recorded for each set to average the emission at each pixel over the 200 images.

6. Finally, the adiabatic film cooling effectiveness in the area of interest (flat plate) can be calculated pixel by pixel to construct an effectiveness distribution, as Figure 9.15. The intensities of four selected locations, from the individual image sets mentioned in Step 5, along with the calculated effectiveness, are listed in Table 9.4.

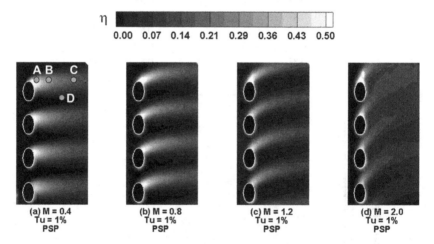

FIGURE 9.15   Sample film cooling effectiveness distributions measured using PSP.

TABLE 9.4

Measured Image Intensities and Film Effectiveness at Select Points on the Cylinder

Location	I_b	I_{ref}	I_{air}	I_{mix}	η
A	49	893	459	1118	0.566
B	53	885	513	752	0.296
C	50	902	560	656	0.133
D	51	897	524	525	0.002

Example 9.3: Film Cooling on a Flat Plate with Anti-Vortex
Film Cooling Holes—PSP Technique

This example is a direct comparison to Example 7.4 from Chapter 7. In the previous example, the detailed film cooling effectiveness distributions were calculated based on the adiabatic wall temperature measured using a wide band liquid crystal. The same geometry, with the same flow conditions, will be examined using the PSP, mass transfer technique. Figure 9.16 provides the details for the film cooling geometry.

The experimental setup for the PSP experiment is similar to that for the steady-state liquid crystal experiment. The wind tunnel is the same. However, the pipe heater is removed from the coolant loop, and a nitrogen line is included. The RGB camera and white lights are replaced with a 14-bit CCD camera and a 400 nm LED light for excitation. Figure 9.17 shows the altered schematic of the

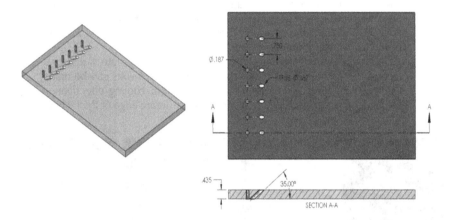

FIGURE 9.16   Anti-vortex hole geometry for the PSP experiment.

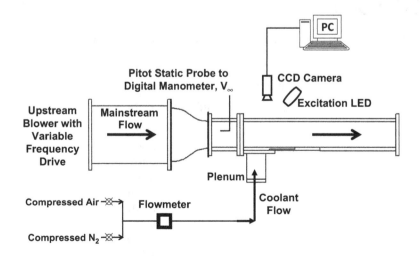

FIGURE 9.17   Overview of the experimental setup for the PSP experiment.

small scale, low speed wind tunnel used for this PSP experiment. The tunnel is used to experimentally investigate the proposed, double hole geometry. The tunnel has a cross section of 6″ × 4″, and the mainstream velocity through the tunnel is approximately 7.2 m/s (as monitored with a pitot-static tube). As measured using a hot wire anemometer, the freestream turbulence intensity in the tunnel is approximately 1.2%.

The following raw data were obtained for this flat plate, film cooling example:

1. Coolant—to—Mainstream Mass Flux Ratio: $M = \frac{\rho_c V_c}{\rho_\infty V_\infty} = 0.5$
2. "Black" intensity distribution the surface (Figure 9.18a)
3. "Reference" intensity distribution the surface (Figure 9.18b)
4. "Air" intensity distribution on the surface (Figure 9.18c)
5. "Nitrogen" intensity distribution on the surface (Figure 9.18d)
6. Raw calibration data (Table 9.5 and Figure 9.19)

   Given the intensities shown in the raw images, the calibration information, the detailed surface pressure distributions (Figure 9.20) can be calculated, and from the pressure, the film cooling effectiveness at every pixel can be determined (Figure 9.21).

   With this geometry being shown for both the steady-state, liquid crystal and PSP techniques, a direct comparison of the results can be made. Figure 9.22 shows the measured, centerline film cooling effectiveness calculated by both techniques. Near the downstream edge of the inclined

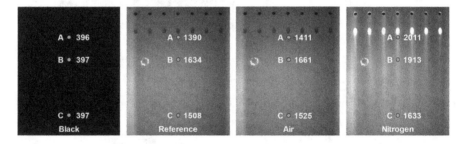

FIGURE 9.18 Average intensity distributions measured on the flat plate.

TABLE 9.5

Selected PSP Calibration Data

P/P_{ref}	I_{ref}/I
0.8695	0.9625
0.7324	0.858
0.5986	0.7539
0.4648	0.6504
0.334	0.5447
0.1973	0.4276
0.0635	0.2759
1	1

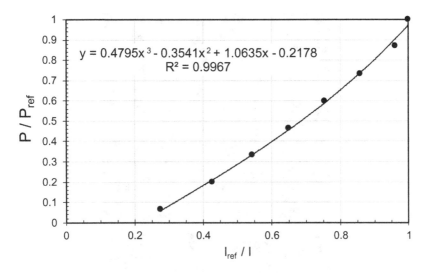

FIGURE 9.19 PSP calibration curve.

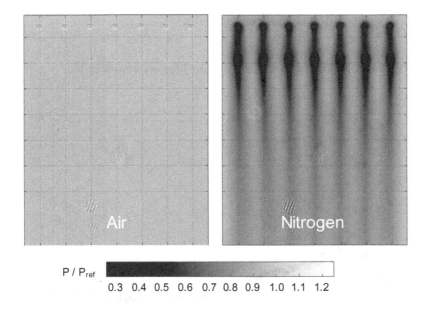

FIGURE 9.20 Measured pressure ratios on the surface with air (left) and nitrogen (right).

hole ($X/D = 0$), the effectiveness measured by the PSP is greater than that from the liquid crystal. With the heat transfer test, the lateral conduction through the plate averages out the high and low effectiveness values expected on the surface. Therefore, in the regions of high heat transfer (immediately downstream of the hole), the measured effectiveness is reduced from the true value. If a laterally averaged effectiveness were to be shown, it is expected, the liquid crystal would be higher than the PSP,

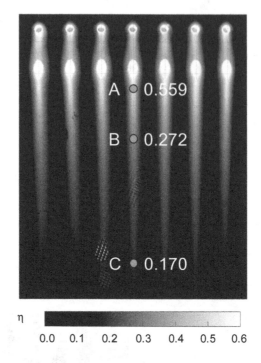

FIGURE 9.21 Detailed film cooling effectiveness distribution on the surface.

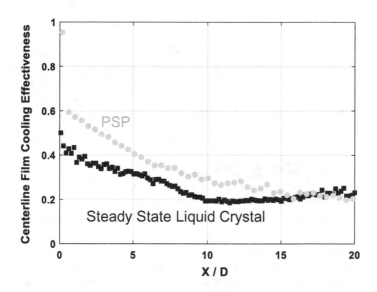

FIGURE 9.22 Centerline comparison between PSP and steady state liquid crystal techniques.

do to conduction between the holes, where the PSP clearly approaches zero. Moving downstream, away from the holes, the film cooling effectiveness from both methods are comparable (in magnitude and trend).

Additional References: The aforementioned PSP measurements have been completed to determine the local film-cooling effectiveness when nitrogen, carbon dioxide or an SF6-argon mixture, is injected through the film-cooling holes. The measured intensities are converted into partial pressures of oxygen over the tested surface. Several selected papers have been published using this methodology for turbine blade tip film cooling by Ahn et al. [5], rotating blade leading-edge film cooling by Ahn et al. [6], the unsteady wake effect on turbine airfoil platform film cooling by Wright et al. [7], turbine blade film cooling with axial shaped holes by Gao et al. [8], rotating blade platform film cooling by Suryanarayanan et al., [9], full coverage turbine blade film cooling by Mhetras et al. [10], turbine cascade blade film cooling by Narzary et al. [11], turbine blade platform film cooling using PSP and TSP methods by Narzary et al. [12], rotating blade tip film cooling by Rezasoltani et al. [13], turbine vane endwall film cooling by Chowdhury et al. [14], full scale turbine vane endwall film cooling by Shiau et al. [15], and transonic turbine vane film cooling by Shiau et al. [16]. Interested readers can read these papers for their detailed flow loop design, test geometry, experimental procedure, data analysis, and results presentation.

9.2 TEMPERATURE-SENSITIVE PAINT (TSP) TECHNIQUE

9.2.1 Fundamental Principle

The first investigations using temperature sensitive paint (TSP) were done by Kolodner et al. [17] in 1982 at Bell Laboratories. They used a europium-based TSP molecule in a polymer binder to examine the surface temperature distributions on an operating integrated circuit. Researchers such as Campbell et al. [18] and Liu et al. [19] further developed this paint and applied it to aerodynamic testing. Over the past decade, the TSP technique has gained popularity with several researchers accepting this technique due to its non-intrusive nature and its ability to give high resolution temperature distributions on the surface. Several different chemical formulations of the TSP paint are available and the correct choice of formulation depends on the temperature range and the application.

Temperature sensitive paint is similar in principle and operation to the Pressure-sensitive paint. When the paint is excited by incident light of a particular wavelength, it emits light at a different wavelength which can then be quantified by capturing images of the test surfaces. Thus, TSP is also a photo-luminescent material that emits light proportional to the temperature of the coated surface. A typical TSP coated surface contains an oxygen impermeable binder and luminescent temperature sensitive molecules. When light is incident to the test surface, these molecules that are initially in a ground state, absorb the incident photons, and their energy level increases. This energy can be dissipated either through the emission of radiation or through

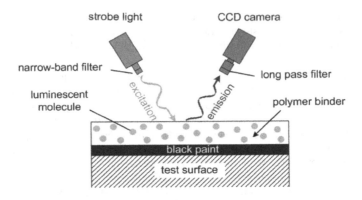

FIGURE 9.23 TSP working principle and basic TSP setup for measurement.

radiationless deactivation. Fluorescence is produced when the molecules lose energy by emitting photons by radiationless deactivation, also known as thermal quenching. Increasing the temperature of the luminescent molecules increases the likelihood of thermal quenching. Photon emission occurs at a higher wavelength, as shown in Figure 9.23, and upon emission, the molecules return to their original ground state.

As with the PSP technique, the test surface is coated with the paint and light intensity from the captured images is calibrated against a known set of temperatures, as opposed to pressures for PSP. The illumination and image capture systems for TSP and PSP techniques are essentially the same. Different optical filters on the illumination source and camera might be needed, depending on the TSP formulation used and their absorption and emission spectra. If a colored illumination source, such as LED light is used, then a filter is not necessary for the light. TSP is generally insensitive to changes in surface pressure as the polymer binder is not permeable to oxygen. For PSP though, the polymer binder is oxygen permeable to allow for oxygen quenching. Hence, PSP needs to be calibrated for pressure as well as temperature.

For TSP measurements, the measured intensity data is normalized using a reference condition. This reference condition will depend on the environment and conditions in which the experiment is performed. For most experiments, the reference condition is usually set as the room temperature before the experiment begins. The primary reason for normalizing the intensity is to eliminate errors due to inconsistencies in spatial illumination, paint thickness and distribution, camera sensitivity, etc. Two images are taken of the test surface under the same lighting and image acquisition conditions. One of the images is at a known temperature (reference), and the second image is at an unknown temperature (during the heat transfer experiment). A ratio is taken of these two images, and the normalization minimizes the errors listed above. The intensity ratio after normalization can be expressed as

$$\frac{I(T) - I_{blk}}{I(T)_{ref} - I_{blk}} = f(T) \tag{9.6}$$

The intensity of the test surface with no lighting is subtracted from the image intensities measured during testing to remove unwanted noise in the data. To determine the functional relationship between intensity and temperature, a calibration must be performed. A small copper piece fitted with a thermocouple to measure its temperature is coated with TSP. A heater is attached to the backside of the copper to regulate the temperature of the TSP coated sample. This copper block is then positioned underneath the camera and the excitation source. The temperature is controlled using the heater over the range in which temperature data will be recorded, and the intensity is recorded through images taken with the camera. A reference condition is set and using the aforementioned equation, the relationship between temperature and intensity is determined. It should be noted that a pressurized chamber, used when calibrating PSP, is not necessary for calibration of TSP as TSP is virtually insensitive to pressure. A typical calibration curve for TSP is shown in Figure 9.24. The raw images from the calibration so the relative behavior of the TSP thermal quenching effect. As the temperature increases, the emission intensity of the paint decreases (becomes darker).

PSP was presented as a mass transfer technique. However, TSP is used in heat transfer experiments. Therefore, it is another means to measure a surface temperature (in comparison to thermocouples, liquid crystals, or infrared thermography). Similar to those measurement methods, TSP can be used in both steady-state and transient experiments. The researcher is cautioned to understand the drawbacks and limitations of these general experiments, with the additional complexity of using the fluorescing paint.

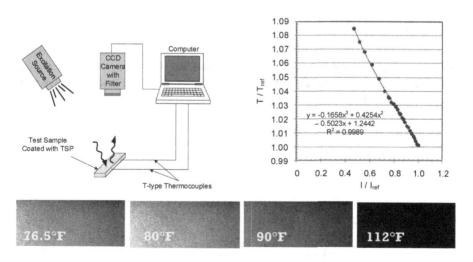

FIGURE 9.24   Typical calibration setup and curve for TSP.

9.2.2 STEADY-STATE TSP TECHNIQUE

The following section describes applications where the TSP technique has been used. Both steady-state and transient experiments involving TSP technique are described.

As shown in Chapter 7 (with the steady-state liquid crystal method), the film cooling effectiveness can be directly measured on a surface if a temperature difference exists between the coolant and mainstream fluids. Based on the measured adiabatic wall temperature (in the present case, with TSP), coolant temperature, and mainstream temperature, the film cooling effectiveness can be calculated:

$$\eta = \frac{T_m - T_{aw}}{T_m - T_c}$$

As the flow temperature (mainstream and coolant) are generally measured by thermocouples, obtaining a detailed distribution of the surface temperature will yield a detailed film cooling effectiveness distribution. When the target test section is made of material with a low thermal conductivity, and the duration of the experiment is controlled to have acceptable thermal penetration depth with respect to material thickness, the surface temperature can be approximated as the adiabatic wall temperature. With the TSP, the adiabatic wall temperature can be determined from the fluorescence intensity from the paint:

$$T_{aw} = f\left(\frac{I - I_b}{I_{\text{ref}} - I_b}\right)$$

Each term in the above equation represents an individual set of recorded images in order to determine the adiabatic wall temperature needed for the calculation of the film cooling effectiveness.

9.2.3 TRANSIENT TSP TECHNIQUE

Transient heat transfer experiments rely on the assumption of "one-dimensional conduction through a semi-infinite solid." By considering conduction through a solid due to a convection boundary condition, it is possible to determine the convection heat transfer coefficients (presented in Chapter 6).

1-Dimensional transient conduction: $\frac{\partial^2 T}{\partial y^2} = \frac{1}{\alpha}\frac{\partial T}{\partial t}$

Initial condition: $t = 0 \; T = T_i$

Boundary conditions: $y = 0 \; -k\frac{\partial T}{\partial y} = h(T_w - T_b) \; \overset{y \to \infty}{} \; T = T_i$

Solving the partial differential equation with the prescribed initial and boundary conditions at the surface ($y = 0$) yields:

$$\frac{T_w(t) - T_i}{T_b - T_i} = 1 - \left[\exp\left(\frac{h^2 \alpha t}{k^2}\right)\right]\left[erfc\left(\frac{h\sqrt{\alpha t}}{k}\right)\right]$$

With knowledge of the solid material properties (α, k), the initial wall temperature (T_i), the bulk air temperature (T_b), and the wall temperature at a known time, $T_w(t)$, the only unknown in the equation is h, the convective heat transfer coefficient. Furthermore, this equation can be applied to every pixel of the domain to yield a detailed heat transfer coefficient distribution.

The measured heat transfer coefficients can be non-dimensionalized using the Nusselt number.

$$Nu = \frac{hD_h}{k}$$

Finally, the Nusselt number can be compared to that for fully developed turbulent flow in a smooth tube (Dittus—Boelter/McAdams Correlation). The comparison is typically made in the form of the Nusselt number ratio.

$$\frac{Nu}{Nu_0} = \frac{Nu}{0.023\,Re^{0.8}\,Pr^{0.4}}$$

9.2.4   EXPERIMENT EXAMPLE

Example 9.4: Film Cooling on a Flat Plate [3]—Steady-State TSP Technique

The following steps detail the preparation and execution of a basic, steady-state, film cooling experiment, utilizing TSP for detailed film cooling effectiveness distributions:

1. First, the test model, with the area of interest, is painted with TSP (white/light pink color). This example utilizes the same test surface and wind tunnel as Example 9.1 earlier in this chapter. Therefore, the details of the geometry can be seen in Figure 9.8 [3].
2. The test model is installed within a wind tunnel. The mainstream flow conditions are set to the desired velocity and monitored with a pitot-static probe. A secondary flow loop is also constructed for delivery of the coolant flow. A heater must be included in the TSP experiment. The pipe heater shown in Figure 9.9 is used to heat the coolant flow; while the heater was shown with the PSP setup, the heater was not utilized. In this example, it is a suction type wind tunnel with an entrance contraction and turbulence grid installed upstream of the film cooling holes, as shown in Figure 9.9.
3. The excitation light source and CCD camera are installed above the wind tunnel, toward the area of interest for TSP activation and to capture the emission intensity from TSP, respectively. In this example, the excitation light source is a strobe light system with a 520 nm wavelength; the camera is equipped with a long pass filter of 610 nm wavelength. To reduce the stray reflection from room light and prevent uncontrolled paint excitation or false signals due to white light's wide wavelength content, the experiment is conducted in a room with the room lights turned off for all images.

4. After determining the location of the light and camera relative to the test surface, a calibration is performed under a similar optical arrangement to generate a correlation between the TSP emission intensity and the surface temperature. A copper test piece is coated with TSP, and attached to a heater and variable transformer. A thermocouple embedded in the copper records the temperature of the surface, and the camera records the TSP emission at this temperature. With the light source, camera, and test piece replicating the real experiment, the TSP emission intensities at different surface temperatures are recorded (Table 9.6) and a calibration curve is generated, as shown in Figure 9.25.

5. Before beginning the test, the reference and black image sets should be recorded. It is necessary to record these images sets before the test surface is exposed to any heated fluids, so the reference condition can be well defined. The first two sets of the required three images should be recorded:

   I_{ref}: Excitation light on, mainstream off, coolant injection off
   I_b: Excitation light off, mainstream off, coolant injection off

6. After the setup and calibration have been completed, actual testing can begin. For this example, the coolant temperature is adjusted to 43.3°C, while the mainstream is at room temperature (24°C). Once the coolant reaches the desired temperature, it is directed into the test section. For this steady-state test, once the heated fluid begins flowing through the film cooling holes, it should remain at the set temperature until the plate reaches steady-state. At this point, the final image set, I, can be recorded.

TABLE 9.6
Selected TSP Calibration Results

I/I_{ref}	T (°C)
1	24
0.91	29.5
0.9	31
0.88	32
0.82	36
0.8	38
0.75	41
0.72	42
0.68	45
0.62	49
0.6	51
0.58	52
0.55	55
0.52	57
0.5	60
0.48	63

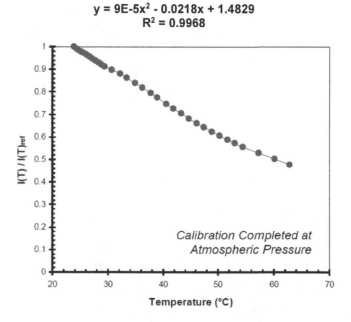

FIGURE 9.25 TSP calibration curve.

I: Excitation light on, mainstream air on, coolant injection on

In practice, the emission intensity of TSP captured by CCD camera will have a slight fluctuation due to nature of the electrical signal that the camera sensor captures (even under the same flow conditions). Therefore, 200 images are recorded for each set to average the emission at each pixel over the 200 images.

7. Finally, the adiabatic film cooling effectiveness in the area of interest (flat plate) can be calculated pixel by pixel to construct an effectiveness distribution, as shown in Figure 9.26. The measured intensities of four

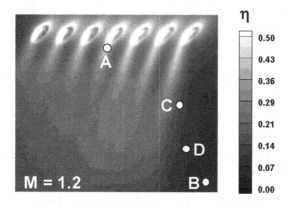

FIGURE 9.26 Sample film cooling effectiveness distribution. Measured using steady state TSP.

TABLE 9.7

Measured Image Intensities and Film Effectiveness at Select Points on the Flat Plate

Location	I_b	I_{ref}	I	η
A	53	887	747	0.557
B	51	892	890	0.047
C	57	896	865	0.147
D	58	890	871	0.106

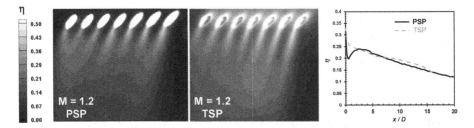

FIGURE 9.27 Comparison of the film cooling effectiveness from PSP (left) and steady state TSP (center).

selected locations, from the individual image sets mentioned in Steps 5 and 6, along with the calculated effectiveness, are listed in Table 9.7.

PSP and TSP are both photo-luminescent paints used to acquire detailed film cooling effectiveness distributions. Viewing the distributions side-by-side (Figure 9.27) highlights the conduction errors associated with the steady-state, heat transfer technique (TSP). Where the test surface material is thin on the upstream side of the film holes, the effectiveness measured by the TSP is greater than zero. PSP correctly measures the effectiveness approaching zero upstream of holes. The effectiveness is averaged across the width of the plate, and this comparison is shown on the right side of Figure 9.27. Near the edge of the holes, the TSP measurements are elevated above the TSP measurements. However, moving further downstream, the film effectiveness measured from the two different techniques converge to the same values. This is a positive result supporting the use of either of these methods to measure the film cooling effectiveness.

Example 9.5: Detailed Heat Transfer Coefficients in a Rectangular Duct with Hemispherical Dimples – Transient TSP Technique

In Section 5.3, a rectangular channel with hemispherical dimples was briefly mentioned. This example builds on the dimples used in that steady-state experiment. Raw data were obtained in a 3:1 rectangular channel. One wall of the channel was lined with a staggered array of hemispherical dimples. Dimples are most notably used to reduce on drag on golf balls. The concept has been introduced as a mechanism for heat transfer enhancement. The periodic depressions effectively trip the boundary layer, resulting in increased heat transfer. With each dimple, there are regions of both increased and decreased heat transfer. Therefore, the general strategy of packing dimples close together is likely not an optimized geometry. In route to creating a more optimized (from a heat transfer standpoint) configuration, a baseline geometry has been chosen for this example.

Data was collected at one flow rate in the 3:1 rectangular channel. Flow through the channel was fully turbulent (Re ≈ 20,000). The test section was composed of a smooth entrance section (to allow hydrodynamic development of the flow) followed by the dimpled test section. Figure 9.28 presents an overview of the experimental setup. Details for the dimple geometry are shown in Figure 9.29. The cross-section of the rectangular duct is 3.063″ × 1.0625″.

A transient experiment was performed to obtain the wall temperature distribution as a function time. Prior to initiating the test, a series of "black" and "reference" images were recorded. In addition, an initial image was recorded, to provide the initial temperature distribution on the surface.

The dimpled surface is initially at room temperature, and air flow through the channel was hot air. The air was routed through a bypass loop while it was heated to approximately 200°F. When the temperature of 200°F was reached, the air was

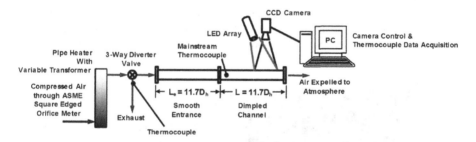

FIGURE 9.28   Overview of the channel flow, transient TSP experiment.

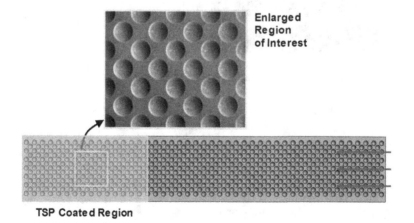

FIGURE 9.29   Overview of the channel flow, transient TSP experiment.

diverted to the test section, and the transient test began. The air temperature was measured with thermocouples placed near the inlet and outlet of the painted section. The hot air was allowed to pass through the test section for approximately 2 minutes.

Detailed heat transfer coefficient (Nusselt number ratio) distributions were obtained in a 3:1 rectangular channel, modeling a typical cooling passage located near the trailing edge of a gas turbine blade.

During the transient test, the following raw data were obtained in the experiment:

1. "Black" intensity distribution on the surface (Figure 9.31a)
2. "Reference" intensity distribution on the surface, and the corresponding reference temperature (Figure 9.30b); the temperature at the reference condition was measured as 74.15°F.
3. "Initial" intensity distribution on the surface (Figure 9.30a)
4. "Wall" intensity distributions at 2 different times [$T_{w1}(t_1)$ and $T_{w2}(t_2)$] (Figure 9.31b and c)
5. Calibration data (intensity ratio, temperature ratio)

The intensity distributions obtained for the "black," "reference," and "initial" conditions represent the average of 100 images.

To begin analyzing any of the images collected for this experiment, first a calibration curve must be developed. The raw data is shown in Table 9.8, and the data is plotted in Figure 9.32.

For this example, it is assumed, there is a true step change in the coolant temperature at $t = 0$. Therefore, the solution to the one-dimensional transient equation does not require modification by Duhamel's Principle of Superposition.

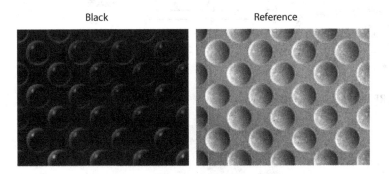

FIGURE 9.30 Black and reference average intensity distributions.

FIGURE 9.31 Intensity distributions at $t = 0$ (initial), $t = 28$ s, and $t = 48$ s.

TABLE 9.8
Selected TSP Calibration Results

I/I_{ref}	T/T_{ref}
1.004	1.001
0.991	1.002
0.98	1.004
0.97	1.006
0.962	1.007
0.947	1.01
0.932	1.012
0.917	1.014
0.903	1.015
0.887	1.018
0.876	1.02
0.865	1.021
0.849	1.024
0.842	1.024
0.826	1.027
0.812	1.029
0.793	1.031
0.772	1.032
0.758	1.035
0.75	1.036
0.72	1.04
0.672	1.049
0.617	1.059
0.558	1.068
0.514	1.075
0.468	1.085

The wall temperatures can be determined from the TSP calibration, and with the detailed wall temperature distributions, the detailed heat transfer coefficient distributions can be numerically calculated. Table 9.9 summarizes the bulk coolant temperature at two different instances in time.

From the raw intensities in Figures 9.30 and 9.31, and the calibration data, the detailed surface temperature distributions can be obtained from the TSP. Figure 9.33 shows the wall temperatures at $t = 0$, 28 s, and 48 s.

Based on these temperature distributions, and the material properties of polycarbonate ($\alpha = 0.1073 \times 10^{-6}$ m²/s, $k = 0.1812$ W/mK), the detailed heat transfer coefficients can be obtained at both instances in time. Pay attention to units through this process as SI and English units are mixed.

Finally, the heat transfer coefficients can be non-dimensionalized and presented in terms of the heat transfer enhancement (Nu/Nu_o).

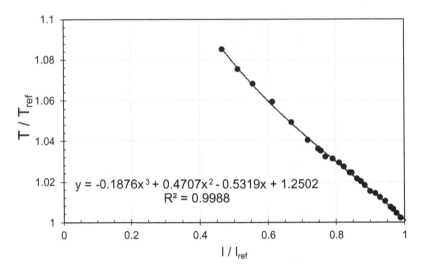

FIGURE 9.32   TSP calibration curve.

TABLE 9.9
Measured Bulk Temperatures during the Transient Test

Time	T_b
28 s	100.7°F
48 s	104.3°F

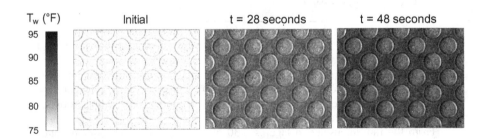

FIGURE 9.33   Wall temperature distributions from the TSP.

Viewing Figures 9.34 and 9.35 show the robustness of this transient TSP experimental method. First, the heat transfer coefficient (Nusselt number ratio) distributions measured at two different times yield comparable values. The heat transfer coefficients should be independent of time, and the results, based on two separate times, so the results are independent of time. In addition, the areas of both high and low heat transfer created by the dimples are clearly captured. With flow from

h
(W/m²K)

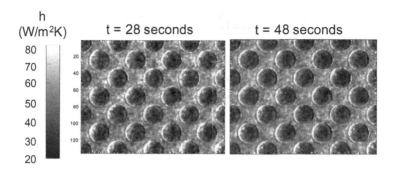

FIGURE 9.34   Detailed heat transfer coefficient distributions at $t = 28$s and $t = 48$s.

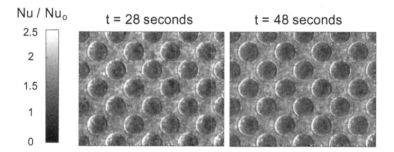

FIGURE 9.35   Detailed Nusselt number ratio distributions at $t = 28$s and $t = 48$ s.

right to left in the images, the flow separates on the upstream (right) side of the dimple, and there is low heat transfer within the dimple. As the fluid reattaches to the surface, downstream of the dimple, the heat transfer coefficients increase. This pattern is repeated for each dimple in the array, and the TSP method is capable of capturing this behavior.

PROBLEMS

1. For industrial applications, researchers are interested in determining the local film cooling or heating distributions on a flat-plate surface through two rows of inclined cooling holes. The required compressed air (25°C), heaters, piping, flow meters, materials, thermocouples, CCD camera, and equipment for calibrations and measurements are available for your use. Two proposed experimental methods, under steady-state flow conditions, are being considered to map the surface concentration or temperature distributions in the region of interest, as shown in the following sketch.

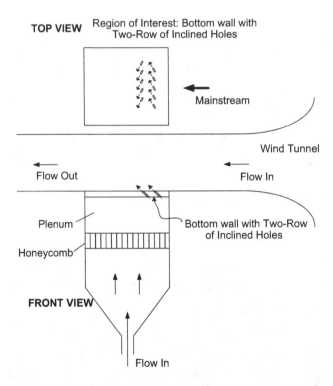

a. Begin by considering the Pressure-sensitive paint method (PSP). Show
    an overall sketch of your experimental test loop and the associated equip-
    ment and instrumentation required to conduct the proposed test, i.e.,
    sketch your overall experimental setup along with a suitable flow measur-
    ing device, detail your test section design with proper materials, describe
    the PSP measurement principle, PSP calibration method, and the detailed
    measurement procedures with the steady-state PSP technique.
b. Based on the wind tunnel flow and cooling flow capability, one
    steady-state blowing ratio (cooling to mainstream mass flux ratio) will
    be tested: $M = 1.0$. Sketch and discuss the expected target surface film
    cooling effectiveness distributions (contours) for one tested blowing
    ratio and estimate the possible uncertainty.
2. Repeat Problem 1 by using the steady-state TSP method.
    a. Show an overall sketch of your experimental test loop and the associated
        equipment and instrumentation required to conduct the proposed test, i.e.,
        sketch your overall experimental setup along with a suitable flow mea-
        suring device and detail your test section design with proper materials,
        describe the TSP measurement principle, TSP calibration method, and the
        detailed measurement procedures with the steady-state TSP technique.
    b. Based on the wind tunnel flow and cooling flow capability, one
        steady-state blowing ratio will be tested: $M = 1.0$. Sketch and discuss the
        expected target surface film cooling effectiveness distributions (contours)

for one tested blowing ratio and estimate the possible uncertainty. Can you compare the film cooling effectiveness contours between (1) and (2) at the same blowing ratio? Are they the same or difference? Why?

3. Repeat Problem 2 by using the transient TSP method.

4. For turbine blade film cooling applications, researchers are interested in measuring the local film cooling or heating distributions on the suction surface of the airfoil with several rows of inclined cooling holes. The required compressed air (25°C), heaters, piping, flow meters, materials, thermocouples, CCD camera, and equipment for calibrations and measurements are available for your use. Two proposed experimental methods, under steady-state flow conditions, are being considered to map the surface concentration or temperature distributions in the region of interest, as shown in below sketch.

   a. Begin by considering the Pressure-sensitive paint method (PSP). Show an overall sketch of your experimental test loop and the associated equipment and instrumentation required to conduct the proposed test, i.e., sketch your overall experimental setup along with a suitable flow measuring device, detail your test section design with proper materials, describe the PSP measurement principle, PSP calibration method, and the detailed measurement procedures with the steady-state PSP technique.

   b. Based on the wind tunnel flow and cooling flow capability, one steady-state blowing ratio (cooling to mainstream mass flux ratio) will be tested: $M = 1.0$. Sketch and discuss the expected target surface film cooling effectiveness distributions (contours) for one tested blowing ratio and estimate the possible uncertainty.

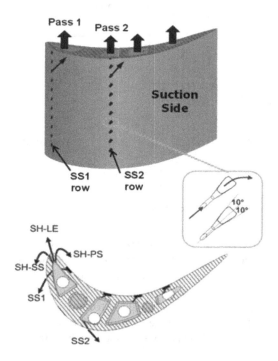

5. Repeat Problem 4 by using the steady-state TSP method.
    a. Show an overall sketch of your experimental test loop and the associated
       equipment and instrumentation required to conduct the proposed test,
       i.e., sketch your overall experimental setup along with a suitable flow mea-
       suring device and detail your test section design with proper materials,
       describe the TSP measurement principle, TSP calibration method, and the
       detailed measurement procedures with the steady-state TSP technique.
    b. Based on the wind tunnel flow and cooling flow capability, one
       steady-state blowing ratio will be tested: $M = 1.0$. Sketch and discuss the
       expected target surface film cooling effectiveness distributions (contours)
       for one tested blowing ratio and estimate the possible uncertainty. Can
       you compare the film cooling effectiveness contours between (1) and (2)
       at the same blowing ratio? Are they the same or difference? Why?
6. Repeat Problem 5 by using the transient TSP method.
7. You are asked to prepare an experimental setup to evaluate the proposed,
   jet impingement configuration for industrial heating/cooling applications.
   The required compressed air (25°C), heaters, piping, flow meters, materials, thermocouples, CCD camera, and equipment for calibrations and measurements are available for your use. Use the transient TSP method to map
   the surface temperature and heat transfer coefficient distributions on the top
   target wall in the region of interest, as shown in below sketch.

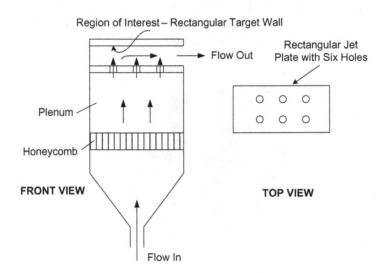

    a. Show an overall sketch of your experimental test loop and the associated
       equipment and instrumentation required to conduct the proposed test,
       i.e., sketch your overall experimental setup along with a suitable flow
       measuring device and detail your test section design with proper mate-
       rials using the transient TSP imaging technique. Based on the jet-hole
       diameter (1 cm), calculate the required mass flow rates for two steady
       flow Reynolds numbers at Re = 10,000 and 30,000.

   b. Describe the TSP measurement principle, TSP calibration method, and detail the TSP measurement procedures by using the transient air heating technique.

   c. Sketch and discuss the expected target surface temperature distributions (contours) as well as heat transfer coefficient distributions (contours) for the two tested Reynolds numbers at time = 50 seconds and 100 seconds. Also, estimate the possible uncertainty.

8. Repeat Problem 7 by using the steady-state TSP method to map the surface temperature and heat transfer coefficient distributions on the top target wall in the region of interest as shown in the sketch with Problem 7. Compare the heat transfer coefficient contours between Problems 7 and 8 at the same flow conditions? Are they the same or difference? Why?

9. You are asked to prepare experimental methods to evaluate surface temperature and heat transfer coefficient distributions for steady, incompressible, turbulent air flow through a two-passage, square, V-shaped ribbed, cooling channel with a 180° turn for heat exchanger applications (as sketched below). Focus on the region of interest shown in the sketch.

   a. Use the steady-state, TSP method to map the surface temperature distributions on the bottom wall as shown in the region of interest.

   b. Sketch the major experimental setup, and describe the measurement principle using the steady-state, uniform heat flux heating technique.

   c. Sketch and discuss the expected experimental results (surface heat transfer coefficient distributions) and perform uncertainty analysis.

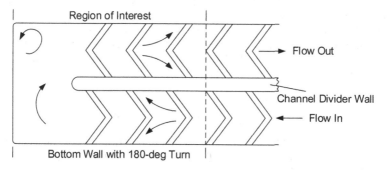

TOP VIEW of Two-Passage Channel Design with V-Shaped Rib

10. Repeat Problem 9 using the transient TSP method to map the surface temperature and heat transfer coefficient distributions on the top target wall in region of interest as shown in the figure with Problem 9. Compare the heat transfer coefficient contours between Problems 9 and 10 at the same flow conditions? Are they the same or difference? Why?

11. You are asked to prepare experimental methods to evaluate surface temperature and heat transfer coefficient distributions for steady, incompressible, turbulent air flow through a square channel with nine circular dimples on the bottom wall, as shown in the following sketch. Focus on the region

of interest shown in the sketch. Use the transient TSP method to map the surface temperature distributions on the bottom wall.

a. Sketch the major experimental setup, and describe the measurement principle using the transient heating technique.

b. Sketch and discuss the expected experimental results (surface heat transfer coefficient distributions) and perform uncertainty analysis.

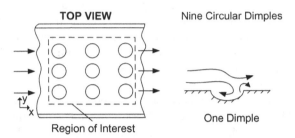

TOP VIEW   Nine Circular Dimples

One Dimple

Region of Interest

12. Repeat Problem 11 using a steady-state TSP method to map the surface temperature and heat transfer coefficient distributions on the bottom target wall in region of interest as shown in the sketch with Problem 11. Can you compare the heat transfer coefficient contours between 11 and 12 at the same flow conditions? Are they the same or difference? Why?

REFERENCES

1. Zhang, L.J. and Fox, M., "Flat Plate Film Cooling Measurement Using PSP and Gas Chromatograph Techniques," *ASME Paper No. AJTE99-6241*, 1999.
2. Han, J.C. and Rallabandi, A.P., "Turbine Blade Film Cooling Suing PSP Technique," *Frontiers in Heat and Mass Transfer*, Vol. 1, Article No. 013001, 2010, p. 21.
3. Wright, L.M., Gao, Z., Varvel, T.A. and Han, J.C., "Assessment of Steady State PSP, TSP, and IR Measurement Techniques for Flat Plate Film Cooling," *ASME Paper No. HT2005-72362*, 2005.
4. Gao, Z., Wright, L.M. and Han, J.C., "Assessment of Steady State PSP and Transient IR Measurement Techniques for Leading Edge Film Cooling," *ASME Paper No. IMECE2005-80146*, 2005.
5. Ahn, J.Y., Mhetras, S.P. and Han, J.C., "Film-Cooling Effectiveness on a Gas Turbine Blade Tip Using Pressure Sensitive Paint," *ASME Journal of Heat Transfer*, Vol. 127, No. 5, 2005, pp. 521–530.
6. Ahn, J.Y., Schobeiri, M.T., Han, J.C. and Moon, H.K., "Film Cooling Effectiveness on the Leading Edge Region of a Rotating Turbine Blade with Two Rows of Film Cooling Holes Using Pressure Sensitive Paint," *ASME Journal of Heat Transfer*, Vol. 128, No. 9, 2006, pp. 879–888.
7. Wright, L.M., Gao, Z., Yang, H. and Han, J.C., "Film Cooling Effectiveness Distribution on a Gas Turbine Blade Platform with Inclined Slot Leakage and Discrete Film Hole Flows," *ASME Journal of Heat Transfer*, Vol. 130, No. 7, 2008, Article No. 071702, p. 11.
8. Gao, Z., Narzary, D.P. and Han, J.C., "Film Cooling on a Gas Turbine Blade Pressure side or Suction Side with Axial Shaped Holes," *International Journal of Heat and Mass Transfer*, Vol. 51, No. 9–10, 2008, pp. 2139–2152.

9. Suryanarayanan, A., Mhetras, S.P., Schobeiri, M.T. and Han, J.C., "Film-Cooling Effectiveness on a Rotating Blade Platform," *ASME J. Turbomachinery*, Vol. 131, No. 1, 2009, Article No. 011014, p. 12.

10. Mhetras, S.P., Han, J.C. and Rudolph, R., "Effect of Flow Parameter Variations on Full Coverage Film-Cooling Effectiveness for a Gas Turbine Blade," *ASME Journal of Turbomachinery*, Vol. 134, No. 1, 2012, Article No. 011004, p. 10.

11. Narzary, D.P., Liu, K.C., Rallabandi, A.P., and Han, J.C., "Influence of Coolant Density on Turbine Blade Film-Cooling using Pressure Sensitive Paint Technique," *ASME Journal of Turbomachinery*, Vol. 134, No. 3, 2012Article No. 031006, p. 10.

12. Narzary, D.P., Liu, K.C. and Han, J.C., "Influence of Coolant Density on Turbine Blade Platform Film-Cooling," *ASME Journal of Thermal Science and Engineering Applications*, Vol. 4, No. 2, 2012Article No. 021002, p. 10.

13. Rezasoltani, M., Lu, K., Schobeiri, M.T. and Han, J.C., "A Combined Experimental and Numerical Study of the Turbine Blade Tip Film Cooling Effectiveness under Rotation Condition," *ASME Journal of Turbomachinery*, Vol. 137, No. 3, 2015, Article No. 051009, p. 12.

14. Chowdhury, N.H.K., Shiau, C.C., Han, J.C., Zhang, L. and Moon, H.K., "Turbine Vane Endwall Film Cooling with Slashface Leakage and Discrete Hole Configuration," *ASME Journal of Turbomachinery*, Vol. 139, 2017, Article No. 061003, p. 11.

15. Shiau, C.C., Chen, A.F., Han, J.C., Lee, C.P. and Azad, S, "Film Effectiveness Comparison on Full-Scale Turbine Vane Endwalls using PSP Technique," *ASME Journal of Turbomachinery*, Vol. 140, No. 2, 2018, Article No. 021009, p. 12.

16. Shiau, C.C., Chowdhury, N.H.K., Han, J.C., MirzaMoghadam, A. and Riahi, A., "Turbine Vane Suction Side Film Cooling Effectiveness with the Presence of Showerhead Cooling using TSP Technique under Transonic Flow Condition," *AIAA Journal of Thermophysics and Heat Transfer*, Vol. 32, No. 3, 2018, pp. 637–647.

17. Kolodner, P. and Tyson, A., "Microscope Fluorescent Imaging of Surface Temperature Profiles with 0.01 C Resolution," *Applied Physical Letters*, Vol. 40, 1982, pp. 782–784.

18. Campbell, B.T., Liu, T. and Sullivan, J.P., "Temperature Sensitive Fluorescent Paint Systems," *AIAA Paper No. 94-2483*, 1994.

19. Liu, T., Campbell, B.T. and Sullivan, J.P., 1992, "Thermal Paints for Shock/Boundary Layer Interaction in Inlet Flows," *AIAA/SAE/ASME/ASEE 28th Joint Propulsion Conference and Exhibit*, AIAA Paper No. 92-3626, Nashville, TN.

10 Mass Transfer Analogy Measurement Techniques

10.1 INTRODUCTION

Mass transfer methods involve the change in concentration of a fluid or solid medium. When a concentration gradient exists between two similar mediums, particle diffusion takes place from the high concentration to the low concentration. Thus, mass transfer can be explained as a phenomenon which occurs due to a concentration gradient between two regions. Momentum transport takes place at the interface of these two regions. This can be compared to the momentum transport which takes place between a heated surface and its cooler surroundings due to the temperature gradient that consequently results in the hot object losing heat. Mass transfer is analogous to heat transfer as both involve momentum transport at the surface. This property has encouraged several researchers to study convective heat transfer using mass transfer methods. There are several techniques which utilize mass transfer for the study of convective heat transfer and film cooling. The more popular techniques using mass transfer are naphthalene sublimation, pressure sensitive paint (discussed in Chapter 9), swollen polymers, and foreign-gas concentration sampling.

One of the primary advantages of using mass transfer over traditional heat transfer methods for the measurement of convective heat transfer coefficients is that no heating of the surface or the fluid is required. Large heat fluxes can cause substantial conduction losses through the substrate when the surface is heated. These conduction losses can result in large uncertainties due to the difficulty associated with predicting them accurately. Heating the fluid instead of the surface does not solve this problem as a substantially larger heater may be required to heat the large volume of fluid. Also, additional heat losses between the ambient surroundings and the hot fluid in the test section can make it difficult to measure the correct bulk temperature. The costs associated with purchasing the heater and other ancillary equipment can be avoided. Another advantage of using a mass transfer method is that most techniques provide a local distribution for heat transfer and film cooling on the surface.

A major disadvantage of these techniques is the high complexity involved in setting up the experiment. The setup time for each test, on average, is more when compared to a typical heat transfer experiment using copper plates and heaters. Also, an experienced experimentalist is required in order to acquire accurate and dependable results.

10.2 NAPHTHALENE SUBLIMATION FOR HEAT/MASS TRANSFER MEASUREMENTS

10.2.1 FUNDAMENTAL PRINCIPAL

This technique utilizes the sublimation of naphthalene to measure the convective heat transfer and film cooling effectiveness. Sublimation is the phase change directly from the solid to the vapor phase. Naphthalene has a relatively low triple point, so sublimation occurs at atmospheric conditions. Therefore, when a surface coated with naphthalene is exposed to forced convection, it sublimates. Figure 10.1 shows the concept of forced convection over a naphthalene coated flat plate. The amount of mass that sublimates can be measured to acquire the local/average mass transfer coefficient. By using the mass transfer analogy, the heat transfer coefficients can then be calculated.

This technique has been used by researchers for several decades and was one of the first techniques to give local distributions of heat transfer coefficients on a surface. A measurement accuracy of ±8% is typical for these experiments. External as well as internal convective heat transfer experiments can be performed using this method. However, this technique is intensive and requires a careful experimentalist due to the brittle nature of the cast naphthalene on the test surface.

By determining the rate of mass lost per unit surface area, $\dot{m}''$, by sublimation of the naphthalene, the mass transfer coefficient, h_m, can be calculated.

$$h_m = \frac{\dot{m}''}{\rho_w - \rho_\infty} \tag{10.1}$$

where ρ_w is the local density of naphthalene at the surface and ρ_∞ is the density of naphthalene in the mainstream. For experiments in a large wind tunnel, $\rho_\infty \sim 0$. However, for internal heat transfer experiments, $\rho_\infty \sim \rho_b$, which is the local, bulk, naphthalene vapor density. The local density of naphthalene vapor at the surface, ρ_w, can be calculated using the ideal gas law.

$$\rho_w = \frac{P_w}{R_v T} \tag{10.2}$$

where R_v is the naphthalene vapor gas constant by Sogin [1] and P_w is the partial pressure of naphthalene at the surface which is given by

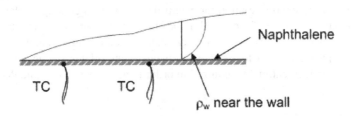

FIGURE 10.1 Concept of forced convection on a naphthalene coated flat plate.

$$P_w = 47.88026 \left(10^{\left(11.884 - \frac{6713}{T(^\circ R)} \right)} \right) \tag{10.3}$$

where T is the surface temperature of naphthalene during the experiment. A thermocouple may be inserted underneath the surface to measure this temperature. The local Sherwood number can then be obtained from

$$Sh = h_m \frac{D_h}{D} \tag{10.4}$$

Here D_h is the hydraulic diameter of the test section in case of internal flow experiments and can be replaced with a suitable length scale for external flow experiments. $\bar{D}$ is the binary diffusion coefficient given as $\bar{D} = \upsilon/Sc$, where υ is the kinematic viscosity of air and Sc is the Schmidt number ($Sc = 2.5$ for naphthalene diffusing in air at ambient temperature). Thus, the Sherwood number can now be expressed as

$$Sh = h_m D_h \frac{Sc}{\left(\mu/\rho \right)} \tag{10.5}$$

A direct correlation exists between the convective heat transfer and mass transfer processes as the governing equations for mass and energy transport are essentially the same with the temperature and the Prandtl number in the energy equation corresponding to the mass concentration and the Schmidt number in the mass transport equation. This analogy will hold true if the differences in properties are taken into account and by assuming that the turbulent transport in the fluid system, as well as, the boundary conditions in the heat transfer and mass transfer processes are identical. Thus, a constant wall temperature boundary condition in a heat transfer process can be considered equivalent to a constant vapor pressure and vapor concentration at the surface in an isothermal air-naphthalene system. Continuous sublimation of naphthalene causes a normal component of velocity to the surface that is absent in most convective heat transfer experiments. However, the effect of this normal velocity component is relatively small when compared to the mainstream velocities, and hence, it can be ignored. The heat—mass transfer analogy has been validated by Sogin [1], and the analogy can be expressed as

$$Nu = \left(\frac{Pr}{Sc} \right)^{0.4} Sh \tag{10.6}$$

When manufacturing the frame (structure) of the test section, a recess should be cut into the test surface to provide a cavity for casting the naphthalene. The molded naphthalene test surface is prepared by first placing the recessed cassette upside down on a polished surface and then pouring molten naphthalene, obtained by

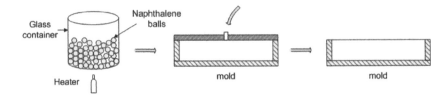

FIGURE 10.2   Procedure of preparing the cast naphthalene test surface.

melting pure (99%) naphthalene crystals, into this gap through a hole on the underside of this recessed cassette. Figure 10.2 shows the procedure of preparing the molded naphthalene test surface. The polished aluminum or stainless steel surface ensures that the cast naphthalene has a smooth finish. Formation of air bubbles in the cast naphthalene during the casting process should be avoided. These bubbles can cause the cast surface to rupture. Pre-heating the polished surface before casting and a ensuring a uniform molten stream of naphthalene while pouring can avoid the formation of bubbles. An air vent should be provided on the cassette to remove the air displaced by naphthalene. The cast plates should rest for at least eight hours to cool to room temperature and solidify. These plates should be placed in a sealed container saturated with naphthalene vapor to prevent sublimation due to natural convection. For example, before running the test, these naphthalene-coated plates can be placed in a sealed plastic container to prevent sublimation due to natural convection. Figure 10.3 shows the basic required tools for manufacturing a naphthalene coated test plate.

The recessed cassette, now filled with naphthalene, should be weighed prior to fitting it in the test section. This measurement provides the overall mass of the naphthalene in this cassette. By measuring the difference between the mass of the

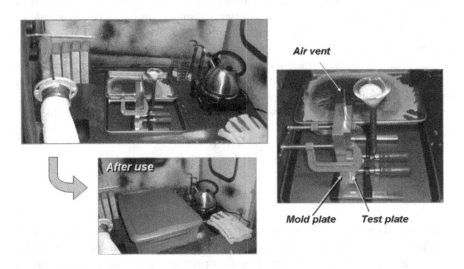

FIGURE 10.3   Basic tools for manufacturing a naphthalene coated test plate.

naphthalene before and after the test, the amount of sublimated naphthalene can be calculated. This direct mass measurement yields the overall (average) mass transfer for the cast test surface. To obtain local mass transfer results, detailed profile distributions of the cast naphthalene surface are needed. By using a profilometer (or high-resolution 3D scanner), the cast surface can be mapped into an x–y grid. It should be ensured that the cast surface suffers minimal damage when the probe is scanning the surface. The test surface is then inserted into the test section and the experiment is started. The naphthalene will start sublimating and the amount of sublimation at a particular grid point will be proportional to the heat transfer coefficient at that point. Figure 10.4 shows the components of the naphthalene measurement technique, they include a LVDT sensor (profilometer), signal conditioner, digital multimeter, motor dive, and x–y coordinate table. A diagram of the naphthalene measurement system is shown in Figure 10.5.

The naphthalene vapor pressure is highly sensitive to its temperature. Hence, a thermocouple embedded in the cast surface should be used to measure the temperature. The test should be run for a sufficient amount of time to have a measurable impact on the cast surface profile. However, the time of the test should be limited, so the surface does not deteriorate to a level that changes the flow over the surface. Upon completion of the experiment, the test section is disassembled, the cast surface is removed and then weighed to calculate the total loss of naphthalene by sublimation. Detailed profile measurements of the cast surface are then carried out at the same grid points established before the experiment by using the profilometer and the difference in height, Δz, from the initial scan is recorded. Losses due to naphthalene sublimation by natural convection during the time needed for assembling

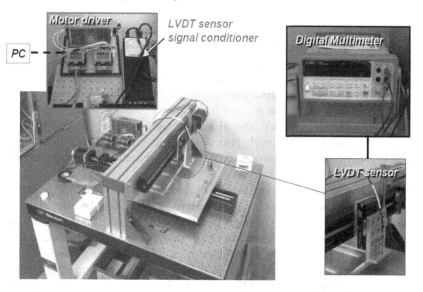

FIGURE 10.4 Components for the local measurement of naphthalene sublimation.

Diagram of Measurement system

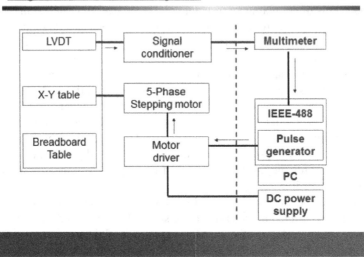

FIGURE 10.5   Diagram of naphthalene measurement system.

and disassembling the test section should be considered. The mass lost from the test surface per unit surface area at each grid point due to naphthalene sublimation can then be calculated using

$$\dot{m}'' = \rho_{\text{solid naph}} \frac{\Delta z}{\Delta t} \tag{10.7}$$

where $\rho_{\text{solidnaph}}$ is the density of solid naphthalene at the testing fluid temperature and Δt is the duration of the experiment. Using the set of equations described earlier, the data can be reduced and mass transfer coefficients can be calculated. The duration of the experiment Δt can be longer for relatively low convective heat transfer but must be shorter for relatively high heat transfer flows. The local naphthalene sublimation depth, Δz, will be too small to measure accurately if Δt is too short, while the local naphthalene sublimation depth Δz will be too large and alter the original surface contour if Δt is too long. Researchers must be aware of the relationship between Δz and Δt in order to obtain reliable data. The typical duration time is approximately 30 minutes.

The aforementioned naphthalene measurements were established by Sogin [1] in 1958. The method was first used to measure the local heat transfer coefficients around film cooling holes by Goldstein and Taylor [2] in 1982, and to determine the local heat transfer coefficients inside a rectangular channel by Sparrow and Cur [3] in 1982. Since the 1980s many papers have been published using this methodology for turbine blade internal cooling as well as film cooling applications. For example, turbine blade internal cooling with pin-fins by Lau et al. [4], turbine blade internal cooling with rib turbulators by Han et al. [5], flat surface with wall-mounted cylinders by Chyu and Goldstein [6], and film cooling by Cho and Goldstein [7] have all been investigated. Interested readers are referred to these papers for details of the experimental setups and procedures.

10.2.2 EXPERIMENT EXAMPLE

Example 10.1: Local Mass/Heat Transfer Coefficients around an Inclined Film Cooling Hole [2]

The naphthalene sublimation technique has been used to measure the mass/heat transfer coefficient distributions around a row of jets entering a cross flow for turbine blade film cooling applications. The naphthalene-coated test plate with one row of inclined film cooling holes and the area of surface measurements are shown in Figure 10.6 [2]. The mass transfer rate is directly related to the heat transfer rate that would occur on a film-cooled surface. Figure 10.7 [2] shows one of many test results at the blowing ratio $M = 1.0$. The results indicate that the heat transfer coefficients are significantly enhanced by the cooling jet interaction with the mainstream flow. The area immediately downstream of the holes is most significantly affected by the jet-to-mainstream interaction, but the region far downstream also experiences heat transfer enhancement (compared to the surface with no film cooling). One of the advantages of using this mass/heat transfer analogy technique is that there is no heat conduction error as normally seen with traditional heat transfer methods. Refer to reference [2] for the detailed flow loop design, naphthalene-coated test plate preparation, test procedure, and data analysis and results presentation.

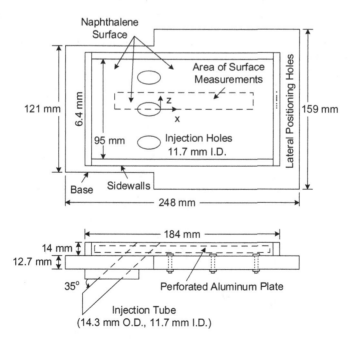

FIGURE 10.6   Naphthalene-coated test plate with one row of inclined film cooling holes.

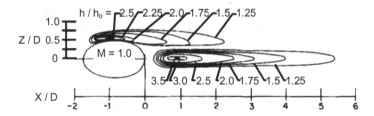

FIGURE 10.7 Heat transfer enhancement around a film cooling hole at the blowing ratio $M = 1.0$.

Example 10.2: Local Mass/Heat Transfer Coefficients around a Circular Pin Array [4]

The naphthalene sublimation technique was used to measure the mass/heat transfer coefficient distributions in the neighborhood of a circular pin array for turbine blade internal cooling applications. This study focused on the effects of the pin configuration, the pin length-to-diameter ratio, and the entrance length on local endwall heat/mass transfer in a rectangular channel with short pin fins (pin length-to-diameter ratios of 0.5 and 1.0). The detailed distributions of the local endwall heat/mass transfer coefficient were obtained for staggered and aligned arrays of pin fins, for the spanwise pin spacing-to-diameter ratio of 2.5, and for streamwise pin spacing-to-diameter ratios of 1.25 and 2.5. Figure 10.8 [4] shows the sketch of the test sector for the tests. The measurement region on the naphthalene-coated test plate around the pin array is shown in Figure 10.9 [4]. The mass transfer rate is directly related to the heat transfer rate that would occur around the pinned surface. Figure 10.10 [4] shows one of many test results for the staggered pin array at the channel Reynolds number Re = 33,000. The results indicate that the heat transfer coefficients (Sherwood numbers) are significantly higher near the pin stagnation region and also high in the pin endwall junction region. Additional details of the experiment can be found in the original paper by Lau et al. [4]

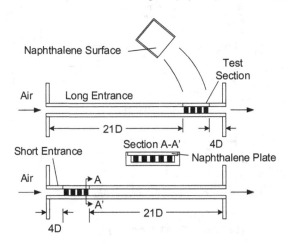

FIGURE 10.8 Schematic of the test channel for the naphthalene sublimation tests.

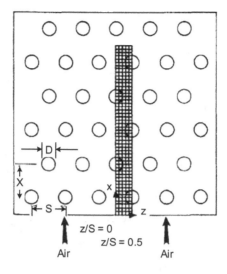

FIGURE 10.9 Measurement region on the naphthalene coated surface around the pin array.

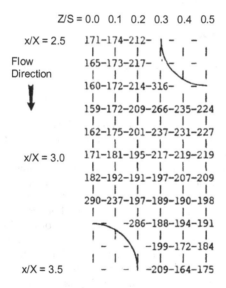

FIGURE 10.10 Sherwood number distribution for the staggered pin array at Re = 33,000.

Example 10.3: Local Mass/Heat Transfer Coefficients inside a Rib Roughened Channel [5]

The naphthalene sublimation technique was used to measure the mass/ heat transfer coefficient distributions inside the two-passage, square channel with repeated ribs, for turbine blade internal cooling applications. This study

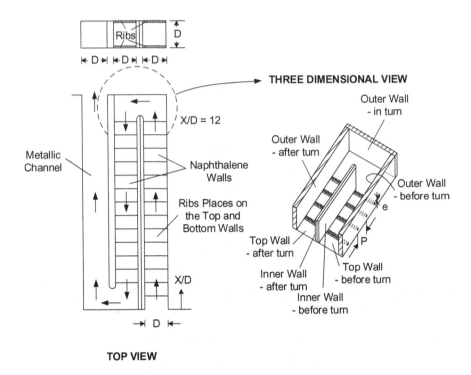

FIGURE 10.11   Two-passage, rib-roughened channel for naphthalene measurements.

focused on the effects of the 180° sharp turn in a two-passage, square channel with smooth walls or rib-roughened walls, for a range of Reynolds numbers. Figure 10.11 [5] shows a sketch of the test section design. The detailed distributions of the local heat/mass-transfer coefficients were obtained along the center line, outer line, and inner line of every naphthalene-coated surface of the two-passage square channel, as shown in Figure 10.12 [5]. The mass transfer rate is directly related to the heat transfer rate that would occur in the ribbed channel. Figure 10.13 [5] shows one of the many test results for the rib-roughened channel at a Reynolds number of Re = 30,000. The results indicate that the heat transfer coefficients (Sherwood numbers) are significantly enhanced on the rib-roughened surface as well as the adjacent smooth surface (2–3 times as compared with the fully developed tube flow with smooth wall). Periodic behavior of heat transfer enhancement is due to the flow over the repeated ribs on the surface. Results also show that the heat/mass transfer enhancement in the second channel (after the 180° sharp turn) is higher than the first channel. Full details of the experimental design and execution are presented in the original publication of this work [5].

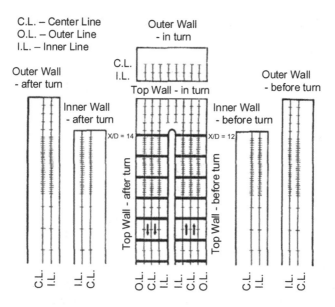

FIGURE 10.12 Measurement points along the centerline, outer line, and inner line within the two-passage channel.

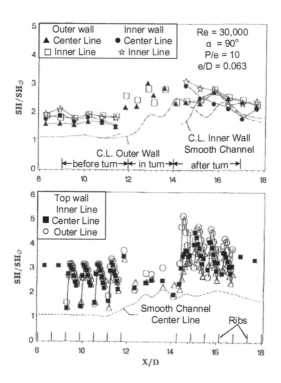

FIGURE 10.13 Local Nusselt Number (Sherwood Number) enhancement for the rib-roughened channel at Re = 30,000.

10.3 FOREIGN GAS CONCENTRATION SAMPLING TECHNIQUE

10.3.1 FUNDAMENTAL PRINCIPLE

The PSP technique presented in Chapter 9 is a fundamentally a foreign gas sampling technique. For this method, a mass transfer analogy is applied for measuring only the film cooling effectiveness. Instead of heating or cooling the injected fluid and measuring surface temperatures, a fluid, at mainstream temperature, containing a foreign gas is injected. The local, impermeable, wall effectiveness, η, is defined as

$$\eta = \frac{T_w - T_\infty}{T_c - T_\infty} = \frac{C_w - C_\infty}{C_{fg} - C_\infty} \tag{10.8}$$

where C_∞ is the concentration of the foreign gas in the mainstream, C_{fg} is the concentration of the foreign gas in the injected fluid, and C_w is the concentration of the foreign gas on the film cooled surface. If there is no foreign gas in the mainstream, then $\eta = C_w/C_{fg}$.

The analogy of mass transfer to heat transfer holds if the turbulent Lewis number and molecular Lewis number are unity. Figure 10.14 shows the concept of a flat plate with compound angle film cooling holes using the foreign gas sampling technique. Many tiny holes are drilled through the test plate downstream of the film cooling holes; these holes (sampling ports) are individually connected to the gas analyzer.

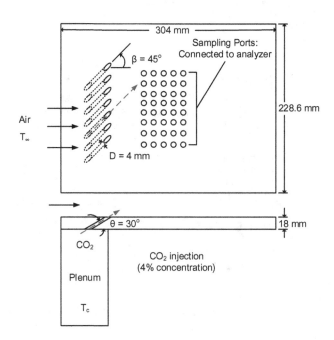

FIGURE 10.14 Conceptual representation of the Foreign gas sampling technique on a film cooled flat plate.

From the outputs of gas analyzer, the local foreign gas concentration and film cooling effectiveness at each sampling port can be determined.

The foreign gas concentration sampling technique was first introduced by Pedersen et al. [8] in 1977. They measured the foreign gas content using gas chromatography. The air was sampled along the test wall at discrete locations, and the film effectiveness values were estimated. This technique provides film effectiveness only. This is a very useful technique as it can be used to simulate actual density ratios that occur inside real gas turbine engines without having to operate at high temperature conditions. The uncertainty of the measured, impermeable, wall film effectiveness is about 6% for effectiveness values greater than 0.1. The uncertainty levels could be higher for lower effectiveness values. As with the naphthalene sublimation, mass transfer technique, there are no heat losses to affect the surface results. Refer to the original work from Pedersen et al. [8] for the details of the experimental setup, methodology, data reduction, and results.

10.4 AMMONIA-DIAZO TECHNIQUE

10.4.1 FUNDAMENTAL PRINCIPLE

The ammonia and diazo technique began as a surface, flow visualization technique. The surface of the test piece is covered with a diazo coated, paper film (film thickness ~ 0.05 mm). When pure ammonia gas is passed over the surface, it reacts with the diazo coating, leaving a trace on the paper as it is transported over the surface. Therefore, coolant air is seeded with ammonia gas and water vapor. Ammonia gas passes over the surface (coated with diazos), and reacts with the coating, a trace is left on the paper as the ammonia moves over the surface. Friedrichs et al. [9], in 1996, were the first to calibrate the traces of the ammonia gas on diazo paper for quantitative measurements. The ammonia-diazo method is a variation of the foreign gas sampling technique. With the diazo paper being used, a specific foreign gas must be used to cause a reaction on the paper. Therefore, the measurement technique depends on the mass transfer principle, which states that the traces on the diazo coated surface are dependent on the surface concentration of ammonia and water vapor in the coolant gas. Prior to the experiment, the test surface is coated with the diazo film. The cooling air is then seeded with ammonia gas and water vapor. The test surface is then exposed to the seeded coolant for 1–2 minutes. The ammonia concentration and relative humidity can be varied to achieve the desired darkness levels. The image is then fixed by exposure to light to prevent further reaction (similar to the traditional film developing process).

10.4.2 EXPERIMENT EXAMPLE

Example 10.4: Film Cooling Effectiveness Distribution on a Turbine Endwall [9]

To determine quantitative data, the relationship between the darkness of the trace and surface concentration of coolant must be determined. Calibration of the trace depends on the ammonia concentration, humidity, and exposure time. Also, an increase in temperature results in a lighter image. A reference experiment is performed parallel to the actual experiment to avoid calibration errors. A calibration strip is produced by mixing the coolant gas mixture with freestream air from the wind tunnel in known ratios. To quantify the darkness distribution, both the calibration strip and the exposed test surface are digitized simultaneously using an optical scanner. The analysis of the calibration strip provides a relationship between the darkness of the trace and the relative concentration of the coolant. Figure 10.15 shows a calibration relating darkness of the trace to the relative concentration of the coolant [9].

Therefore, the local film cooling effectiveness distributions can be determined from the darkness of the trace on the surface, i.e., from the relative concentration C_{rel} [%] on the surface. Figure 10.16 shows the diazo traces on the film-cooled turbine endwall [9]. Since only the coolant is seeded with ammonia and water vapor, the adiabatic film cooling effectiveness can be determined from the relative concentration (trace darkness-relative concentration calibration). Figure 10.17 shows the detailed film cooling effectiveness distributions on the turbine endwall [9]. This technique is an excellent surface flow visualization tool and also a quantitative method for measuring the film effectiveness. However, like other foreign gas sampling techniques, it provides only film effectiveness measurements. The complete details for this experiment have been provided by Friedrichs et al. [9]. In addition, Han et al. [10] have also summarized other investigations using this technique along with other mass transfer studies.

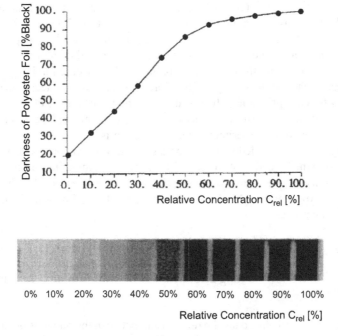

FIGURE 10.15 Calibration relating darkness of the trace to the relative concentration of the coolant.

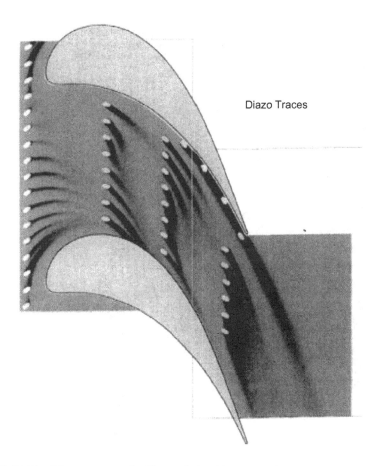

Diazo Traces

FIGURE 10.16 Diazo traces on the film cooled turbine endwall.

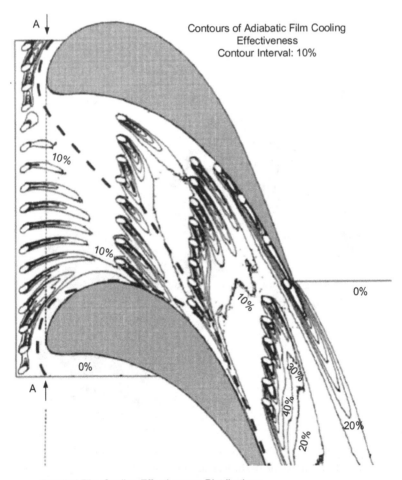

Detailed Film Cooling Effectiveness Distributions

FIGURE 10.17   Detailed film cooling effectiveness distributions on the turbine endwall.

PROBLEMS

1. For industrial applications, researchers are interested in determining the local heat transfer coefficient distributions on a flat plate surface through two rows of inclined cooling holes. The required compressed air (25°C), heaters, piping, flow meters, materials, thermocouples, and equipment for calibrations and measurements are available for your use. The naphthalene sublimation method should be used to obtain detailed heat transfer coefficient distributions on the film cooled surface.

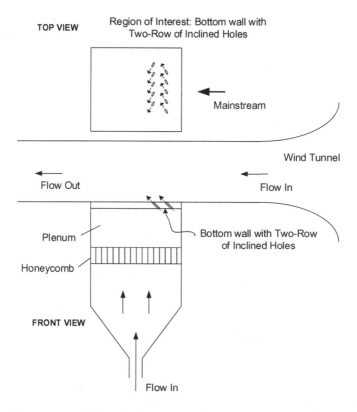

a. Show an overall sketch of your experimental test loop and the associated equipment and instrumentation required to conduct the proposed test, i.e., sketch your overall experimental setup along with a suitable flow measuring device, detail your test section design with proper materials, describe the naphthalene sublimation measurement principle using the heat and mass transfer analogy, including the process for casting the naphthalene surface and using a LVDT for local depth measurements, and the detailed measurement procedures at the steady-state condition.
b. Based on the wind tunnel flow and cooling flow capability, one steady-state blowing ratio (cooling to mainstream mass flux ratio) will be tested: $M = 1.0$. Sketch and discuss the expected target surface mass transfer/heat transfer coefficient distributions (contours) for one tested blowing ratio and estimate the possible uncertainty.
2. Repeat Problem 1 by using the steady-state, ammonia and diazo method.
a. Show an overall sketch of your experimental test loop and the associated equipment and instrumentation required to conduct the proposed test, i.e., sketch your overall experimental setup along with a suitable flow measuring device and detail your test section design with proper materials, describe the ammonia-diazo measurement principle (including the dark trace vs relative concentration calibration method), and the detailed measurement procedures at the steady-state condition.

b. Based on the wind tunnel flow and cooling flow capability, one steady-state blowing ratio will be tested: $M = 1.0$. Sketch and discuss the expected target surface film cooling effectiveness distributions (contours) for one tested blowing ratio and estimate the possible uncertainty.

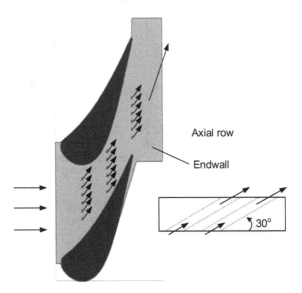

3. For turbine blade, film cooling applications, researchers are interested in measuring the local heat transfer coefficient distributions on the endwall surface through three rows of 30° inclined cooling holes. The required compressed air (25°C), heaters, piping, flow meters, materials, thermocouples, and equipment for calibrations and measurements are available for your use. The naphthalene sublimation method should be used to obtain detailed heat transfer coefficient distributions on the film cooled endwall, as shown in the following sketch.

a. Show an overall sketch of your experimental test loop and the associated equipment and instrumentation required to conduct the proposed test, i.e., sketch your overall experimental setup along with a suitable flow measuring device, detail your test section design with proper materials, describe the naphthalene sublimation measurement principle using the heat and mass transfer analogy, including the process for casting the naphthalene surface and using a LVDT for local depth measurements, and the detailed measurement procedures at the steady-state condition.

b. Based on the wind tunnel flow and cooling flow capability, one steady-state blowing ratio (cooling to mainstream mass flux ratio) will be tested: $M = 1.0$. Sketch and discuss the expected target surface mass transfer/heat transfer coefficient distributions (contours) for one tested blowing ratio and estimate the possible uncertainty.

4. Repeat Problem 3 by using the steady-state ammonia and diazo technique method.

a. Show an overall sketch of your experimental test loop and the associated equipment and instrumentation required to conduct the proposed test, i.e., sketch your overall experimental setup along with a suitable flow measuring device and detail your test section design with proper materials, describe the ammonia-diazo measurement principle (including the dark trace vs relative concentration calibration method), and the detailed measurement procedures at the steady-state condition.

b. Based on the wind tunnel flow and cooling flow capability, one steady-state blowing ratio will be tested: $M = 1.0$. Sketch and discuss the expected target surface film cooling effectiveness distributions (contours) for one tested blowing ratio and estimate the possible uncertainty.

5. You are asked to prepare an experimental setup to evaluate the proposed, jet impingement configuration. Heat transfer coefficients will be measured on the upper target surface, as shown in the below sketch. The required compressed air (25°C), heaters, piping, flow meters, materials, thermocouples, and equipment for calibrations and measurements are available for your use. Use the naphthalene sublimation method to map the surface mass and heat transfer coefficient distributions on the top target wall.

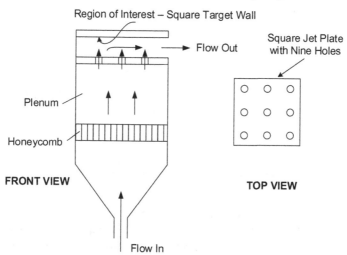

a. Show an overall sketch of your experimental test loop and the associated equipment and instrumentation required to conduct the proposed test, i.e., sketch your overall experimental setup along with a suitable flow measuring device and detail your test section design with proper materials using the naphthalene sublimation technique. Based on jet hole diameter (1 cm), calculate the required mass flow rates for two steady flow Reynolds numbers at Re = 10,000 and 30,000.

b. Describe the measurement principle, including the process for casting the naphthalene surface and using a LVDT for local depth measurements, and detail the measurement procedures by using the naphthalene sublimation technique.

   c.   Sketch and discuss the expected target surface mass/heat transfer coefficient distributions (contours) for two tested Reynolds numbers and estimate the possible uncertainty.

6. Prepare experimental methods to evaluate the surface heat transfer coefficient distributions for steady incompressible turbulent air flow through a two-passage, rectangular, cooling channel with a 180° turn for heat exchanger applications, as sketched below. Focus on the region of interest shown in the sketch.

   a.   Use the steady-state, naphthalene sublimation method to map the surface heat transfer coefficient distributions on the bottom wall as shown in the region of interest.

   b.   Sketch the major experimental setup, describe the measurement principle, including the process for casting the naphthalene surface and using a LVDT for local depth measurements, and detail the measurement procedures by using the naphthalene sublimation technique.

   c.   Sketch and discuss the expected experimental results (surface heat transfer coefficient distributions), and perform uncertainty analysis.

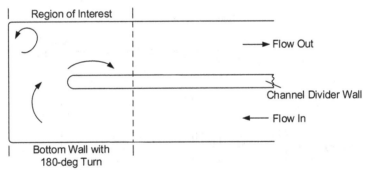

TOP VIEW of Two-Passage Rectangular Channel Design

7. You are asked to prepare experimental methods to evaluate surface heat transfer coefficient distributions for steady, incompressible, turbulent air flow over a surface with two repeated ribs for heat exchanger applications (as shown in the following sketch). The naphthalene sublimation method will be used to map the surface mass/heat transfer coefficient distributions on the bottom surface between the ribs as well as on the rib surface.

   a.   Sketch the major experimental setup, describe the measurement principle, including the process for casting the naphthalene surface and using a LVDT for local depth measurements, and detail the measurement procedures of the naphthalene sublimation technique.

   b.   Sketch and discuss the expected experimental results (surface mass and heat transfer coefficient distributions) and perform uncertainty analysis.

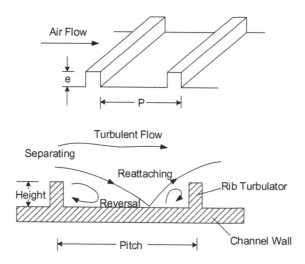

8. You are asked to prepare experimental methods to evaluate surface heat transfer coefficient distributions for steady, incompressible, turbulent, air flow over a surface with three repeated wedge-shaped ribs for heat exchanger applications, as sketched below. The naphthalene sublimation method will be used to map the surface mass/heat transfer coefficient distributions on the bottom surface between the ribs as well as on the rib surface.

   a. Sketch the major experimental setup, describe the measurement principle, including the process for casting the naphthalene surface and using a LVDT for local depth measurements, and detail the measurement procedures of the naphthalene sublimation technique.

   b. Sketch and discuss the expected experimental results (surface mass and heat transfer coefficient distributions) and perform uncertainty analysis.

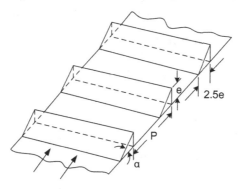

Wedge-Shaped Continuous Rib

9. Prepare an experiment to evaluate surface heat transfer coefficient distributions for steady, incompressible, turbulent air flow inside a rectangular channel with nine square pins, for heat exchanger applications. Focus on the region of interest identified in the sketch. The naphthalene sublimation

method will be used to map the surface mass/heat transfer coefficient distributions on the bottom surface around the pins and the pins themselves.

a. Sketch the major experimental setup, describe the measurement principle, including the process for casting the naphthalene surface and using a LVDT for local depth measurements, and detail the measurement procedures of the naphthalene sublimation technique.

b. Sketch and discuss the expected experimental results (surface mass and heat transfer coefficient distributions) and perform uncertainty analysis.

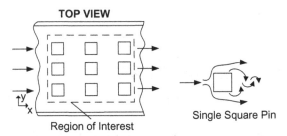

10. Researchers would like to investigate the heat/mass transfer in a rectangular channel with nine strip fins for potential heat exchanger applications (as shown in the sketch). Airflow through the channel, is steady, turbulent, and incompressible. Focusing on the center row of strip fins, use the naphthalene sublimation technique to evaluate the thermal performance of the strip fins.

a. Sketch the major experimental setup, describe the measurement principle, including the process for casting the naphthalene surface and using a LVDT for local depth measurements, and detail the measurement procedures of the naphthalene sublimation technique.

b. Sketch and discuss the expected experimental results (surface mass and heat transfer coefficient distributions) and perform uncertainty analysis.

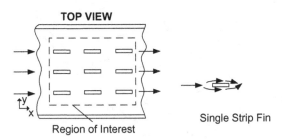

REFERENCES

1. Sogin, H.H., "Sublimation from Disks to Air Streams Flowing Normal to Their Surfaces," *Transactions of the American Society of Mechanical Engineers*, Vol. 80, 1958, pp. 61–69.

2. Goldstein, R.J. and Taylor, J.R., 1982, "Mass Transfer in the Neighborhood of Jets Entering a Crossflow," *The ASME Journal of Heat Transfer*, Vol. 104, 1982, pp. 715–721.

3. Sparrow, E.M. and Cur, N., "Turbulent Heat Transfer in a Symmetrically or Asymmetrically Heated Flat Rectangular Duct with Flow Separation at Inlet," *ASME Journal of Heat Transfer*, Vol. 104, 1982, pp. 82–89.

4. Lau, S.C., Kim, Y.S. and Han, J.C., "Local Endwall Heat/Mass Transfer Distributions in Pin Fin Channels," *AIAA Journal of Thermophysics and Heat Transfer*, Vol. 1, No. 4, 1987, pp. 365–372.

5. Han, J.C., Chandra, P.R. and Lau, S.C., "Local Heat/Mass Transfer Distributions around Sharp 180E Turns in Two-Pass Smooth and Rib-Roughened Channels," *ASME Journal of Heat Transfer*, Vol. 110, No. 1, 1988, pp. 91–98.

6. Chyu, M.K. and Goldstein, R.J., "Influence of an Array of Wall-mounted Cylinders on the Mass Transfer from a Flat Surface," *International Journal of Heat and Mass Transfer*, Vol. 34, No. 9, 1991, pp. 2175–2186.

7. Cho, H.H. and Goldstein, R.J., "Heat (Mass) Transfer and Film Cooling Effectiveness with Injection through Discrete Holes Part I: Within Holes and on the Back Surface," *ASME Journal of Turbomachinery*, Vol. 117, 1995, pp. 451–460.

8. Pedersen, D.R., Eckert, E.R.G. and Goldstein, R.J., "Film Cooling with Large Density Differences between the Mainstream and Secondary Fluid Measured by the Heat-Mass Transfer Analogy," *ASME Journal of Heat Transfer*, Vol. 99, 1977, pp. 620–627.

9. Friedrichs, S., Hodson, H.P. and Dawes, W.N., "Distribution of Film-Cooling Effectiveness on a Turbine Endwall Measured using the Ammonia and Diazo Technique," *ASME Journal of Turbomachinery*, Vol. 118, 1996, pp. 613–621.

10. Han, J.C., Dutta, S. and Ekkad, S.V., "Experimental Methods" in *Gas Turbine Heat Transfer and Cooling Technology*, Taylor & Francis Group, New York, December, 2000.

11 Flow and Thermal Field Measurement Techniques

Up to this point, each chapter has focused on surface measurements, primarily surface temperatures. While detailed surface measurements can provide researchers with valuable insight of how the fluid is interacting with the surface, experimentalists are often left to conjecture about the actual flow behavior near the test surface. Experimental hardware and methods exist to allow researchers to directly investigate fluid flow behavior. These techniques may use either discrete probes or optical methods where the fluid is seeded with suitable "tracer" particles. Depending on the measurement method, researchers can obtain velocity (one-, two-, or three-dimensional), turbulence (Reynolds stress, vorticity, turbulence intensity), and flow direction (yaw and pitch). In addition, thermal probes can be traversed through a fluid to obtain temperature distributions for the fluid. While the techniques mentioned in this chapter are not an exhaustive list for flow measurements, they do provide insight into a variety of methods available to researchers.

11.1 MINIATURE FIVE-HOLE AND SEVEN-HOLE PROBES

Miniature five-hole probes have been used for measuring 3D, mean velocity components at a single location. Figure 11.1 shows a general schematic of the miniature five-hole probe. The diameter of the probe tip is around 1.22 mm, where the central tube is surrounded by four outer tubes. The end of each tube is tapered at a 45° angle, with respect to the mouth of the central tube. Five-hole probes can measure velocity vectors inclined to about 45° with respect to the probe axis. The probe needs to be calibrated in a wind tunnel before actual application. Because the outer holes are angled to the flow, they do not provide a direct measure of the static pressure. Therefore, in order to obtain the velocity, the probe must be calibrated to relate the measured pressure to the actual, static pressure. A traversing mechanism is required to place the probe at the specified location for measurements. The three-dimensional flow field can be measured by traversing the miniature five-hole probe. In 1989, Ligrani et al. [1] provided a detailed description of the usage of five-hole probes. From calibration, the yaw, pitch, total, and total minus static pressure coefficients can be calculated. Morrison et al. [2] presented a five-hole pressure probe analysis technique in 1998. Recently, Paul et al. [3] reported a novel calibration algorithm for five-hole pressure probes in 2011.

Increasing the number of measurement points on the probe increases the amount of data available with the probe. In 1997, Takahashi [4] used a seven-hole, cone probe

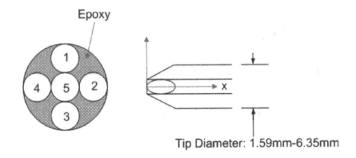

FIGURE 11.1   Schematic of the miniature five-hole probe.

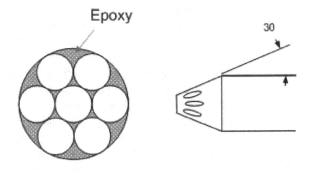

FIGURE 11.2   Schematic of the miniature seven-hole probe.

for 3D, mean velocity distributions, including high flow angularity measurements. Figure 11.2 shows the general schematic of the seven-hole probe. Seven-hole probes can measure velocity vectors inclined to about 70° with respect to the probe axis. Takahashi developed a simple, very efficient algorithm to compute flow properties from the measured probe tip pressures. Disadvantages of using miniature five-hole or seven-hole probes are its low frequency response (<50 Hz) and flow interference. In addition, dirty particles may partially block the pressure lines which will cause inaccurate pressure measurements.

11.1.1   Fundamental Principles of Multiple-Hole Probes

Two-Hole Probes for Measurement of Flow Direction: Figure 11.3 shows a two-hole "yaw" probe which can be used to determine the flow direction of 2D flow by rotating the probe until $p_A = p_B$. The probe is placed in the flow, and it rotated until the pressures are equal (or $\Delta P = 0$). At this point, the angle between the wind tunnel axis and the axis of the instrument is measured, yielding the angularity of the flow. Because the pressure taps are angled to the flow, they do not provide a direct measurement of the static or total pressures. Therefore, only the flow direction (in a 2D flow) is measured; no velocity information is generated.

Three-Hole Probes for Measurement of Flow Direction and Stagnation Pressure: Increasing the number of pressure measurements with the probe,

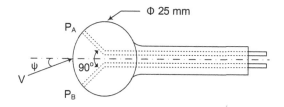

FIGURE 11.3 A two-hole yaw probe.

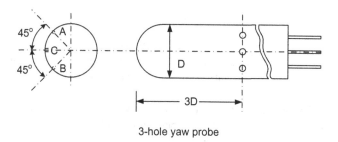

3-hole yaw probe

FIGURE 11.4 Three-hole yaw probe.

increases the amount of information collected with the probe. Moving from two holes to three holes allows for the measurement of the total pressure of the fluid. Figure 11.4 shows a three-hole, yaw probe which can be used to determine the flow direction of a 2D flow by rotating the probe until $p_A = p_B$. Because the probe is now oriented with the flow, the center port is measuring the stagnation pressure of the fluid. Again, the outer pressures (p_A and p_B) are not equal to the static pressure, but through a calibration, these pressures can be related to the static pressure, and from the stagnation and calibrated static pressure, the velocity can be determined.

Another three-hole probe is the Cobra probe. As shown in Figure 11.5, the head of the probe has the appearance of the head of a cobra snake. Like the three-hole, yaw probe, the Cobra probe is rotated until the outer pressures balance. At this point the flow direction is known. The center port is now pointed directly into the flow, so the stagnation pressure is

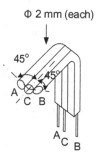

FIGURE 11.5 Three-hole Cobra probe.

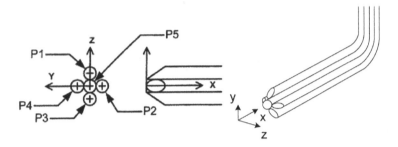

FIGURE 11.6 Five-hole probes.

directly measured. With a rigorous calibration, the outer pressure can be related to the true static pressure, and with the corrected pressure (along with the measured stagnation pressure), the velocity of the fluid can be determined.

Five-Hole Probes for Measurement of Flow Direction and Velocity: For three-dimensional flows, probes with additional pressure ports are required. Increasing the number of pressure ports from three to five allows for both the yaw and pitch of the fluid to be measured. Figure 11.6 shows a typical five-hole probe [1]. In this case, the probe is inserted into the flow, and rotated until all the outer pressures balance ($p_1 = p_2 = p_3 = p_4$). The flow direction and stagnation pressure (p_5) are measured. Through an extensive calibration, additional information can be obtained: stagnation pressure, flow direction, static pressure, velocity magnitude, individual velocity components.

11.1.2 Methods for Using the Five-Hole Probe

There are two methods for using the five-hole probe: Nulling Arrangement and Non-Nulling Arrangement.

Nulling Arrangement: The miniature five-hole probe is mounted to a traversing system. The probe is rotated in the pitch and yaw directions at a set location. As shown in Figure 11.7, the pitch and yaw angles can be determined by rotating the probe until $P1 = P2 = P3 = P4$. The stagnation pressure (total pressure), $P5$, is directly measured, when the probe is oriented directly into the flow. The outer pressures ($P1$, $P2$, $P3$, $P4$) are not equal to the static pressure of the fluid. However, the probe can be calibrated in a clean, streamline wind tunnel. At a known static pressure, the pressures at ports 1, 2, 3, and 4 are recorded, and with this calibration, the static pressure can be determined from the actual experiments. The disadvantage of this Nulling Arrangement is that it requires the traversing system which can be rotated in pitch and yaw directions.

Non-Nulling Arrangement: The miniature five-hole probe is mounted to a linear traversing system (its axis is not rotated). $P1$, $P2$, $P3$, $P4$, and $P5$ are measured directly at each location. With an extensive calibration, pitch and yaw angles as

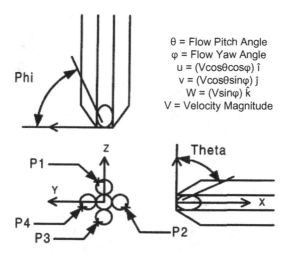

FIGURE 11.7 Miniature five-hole probe for flow direction and velocity components.

well as the three components of the mean velocity (magnitude and direction) are determined, as shown in Figure 11.7. This is a simple traversing system; however, it requires detailed calibration data and tedious post processing [2].

11.1.3 EXPERIMENT EXAMPLE

The miniature five-hole probe needs to be calibrated in a nozzle (streamline flow) before using in the desired application. A traversing mechanism is used to place the probe at the required location for measurements. The five-hole probe for calibration is designed in such a way that its pitch (theta) and yaw (phi) angles typically can be rotated from +30° to −30°, with 3°–5° increments. Measured pressures from each probe port are stored through pressure transducers into the computer at each traverse location. The following calibration procedures with figures and equations are obtained from reference [2]. Figure 11.8 shows typical calibration data of $P1$, $P2$, $P3$, $P4$, and $P5$ over a range of theta and phi angles [2]. $P1$ and $P3$ are for the yaw angle, phi variation; $P2$ and $P4$ are for the pitch angle, theta variation; and $P5$ is for the stagnation pressure. The complete information is required for the pressure coefficient calculation.

As seen from Figure 11.9 [2], four pressure coefficients, $C_{p\theta}$, $C_{p\phi}$, C_{p5}, and C_{pave} can be calculated from the above-measured $P1$, $P2$, $P3$, $P4$, and $P5$ with additional information of the local static pressure (P), local stagnation pressure (P_o), and average pressure (P_{ave}). Note that local static pressure (P) and local stagnation pressure (P_o) are part of the solutions that will be determined later, and $P_{ave} = \frac{1}{4}\,(P1 + P2 + P3 + P4)$.

$$C_{p\theta} = \frac{P3 - P1}{P5 - P_{ave}} \qquad C_{p\phi} = \frac{P2 - P4}{P5 - P_{ave}}$$

$$C_{p5} = \frac{P5 - P}{P_o - P} \qquad C_{pave} = \frac{P_{ave} - P}{P_o - P}$$

(11.1)

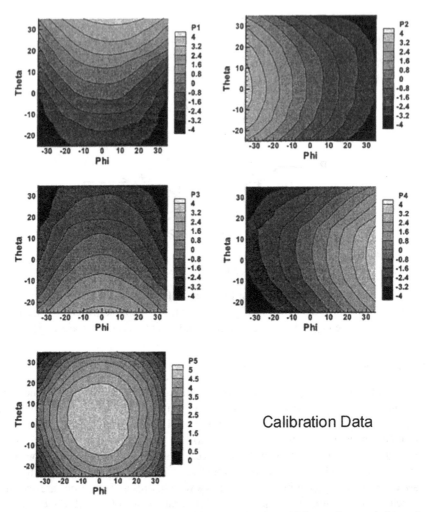

Calibration Data

FIGURE 11.8 Sample calibration data of $P1$, $P2$, $P3$, $P4$, and $P5$ over theta and phi angles.

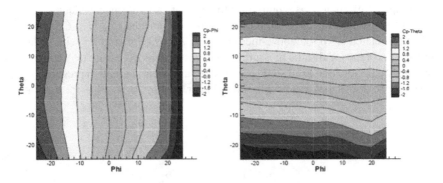

FIGURE 11.9 Calibration data for pressure coefficients $C_{p\theta}$ and $C_{p\phi}$.

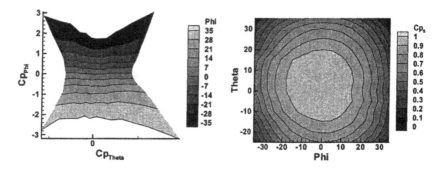

FIGURE 11.10   3D surface fitting for ϕ, θ, C_{p5}, and C_{pave}.

From the above-calculated $C_{p\phi}$ and $C_{p\theta}$ data, 3D surface fitting can be used to obtain ϕ (coefficients a, b, c, d, e, f, g, h, and i can be determined by curve fitting) and θ (coefficients A, B, C, D, E, F, G, H, I, J, and K can be determined by curve fitting) as shown in Figure 11.10. The general forms of these surface equations are shown in Equation (11.2). Similarly, from this ϕ and θ information, C_{p5} and C_{pave} can be determined using the same 3D surface fitting technique [2].

$$\theta = \frac{a + bC_{p\theta} + cC_{p\phi} + dC_{p\phi}^2 + eC_{p\phi}^3}{1 + fC_{p\theta} + gC_{p\phi} + hC_{p\phi}^2 + iC_{p\phi}^3} \qquad \phi = \frac{A + BC_{p\theta} + CC_{p\phi} + DC_{p\theta}^2 + EC_{p\phi}^3 + FC_{p\theta}C_{p\phi}}{1 + GC_{p\theta} + HC_{p\phi} + IC_{p\theta}^2 + JC_{p\phi}^3 + KC_{p\theta}C_{p\phi}}$$

$$C_{p5} = \frac{a + b\phi + c\phi^2 + e\theta + f\theta^2}{1 + g\phi + h\theta + i\theta^2 + i\theta^3}$$

$$C_{Pave} = A + B\phi + C\theta + D\phi^2 + E\theta^2 + F\phi\theta + G\phi^3 + H\theta^3 + I\phi\theta^2 + J\phi^2\theta \qquad (11.2)$$

Finally, the local static pressure can be related to the measured pressure, $P5$, and the above-calculated pressure coefficient C_{p5}, as well as measured average pressure P_{ave} and the above-calculated average pressure coefficient C_{pave}. The local stagnation pressure can be related to calculated local static pressure, measured pressure $P5$, and the above-calculated pressure coefficient C_{p5}. By applying Bernoulli's equation, mentioned in Chapter 2, the local velocity, V, can be calculated from the calculated stagnation pressure, P_o, and the static pressure, P. Therefore, three velocity components, u, v, and w can be determined according to vector relationship in x, y, and z directions, i.e., u, v, and w are related to local calculated velocity V and local calculated θ (pitch angle) and ϕ (yaw angle), respectively.

$$\text{Static Pressure } P = \frac{C_{p5}P_{ave} - C_{pave}P_5}{C_{p5} - C_{pave}} \qquad (11.3)$$

$$\text{Stagnation Pressure } P_o = P + \frac{P_5 - P}{C_{p5}} \qquad (11.4)$$

$$\text{Velocity } V = \sqrt{\frac{2(P_o - P)}{\rho}} \qquad (11.5)$$

$$\text{Velocity Components } u = (V \cos\theta \cdot \cos\phi)\vec{i}$$

$$v = (V \cos\theta \cdot \sin\phi)\vec{j} \qquad (11.6)$$

$$w = (V \sin\theta)\vec{k}$$

11.1.3.1 Fast Response Probes for Flow Measurements in Turbomachinery

The above-mentioned five-hole probe can be used to measure the steady, time-averaged, three component velocities. For gas turbine applications, typical flows are unsteady, under high velocity, high pressure, and high temperature conditions. In order to measure unsteady flow conditions, in 1995, Gossweiier et al. [5] developed a fast response, four-hole probe using four miniature, piezoresistive, pressure transducers placed in the head of a probe. They used this probe to determine the velocity fluctuations, as characterized by yaw, pitch, total pressure, and static pressure and to derive mean values and spectral, or turbulence parameters. Meanwhile, Ainsworth, et al. [6] developed two- and three-dimensional fast response aerodynamic probes by using small-scale, semiconductor pressure sensors, which were directly mounted on the surface of the probes. They used these probes to measure Mach number and flow direction in compressible, unsteady flow regimes. Interested readers can read these papers for their detailed fast-response probe design, test geometry, experimental procedures, data analysis, and results presentation.

11.2 HOT WIRE ANEMOMETRY

Hot wire anemometry can be used for 1D (single-wire probe), 2D (two-wire probe), 3D (three-wire probe) mean and fluctuating velocity measurements. A traversing mechanism is used to place the probe at the required location for measurements. 1D, 2D, or 3D flow fields can be measured by traversing the single-wire, two-wire, or three-wire probes, respectively. The hot wire probe is calibrated in a wind tunnel before it is used to measure an unknown velocity. After calibration, the 3D mean velocity (u, v, w) and the 3D fluctuating velocity (u', v', w') can be calculated from the three-wire probe. Flow properties, such as Reynolds stress ($u'v'$, $v'w'$, $w'u'$, u'^2, v'^2, w'^2), can be subsequently obtained.

Figure 11.11 shows the typical dimensions of a hot wire sensor: sensor diameter: 0.00015–0.0002 inches (0.0038–0.005 mm); sensor length: 0.040–0.080 inches (1.0–2.0 mm); sensor material: tungsten, platinum, or platinum-iridium alloy. Advantages of hot wire probes include high frequency response (~500 kHz), low flow interference, and good spatial resolution. In regions where a hot-wire probe would quickly break, for example, for water flow measurements, a hot film probe (diameters: 0.020–0.050 mm) is used, as shown in Figure 11.12. Hot wire anemometry provide accurate, quick, and repeatable measurements in a relatively inexpensive manner.

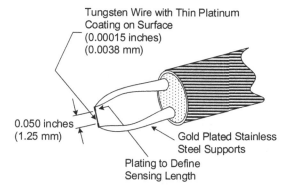

FIGURE 11.11   Schematic of a single hot wire probe.

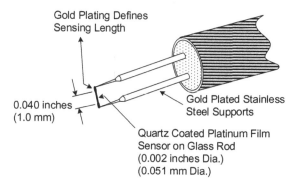

FIGURE 11.12   Schematic of a single hot-film probe.

Cold-wire (single-wire probe) anemometry can be used for mean temperature (T) and fluctuating temperature (T') measurements. A traversing mechanism is used to place the probe at the required location for measurements. The temperature field can be measured by traversing the single cold wire probe (similar to the single thermo-couple probe traversing). The cold wire probe is calibrated in a wind tunnel, with known and controllable temperatures, before it is used in the specific experiment. After calibration, the mean temperature T and the temperature fluctuation T' can be calculated accordingly. By combining T' with the fluctuating velocity (u', v', w') from the hot wire probe, thermal properties such as Reynolds flux ($u'T'$, $v'T'$, $w'T'$) can be obtained.

11.2.1   Fundamental Principles of Hot Wire Probes

A single hot wire probe measures streamwise velocity, a two-wire probe (X wire) mea-sures two velocity components that are in the plane formed by the two sensors, and three-wire probes (triple wire) measure three velocity components. The discussion begins with a single wire, as sketched in Figure 11.13. Thermal anemometers mea-sure flow velocity by sensing the convective heat loss from an electrically heated thin sensor wire (or film) exposed to the fluid. For the Constant Temperature Anemometer

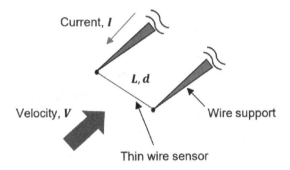

FIGURE 11.13   Concept of a single hot wire anemometer.

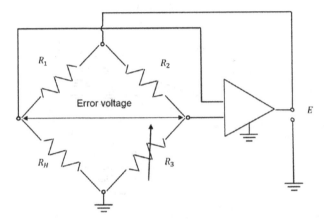

FIGURE 11.14   Single hot wire probe operating in constant temperature mode.

(CTA), shown in Figure 11.14, the cooling effect caused by the flow passing the sensor is balanced by the electric current to the sensor, so the sensor is held at a constant temperature. In other words, the electrical circuit maintains a constant temperature (resistance) across the sensor wire [7]. The change in current due to a change in velocity shows up as a voltage at the anemometer output. When there is a fluctuation in the flow velocity, the convective heat transfer coefficient changes, a corresponding current and voltage change will lead to a new thermal equilibrium condition.

At thermal equilibrium, the rate of heat loss from the hot wire sensor to the fluid equals the rate of heat generated by current through the hot wire sensor. The hot wire is assumed to be an infinitely long, straight cylinder in crossflow; with a diameter of 4–5 µm, and a length of 1–2 mm, the wire length-to-diameter ratio ranges from 100 to 600. Therefore, assuming the cylinder is infinitely long is justified.

Considering, fundamental flow around a cylinder, the heat transfer coefficient (Nusselt number) on the surface of the cylinder can be calculated from a First Law, Energy Balance:

Heat transfer from the wire to the fluid $\dot{Q} = hA_s\left(T_w - T_g\right) = h\left(\pi DL\right)\left(T_w - T_g\right)$ (11.7)

Heat transfer generated by current through the wire $\dot{Q} = I^2 R$ (11.8)

Energy balance for the wire $h(\pi DL)(T_w - T_g) = I^2 R$ (11.9)

Substituting Nusselt number definition $\mathrm{Nu}(\pi kL)(T_w - T_g) = I^2 R$ (11.10)

The average Nusselt number for a cylinder in crossflow

$$\mathrm{Nu} = 0.42\,\mathrm{Pr}^{0.2} + 0.57\,\mathrm{Pr}^{0.33}\,\mathrm{Re}^{0.5}$$

(11.11)

Wire resistance as a function of wire temperature

$$\left[0.42\,\mathrm{Pr}^{0.2} + 0.57\,\mathrm{Pr}^{0.33}\,\mathrm{Re}^{0.5}\right](\pi kL)(T_w - T_g) = I^2 R$$

(11.12)

In the above equations, I is the current, R is the wire resistance, D is the hot wire diameter, h is the convection heat transfer coefficient, Nu is the Nusselt number, Pr is the Prandtl number, and Re is the Reynolds number. All fluid properties are evaluated at the film temperature. From energy balance, the electric heat generation in the hot wire equals heat loss by convection to the fluid; from the heat transfer correlation, the Nusselt number can be determined by the Reynolds number and Prandtl number. Therefore, a relationship between flow velocity and anemometer voltage output can be determined.

The temperature dependence of the electric resistance of the wire gives an effect of first-order; it is on this effect that the use of a hot wire is based. The temperature coefficient of resistivity allows the interpretation of voltage fluctuations in terms of velocity fluctuations. The wire resistance is related to the wire temperature, as shown below:

$$R = R_o\left[1 + C(T_w - T_o) + C(T_w - T_o)^2 + \ldots\right]$$

(11.13)

where R is the wire resistance at T_w, and R_o is the wire resistance at a reference temperature, T_o. C and C_1 are temperature coefficients of electrical resistivity and vary depending on the wire material:

Platinum wire: $C = 3.5 \times 10^{-3}\mathrm{K}^{-1}$, $C_1 = -5.5 \times 10^{-7}\mathrm{K}^{-2}$

Tungsten wire : $C = 5.2 \times 10^{-3}\mathrm{K}^{-1}$, $C_1 = 7.0 \times 10^{-7}\mathrm{K}^{-2}$

In this type of CTA, shown in Figure 11.14, the sensor (wire) temperature is kept constant (shown as the resistance R_H). This is realized by keeping the Wheatstone bridge in balance (i.e., Error voltage = zero), by adjusting the overall bridge voltage using a servo (feedback) amplifier. From the balanced Wheatstone bridge, $R_1 \times R_3 = R_H \times R_2$, the wire resistance R_H can be fixed, then the wire temperature, T_w, can be kept constant. The voltage output must be calibrated with the corresponding fluid velocity as described in the following section. The concept of Wheatstone bridge is shown

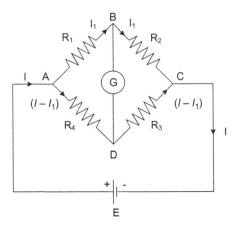

FIGURE 11.15   Concept of Wheatstone bridge.

in Figure 11.15. No current flows through the galvanometer (G) when the circuit is balanced. That means $I_1 \times R_1 = (I-I_1) \times R_4$ and $I_1 \times R_2 = (I-I_1) \times R_3$. Therefore, $R_1/R_2 = R_4/R_3$, or $R_1 \times R_3 = R_4 \times R_2$. This is required to balance the Wheatstone bridge.

11.2.2   Hot Wire Calibration—Constant Temperature Anemometer (CTA)

The following provides the basic concept of the hot wire calibration and the voltage output of the Constant Temperature Anemometer. Air enters a plenum and exits through a converging nozzle to the atmosphere. The pressure difference between air in the plenum and atmospheric air can be measured through a manometer or a transducer. The velocity can be calculated by this measured pressure difference. The corresponding voltage from the hot wire circuit can be obtained from a multimeter. The relationship between air velocity and multimeter voltage output can be constructed, as sketched in Figure 11.16. Therefore, in the actual experiment, air velocity can be determined by the voltage output of the CTA.

> **Measuring V and v' in the Calibration:** The following provides the simple concept of determining velocity and velocity fluctuations. Figure 11.17 shows a simplified electric circuit through a bridge and hot wire. The voltage drop occurs, due to resistances of bridge and wire, when the current passes through the circuit. The heat loss (Nu, Nusselt number) can be related to voltage drop (e) and resistance of bridge (R_b) and hot wire (R_H). The Nusselt number is also related to the fluid velocity (Re, Reynolds number). The voltage $(e,$ or D.C.) and voltage fluctuation (de, RMS) can be related to the velocity (V) and velocity fluctuation $(dV/V,$ or $v')$ as shown in the following equations. Turbulence intensity (Tu) can also be related to the voltage fluctuation (de/e). Figure 11.18 shows a typical relationship between Nu and Re as well as the measured instantaneous velocity output, $V = \bar{V} + v'$. Figures 11.19 and 11.20 show typical calibration data and instantaneous velocity obtained from a hot wire (50 KHz sampling). The results

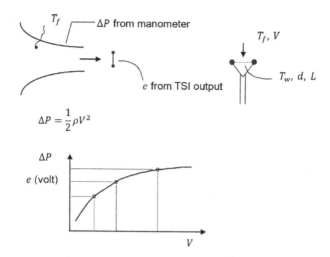

$$\Delta P = \frac{1}{2}\rho V^2$$

FIGURE 11.16   Hot wire calibration setup and corresponding velocity–voltage relationship.

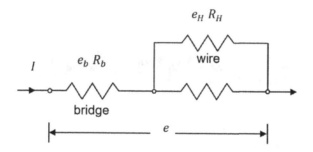

FIGURE 11.17   Simplified electric circuit through bridge and hot wire.

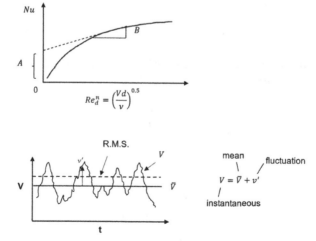

FIGURE 11.18   Typical relationship between Nusselt Number and Reynolds Number and the measured, instantaneous velocity as a function of time.

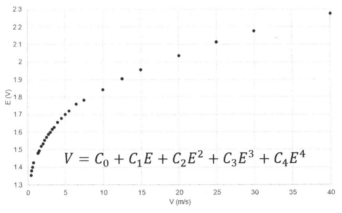

$$V = C_0 + C_1 E + C_2 E^2 + C_3 E^3 + C_4 E^4$$

$$V = -98.729 + 232.65E - 194.98E^2 + 63.617E^3 - 4.882E^4$$

FIGURE 11.19 Typical calibration data for a hot wire (50 KHz sampling).

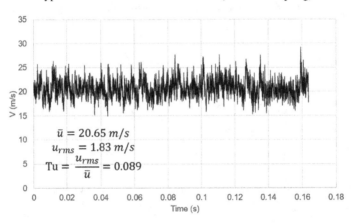

$$\bar{u} = 20.65 \, m/s$$
$$u_{rms} = 1.83 \, m/s$$
$$Tu = \frac{u_{rms}}{\bar{u}} = 0.089$$

FIGURE 11.20 Typical hot wire instantaneous velocity versus time (50 KHz sampling)

indicate that the mean velocity is 20.65 m/s, root mean square fluctuation is 1.83 m/s, and turbulence intensity is 8.9%.

$$Nu = \frac{hd}{k} = \frac{q}{\Delta T A_s} \cdot \frac{d}{k}$$

$$= \frac{I e_H}{\Delta T A_s} \cdot \frac{d}{k}$$

$$= \frac{e^2 R_H}{\left(R_b + R_H\right)^2} \cdot \frac{1}{\pi L K \left(T_w - T_f\right)}$$

$$Nu = A + B Re_d^n$$

$$\sim e^2 = C e^2$$

$$2Ce\frac{de}{dt} = Bn\left(\frac{Vd}{\nu}\right)^{n-1}\frac{d}{\nu}\frac{dV}{dt}$$

$$2eCde = Bn\left(\frac{d}{\nu}\right)^n V^n\frac{dV}{V}$$

$$T_u = \frac{v'}{V} = \frac{dV}{V}$$

$$= \frac{de}{e}\cdot\frac{2}{n}\left(1+\frac{A}{BRe_d^n}\right)$$

$$\sim \frac{de}{e}$$

Single-Wire Probes: Single-wire probes are used for the majority of instantaneous flow measurements. However, for accurate results, the wire should be oriented perpendicular to the flow. If the wire is not properly oriented, the off-components of velocity will effect the results, as shown in Figure 11.21. As shown below, with a proper setup, reasonable accuracy can be maintained without accounting for the wire orientation effects [8].

$$U_{\text{eff}} = \sqrt{U_N^2 + k_T^2 U_T^2 + k_N^2 U_{BN}^2} \tag{11.14}$$

where k_T and k_N are empirically determined correction constants. k_T generally ranges from 0 to 0.2, and decreases as the zero as L/D increases. Therefore, having a length-to-diameter ratio for the wire of 600–800 is desirable. k_N is generally in the range of 1–1.2. However, when inserting the probe into the flow, most often U_{BN} approaches zero. Therefore, with $k_T \to 0$ and $U_{BN} \to 0$, the effective velocity cooling the wire is well represented by the velocity component of interest, U_N.

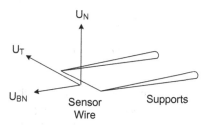

FIGURE 11.21 Hot wire orientation effect.

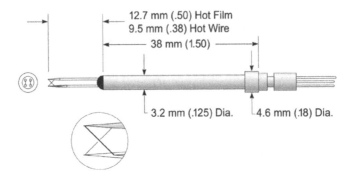

FIGURE 11.22 Typical X wire probe.

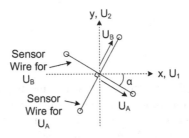

FIGURE 11.23 Typical X wire orientation for two component velocity measurements.

Two-Wire (X Wire) Probes: The X wire (cross wire) probe can be used to measure instantaneous velocity in two directions. Figure 11.22 shows a typical X wire probe. Figure 11.23 shows the calculated, two component velocity from the X wire probe [8]. Assuming $k_T = 0$ and $k_N = 1$, the two velocity components (U_1 and U_2) can be determined from the effective velocity components. The calibration procedures of voltage versus velocity for the X wire are much more tedious than the single-wire calibration as described in the above section.

$$U_{A,\text{eff}}^2 = \left(U_1\cos a - U_2\sin a\right)^2 + k_T^2\left(U_1\sin a + U_2\cos a\right)^2 + k_N^2 U_3^2 \qquad (11.15)$$

$$U_{B,\text{eff}}^2 = \left(U_1\sin a + U_2\cos a\right)^2 + k_T^2\left(U_1\cos a - U_2\sin a\right)^2 + k_N^2 U_3^2 \qquad (11.16)$$

$$U_1 = \frac{1}{\sqrt{2}}\left(U_{A,\text{eff}} + U_{B,\text{eff}}\right) \qquad (11.17)$$

$$U_1 = \frac{1}{\sqrt{2}}\left(U_{A,\text{eff}} - U_{B,\text{eff}}\right) \qquad (11.18)$$

Three-Wire Probes: The three-wire (triple wire) probe can be used to measure instantaneous velocity in three directions. Figure 11.24 shows a typical triple-wire probe. The calibration procedures for voltage versus velocity for triple wire are much more tedious than the X wire.

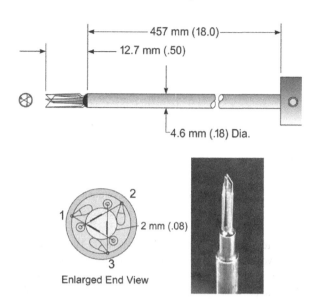

FIGURE 11.24   Photos of typical triple wire probe (from TSI Model 1299).

Turbulence Intensity Measurement: The hot wire (or hot film) outputs give instantaneous velocity, i.e., velocity varies with time. For a single hot wire, mean velocity and root mean square velocity fluctuation can be calculated. Then, turbulence intensity, Tu, can be determined. The following provides the mean velocity across a boundary layer flow. In addition, the turbulence intensity and turbulent length scales are also calculated from the hot wire output. Figure 11.25 shows a typical hot film probe inserted into a turbine

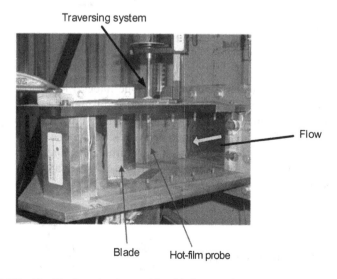

FIGURE 11.25   Hot film location in a turbine blade cascade.

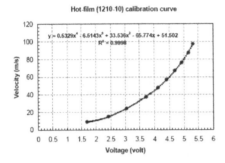
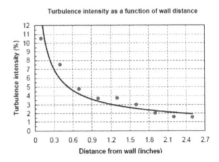

FIGURE 11.26   Hot film calibration and turbulence intensity decay across the boundary layer.

blade cascade test section. Figure 11.26 shows a typical hot film calibration curve of voltage versus velocity. Based on the following equations, Figure 11.25 also shows the calculated turbulence intensity profile inside the boundary layer. The turbulence intensity, Tu, decreases from near the wall (around 10%) to outside the boundary layer (around 1.5%).

$$\text{Mean velocity } \bar{u} = \frac{1}{N} \sum_{1}^{N} u_i \tag{11.19}$$

$$\text{Instantaneous velocity fluctuation } u_i' = u_i - \bar{u} \tag{11.20}$$

$$\text{Mean velocity fluctuation } u_{\text{rms}} = \sqrt{\frac{1}{N} \sum_{1}^{N} (u_i')^2} \tag{11.21}$$

$$\text{Turbulence intensity } Tu = \frac{u_{\text{rms}}}{\bar{u}} \tag{11.22}$$

Turbulence Length Scale Measurement: The integral length scale is representative of the size of the most energetic eddies in the flow. It can be determined through analysis of the autocorrelation of the fluctuating velocity time signal from hot wire anemometry, as shown in the following equations. Figure 11.27 shows the calculated turbulence length scale profile inside the boundary-layer, it increases from near wall (around 0.7 cm) to outside the boundary-layer (around 2.0 cm).

$$\text{Autocorrelation of } u_i' \; R_{uu}(t) = \left(\frac{\overline{u'(t) \cdot u'(t+\tau)}}{u_{\text{RMS}}^2} \right) = \frac{1}{N} \left(\frac{\sum_{1}^{N} u_i' \cdot u_{i+j}'}{u_{\text{RMS}}^2} \right) \tag{11.23}$$

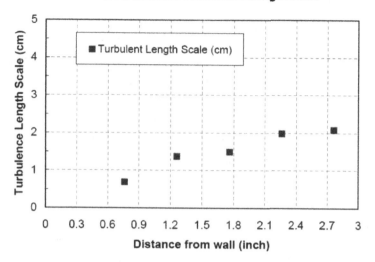

FIGURE 11.27 Turbulent length scale across a boundary layer.

$$T = \sum_{1}^{N_o} R_{uui} \cdot \Delta \tau \qquad (11.24)$$

$$\tau = j\Delta t \qquad (11.25)$$

$$\text{Turbulent length scale } \Lambda_X = \bar{U} \cdot T \qquad (11.26)$$

11.2.3 EXPERIMENT EXAMPLE

Example 11.1: Influence of Jet-Grid Turbulence on Flat Plate, Turbulent, Boundary Layer Flow and Heat Transfer [9]

A low-speed wind tunnel with a jet grid system was used to study the effect of jet grid, turbulence on flat plate, turbulent, boundary layer flow and heat transfer [9]. Figure 11.28 shows the low-speed wind tunnel with a jet grid system. The high mainstream turbulence was produced by a round tube grid with uniform jet injection. Injected air was blown in either an upwind or downwind direction at a controllable flow rate. The jet grid was placed between the nozzle and the downstream test plate. A flat plate test section, instrumented with foil thermocouples was located downstream from the jet grid. A four-channel TSI IFA 100, Constant Temperature Anemometer (CTA) with a four-channel TSI IFA 200, high-speed digitizer were connected to an IBM XT personal computer through a TSI DMA connector for hot film data recording and processing. The hot film was calibrated in the air jet from a third-order polynomial nozzle (similar to Figure 11.15). The local streamwise, instantaneous velocity fluctuations were based on 5160 readings from the hot film anemometer. The total digitizing time for gathering the 5160 readings was about 0.1–0.5 s.

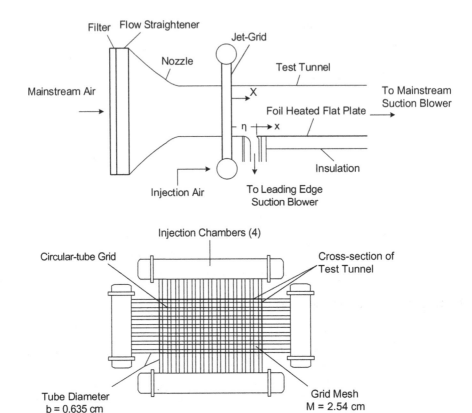

FIGURE 11.28 Low-speed wind tunnel with jet grid system.

The streamwise turbulence intensity decay and streamwise length scale growth along the test plate, the mean velocity and temperature profiles across the boundary layer, and the surface heat transfer distribution were measured. Figure 11.29 shows that the grid with downwind injection produces a slightly higher turbulence intensity and a smaller length scale than the grid with upwind injection, while the passive grid (with no air injection) has the lower turbulence intensity and a smaller length scale than the jet grids. Of course, the clean wind tunnel (with no grid) provides the lowest turbulence intensity along the test plate. Note that the length scale is proportional to the grid size (diameter). A higher turbulence intensity and a smaller length scale enhance the surface heat transfer coefficient as shown in Figure 11.30. The jet grid with downwind blowing provides higher Stanton numbers (proportional to heat transfer coefficient) than the jet grid with upwind blowing, and subsequently higher than the passive grid and no grid cases, respectively.

In order to measure the mean velocity and temperature profiles across the boundary layer, a computer-controlled traversing device on the top of the tunnel held the probe (hot film probe, and total pressure or temperature probe) for data acquisition, as shown in Figure 11.31. It was driven by a superior electric stepping motor (200 steps/ revolution) yielding 0.00254 cm per step. Two surface static pressure taps were installed at $X/b = 28.5$ and $X/b = 117.9$. These taps

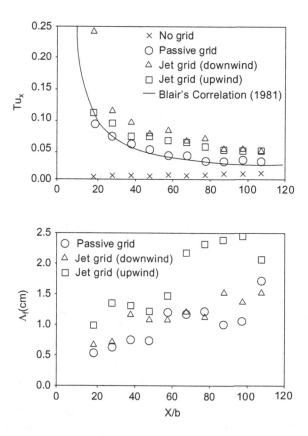

FIGURE 11.29 Turbulence intensity decay and length scale growth along the test plate.

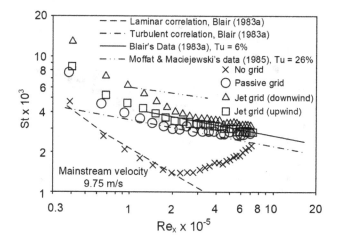

FIGURE 11.30 Effect of grid injection on heat transfer along the test plate.

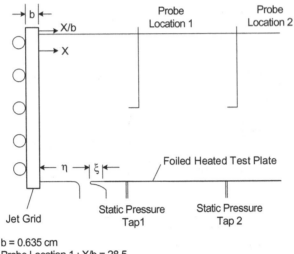

b = 0.635 cm
Probe Location 1 : X/b = 28.5
Probe Location 2 : X/b = 117.9

FIGURE 11.31   Locations of the traversing probes relative to the jet grid and test plate

were incorporated with the total pressure probe for boundary layer velocity
profile measurements. A total pressure probe (United Sensor) with a flattened
tip size of 0.0508 × 0.1016 cm measured the boundary layer velocity profiles.
This probe was moved manually to touch the wall while viewing through a 3x
magnifying lens. The calculated y+ value of this setup is around 10. In addition, a special miniature thermocouple probe (United Sensor) made of 36 gauge
copper-constantan thermocouple wires was employed to measure the boundary
layer temperature profiles at the same locations. The thermocouple bead size
was measured as 0.0508 cm. Two insulated leads were threaded through two
separate 0.05588 cm (ID), U-shaped, hypodermic, stainless steel tubes to minimize the flow disturbance while providing better support. Figure 11.32 shows

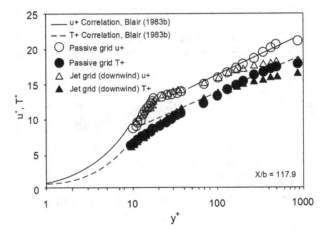

FIGURE 11.32   Universal velocity and temperature profiles across the boundary layer.

the measured universal velocity and temperature profiles across the boundary layer at X/b = 117.9. Results indicate that the effect of the jet grid (downwind blowing) on boundary layer velocity and temperature profiles at the downstream locations is negligible, except in the wake region, as compared with the passive grid. The same conclusion held for the jet grid with upwind blowing (but not shown here).

y^+ = dimensionless distance from the flat plate surface = $(yu^*)/\nu$
u^* = dimensionless local friction velocity = $(\tau_w/\rho)^{1/2}$
u^+ = dimensionless local streamwise velocity = u/u^*
T^+ = dimensionless local temperature = $(T_w-T)/[q''/(\rho u^* C_p)]$

Example 11.2: Influence of High Mainstream Turbulence on Leading-Edge Heat Transfer [10]

A low-speed wind tunnel with a jet grid system was used to study the effect of jet grid turbulence on leading-edge region, boundary layer flow, and heat transfer [10]. Experiments were performed using a blunt body with a semi-cylinder, leading-edge, and flat sidewalls as shown in Figure 11.33. The mainstream Reynolds numbers based on the leading-edge diameter were 25,000, 40,000, and 100,000. Three turbulence grids were fabricated to generate different levels of turbulence intensities. The high mainstream turbulence was produced by a square-tube grid with uniform jet injection in the downwind direction at a controllable flow rate. The turbulence grid was placed between the nozzle and the downstream semi-cylinder test section. The semi-cylinder test section, instrumented with foil thermocouples, was located downstream from the jet grid. A four-channel TSI IFA 100 Constant Temperature Anemometer (CTA) with a four-channel TSI IFA 200 high-speed digitizer was connected to an IBM XT personal computer through a TSI DMA connector for hot film data recording and processing. The hot film was calibrated in the air jet from a third-order polynomial nozzle. The local, streamwise, instantaneous velocity fluctuations were based on 5160 readings from the hot film anemometer. An HP 3478A multimeter monitored true rms fluctuations in the CTA output. The total digitizing time for gathering the 5160 readings was about 0.1–0.5 s.

Velocity and turbulence intensity distributions in the test channel were measured to check the oncoming mainstream flow conditions. Figure 11.34 shows local velocity and turbulence intensity distributions along the centerline and right-side line of the test channel for a Reynolds number of 100,000. It shows that the ratio (local mainstream velocity to incoming mainstream velocity) is around 1.0 for X/b from 10 to 20 downstream of the grid (around 1.5 diameters upstream of the leading edge of the test model). The incoming mainstream velocity is 10 m/s at X/b = 20 with the no-grid condition. Downstream of this point, the local mainstream velocities along the centerline decrease sharply as the flow approaches the leading edge, stagnation region of the cylinder. However, they increase along the right-side line because of the test model blockage effect. Then these curves taper off, but a little increase is due to the boundary layer growing along the flat, side wall of the test model. Figure 11.34 also shows that for the bar, passive, and jet-grid cases, the streamwise turbulence intensity along the right side decreases monotonically with

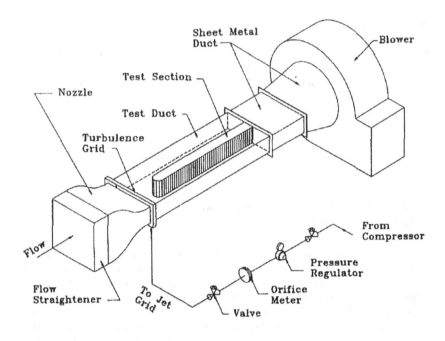

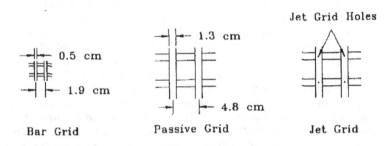

FIGURE 11.33   Sketches of the semi-circular test section and the three turbulence grids.

increasing distance from the grid. However, the streamwise turbulence intensity along the centerline also decreases and then starts increasing as the mainstream velocity starts decreasing because of approaching the semi-cylinder stagnation region. As expected, the jet grid produces higher turbulence intensity than the passive grid and bar grid, the no grid case has the lowest turbulence intensity. Figure 11.35 shows the effect of different grid turbulence on the semi-cylinder surface heat transfer distributions. As expected, the jet grid gives the higher Nusselt number around the semi-cylinder than the passive grid and bar grid; the no grid case has the lowest Nusselt numbers. However, the effect diminishes downstream along the flat-side wall of test section.

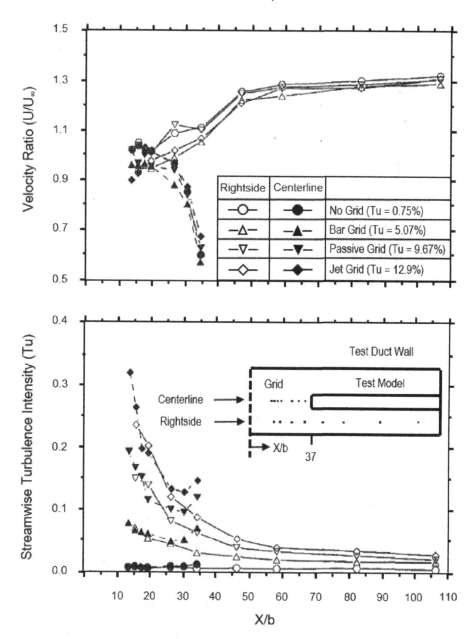

FIGURE 11.34   Streamwise velocity and turbulence intensity distributions.

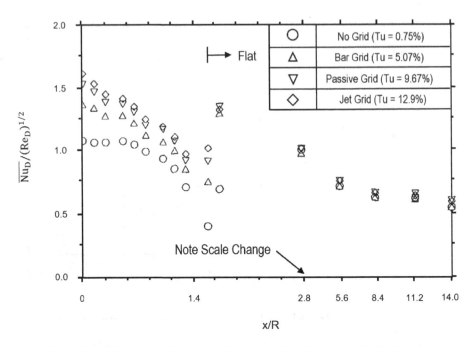

FIGURE 11.35 Effect of turbulence intensity on the Nusselt number distributions.

Example 11.3: Influence of Unsteady Wake on Heat Transfer Coefficients on a Gas Turbine Blade [11]

The effect of unsteady wake on surface heat transfer coefficients of a gas turbine blade was experimentally determined using a spoked wheel type wake generator [11]. The experiments were performed with a five airfoil, linear cascade in a low-speed wind tunnel facility, as shown in Figure 11.36. The cascade inlet Reynolds number based on the blade chord varied from 100,000 to 300,000. The wake Strouhal number (S) varied between 0 and 1.6 by changing the rotating wake passing frequency (rod speed, N, and rod number, n), rod diameter d, and cascade inlet velocity V_1 ($S = 2\pi Ndn/(60V_1)$).

Figure 11.37 shows the wake generator geometry, and a conceptual view of the effect of the unsteady wake on the blade model is shown in Figure 11.38. A hot wire anemometer system was located at the cascade inlet to detect the instantaneous velocity, phase-averaged mean velocity, and turbulence intensity induced by the passing wake. A blade instrumented with thin foil heaters and thermocouples was used to measure the surface heat transfer coefficients on the blade surface.

The hot wire probe was located near the blade leading edge and in the middle of flow passages to measure the instantaneous velocities in unsteady, wake flow conditions. A calibrated single hot wire anemometer, TSI IFA 100, was used to measure the unsteady (instantaneous) velocity profile of the passing wake. The hotwire anemometer was connected to an IBM PC through a 100 kHz A/D converter in order to get sufficient sampling data for analysis. The anemometer was also connected to a NICOLET 446A Spectrum Analyzer that displays the instantaneous wake profile and frequency distribution. The hot wire sensor was located 8.82 cm

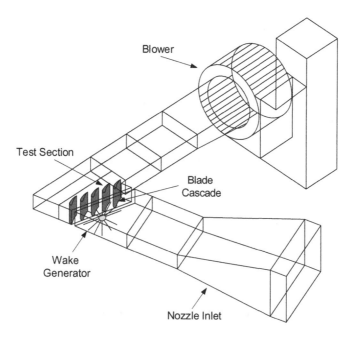

FIGURE 11.36   Overview of the linear turbine blade cascade with a rotating wake generator.

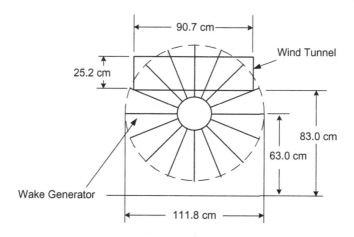

FIGURE 11.37   Unsteady wake generator geometry.

downstream of the rotating rods. Figure 11.39 shows typical, instantaneous velocity profiles, and Figure 11.40 shows typical phase-averaged mean velocity profiles and phase-averaged turbulent intensity profiles for S = 0.1, 0.2, and 0.4, respectively. The Strouhal Number (S) was varied by increasing the rotating rod speed (N) for a given rod diameter (d = 0.63 cm), rod number (n = 16), and cascade inlet velocity (V_i = 21 m/s or Re = 3 × 10^5). The instantaneous velocity profile shows the periodic unsteady fluctuations caused by the upstream passing wake, and the periodic fluctuations increase with the Strouhal number. The phase-averaged profile shows the time dependent, mean velocity defect caused by the upstream passing

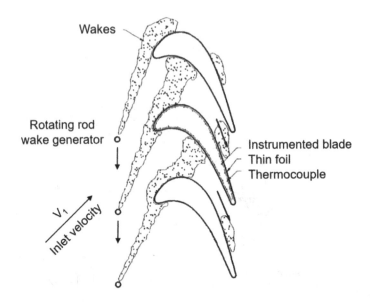

FIGURE 11.38 Conceptual view of the effect of the unsteady wake on the Blade Model.

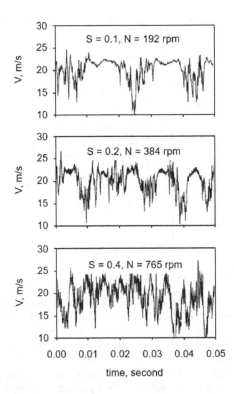

FIGURE 11.39 Typical instantaneous velocity profiles caused by the upstream, unsteady wake.

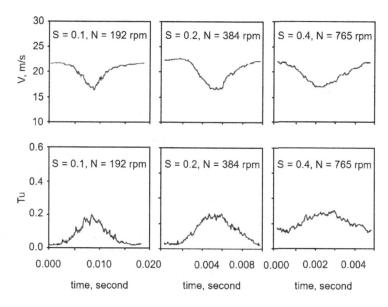

FIGURE 11.40   Phase-averaged mean velocity profiles and turbulence intensity profiles by the upstream, unsteady wake.

wake, whereas the wake width increases with the Strouhal number. The phase-averaged turbulence intensity reaches 20% inside the wake. The time-averaged turbulence intensity is about 8%, 10%, and 15%, for S = 0.1, 0.2, and 0.4, respectively. The background turbulence intensity is only about 0.75% for the case of no rotating rods in the wind tunnel.

From the results of the instantaneous velocity, phase-averaged mean velocity, and turbulence intensity profiles, it is expected that the unsteady wake, with higher Strouhal numbers, will produce a larger impact on the downstream blade. Figure 11.41 shows the unsteady, passing wake promotes earlier and broader

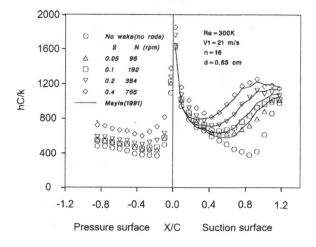

FIGURE 11.41   Unsteady wake effect on turbine blade surface heat transfer.

boundary layer transition and causes much higher heat transfer coefficients on the suction surface, whereas the passing wake also significantly enhances heat transfer coefficients on the pressure surface.

11.3   COLD-WIRE ANEMOMETRY

11.3.1   FUNDAMENTAL PRINCIPLES OF COLD WIRE PROBES

Figure 11.42 shows the electric circuit for the Constant Current Anemometer (CCA). In the constant current mode, the operation is very similar to the resistance measurement for RTDs (see Chapter 3) and thin film heat flux gauges (see Chapter 6). The resistance, $R4$, is adjusted until the Wheatstone bridge is balanced, i.e., no current or voltage reads across Vout. This means $R2/R4 = R1/Rw$. This sets up the Wheatstone resistance relationship for the wire resistance during operation. For calibration of a CCA, the current is kept constant at different levels of known fluid velocities. The Wheatstone bridge is balanced by adjusting $R2$ and $R4$; therefore, Rw can be determined.

$$R_W = R_o\left[1 + C\left(T_W - T_o\right)\right] \tag{11.27}$$

In the above equation, R_w is the wire resistance at the temperature T_w, R_o is the wire resistance at a reference temperature T_o, C is the temperature coefficient of electrical resistivity. The wire resistance fluctuations are related to voltage fluctuations by passing a constant current through the wire. The wire temperature can be obtained from the following equations.

$$\text{Output voltage } V_{\text{out}} = I_o \frac{R3\left(R2 + R4\right) - R4\left(R1 + R3\right)}{\left(R1 + R3\right) + \left(R2 + R4\right)} \tag{11.28}$$

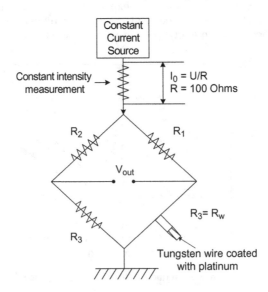

FIGURE 11.42   Single cold wire probe for a constant current anemometer.

Joule heating from the current passing through the wire

$$\dot{Q}_J = R_W I_W^2 = R_o \left[1 + C \left(T_W - T_o \right) \right] I_W^2$$
(11.29)

Joule heating is balanced by forced convection $\left(\text{neglecting conduction effects} \right)$

$$\dot{Q}_J + \dot{Q}_h = 0 \qquad \dot{Q}_h = -hA_s \left(T_W - T_\infty \right)$$
(11.30)

Solving for the wire temperature

$$T_W = \frac{h\pi^2 d_W^3 T_\infty + 4\sigma_o^{-1} I_W^2 \left(1 - T_o C \right)}{h\pi^2 d_W^3 - 4\sigma_o^{-1} I_W^2 C}$$
(11.31)

11.3.2 Cold Wire Calibration—Constant Current Anemometer (CCA)

Figure 11.43 shows the concept of a simplified electric circuit through the bridge and cold wire. In general, the cold wire system includes a tungsten wire probe, 5 μm in diameter and 1.5 mm in length, and a temperature bridge. The temperature bridge is designed to restrict the current applied to the wire at less than 1 mA to ensure negligible sensitivity to velocity. The wire current is typically maintained at approximately 70 mA. The frequency response of the cold wire is about 800 Hz. The signal from the cold wire system is directed to an A/D converter installed on a PC. The A/D converter has 12-bit resolution and maximum gain of 8. The combination provides a temperature measurement accuracy of approximately 0.1°C.

Measuring T and T' in the Calibration: The following provides the simple concept of determining temperature and temperature fluctuations. The voltage drop occurs due to resistances of the bridge and wire when current passes through the circuit. The heat loss (Nu, Nusselt number) can be related to this voltage drop $\left(e_C \right)$, the resistance of bridge (R_b), and the resistance of the cold wire $\left(R_C \right)$. The Nusselt number can also be related to the flow velocity (Re, Reynolds number). The voltage (e, or D.C.) and voltage fluctuation (de, RMS) can be related to the flow temperature $\left(T_w - T_f \right)$ and flow temperature fluctuation $\frac{d\left(T_w - T_f\right)}{dt}$ as shown in the following equations. Figure 11.44 shows the typical calibration between voltage output,

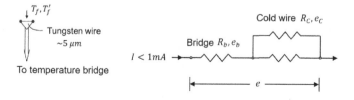

FIGURE 11.43 Simplified electric circuit through bride and cold wire.

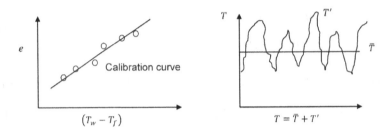

FIGURE 11.44 Typical calibration between voltage and flow temperature along with the measured, instantaneous flow temperature.

e, and flow temperature, $(T_w - T_f)$. The instantaneous fluid temperature, $T = \bar{T} + T'$, is also shown. The term $(T_w - T_f)$ represents the temperature difference between the cold wire and the working fluid.

$$\text{Nu} = A + B Re_D^n$$

$$= \frac{hD}{k} = \frac{q}{A_s \Delta T} \cdot \frac{D}{k}$$

$$= \frac{Ie_C}{(T_w - T_f)} \cdot \frac{1}{\pi DL} \cdot \frac{D}{k}$$

$$= \frac{e^2 R_C}{(R_b + R_C)^2} \cdot \frac{D}{k} \cdot \frac{1}{(T_w - T_f)\pi DL}$$

$$(T_w - T_f) = \frac{R_C}{(R_b + R_c)^2} \cdot \frac{1}{\pi kL} \cdot \frac{e^2}{A + B Re_D^n} \sim e^2$$

Mean temperature D.C. from cold wire measurement

$$\frac{d(T_w - T_f)}{dt} \sim 2e\frac{de}{dt}$$

$$\frac{d(T_w - T_f)}{(T_w - T_f)} \sim \frac{de}{e}$$

Temperature fluctuation A.C. (RMS) output from cold wire measurement

11.3.3 Experiment Example

Example 11.4: Unsteady Wake Effect on Film Temperature and Effectiveness Distributions for a Gas Turbine Blade [12]

The unsteady wake effect on the coolant jet temperature profile on the suction side of a gas turbine blade was measured using the cold wire technique [12]. As shown in Figure 11.45, the blade had only one row of film holes near the gill hole portion, on the suction side of the blade. Tests were performed in a five blade, linear cascade in a low-speed wind tunnel, as shown in Figure 11.36. The mainstream Reynolds number, based on cascade exit velocity, was 500,000. Upstream unsteady wakes were simulated using a spoke-wheel type, wake generator, as shown in Figure 11.46. The coolant blowing ratio varied from 0.6 to 1.2. The wake Strouhal number was kept at $S = 0$ and 0.1. Figure 11.47 shows the measurement planes at selected locations with a cold wire probe traversing system. Figure 11.48 shows the coolant jet temperature

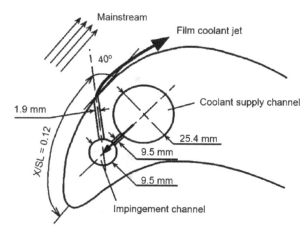

FIGURE 11.45 Film cooling arrangement on the suction side of the turbine blade.

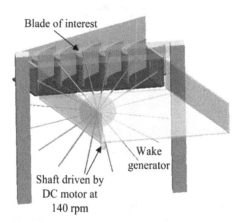

FIGURE 11.46 Spoke-wheel type generator upstream of the turbine blades.

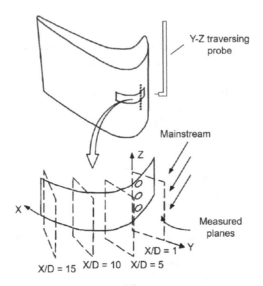

FIGURE 11.47 Measurement planes at selected locations for the cold wire probe.

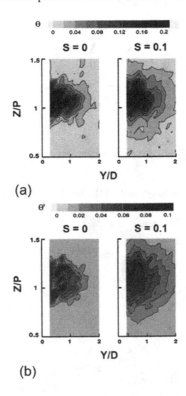

FIGURE 11.48 Detailed film temperature distributions at X/D = 10 and M = 0.8, without and with the unsteady wake effect: (a) time averaged temperature and (b) temperature fluctuation.

contours at X/D = 10 for both steady and unsteady cases of a blowing ratio M = 0.8. For the case with the unsteady wake effect, both the coolant jet mean temperature and temperature fluctuation contours have been expanded to a larger area. The unsteady case has a much larger temperature fluctuation area. This implies that the unsteady wake enhances the mixing between the heated coolant jet and the cold free-stream, and more heated jet is diffused into the free-stream, caused lower effectiveness.

Example 11.5: Coolant Film Temperature Distributions Using a Cold Wire [13]

Kohli and Bogard [13], in 1998, provided temperature measurements using the cold-wire probe inside flows that are highly turbulent, such as film cooling jets. This study provides detailed temperature measurements in mixing flows and can resolve mean and fluctuating components of the temperature field. The near-wall temperature distributions can be used to estimate the film effectiveness for an adiabatic test surface. Figure 11.49 shows the non-dimensional mean temperature and non-dimensional temperature fluctuation distributions along the centerline of film jet [13]. It shows that the coolant jet mean and fluctuating temperatures decrease along the jet trajectory due to the turbulent mixing between coolant jet and oncoming mainstream flow.

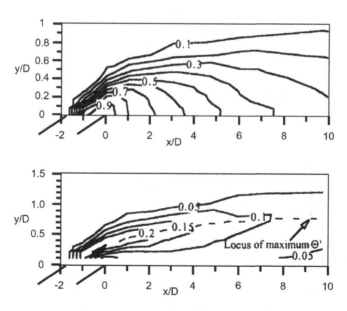

FIGURE 11.49 Non-dimensional mean and fluctuating temperature distributions along the centerline of a film cooling jet.

Example 11.6: Velocity and Temperature Distributions Using an X-Wire with a Cold Wire Probe [14]

Hot wire anemometry systems are an effective tool for measuring fluctuating velocity and temperature components in non-isothermal boundary layer flows. The technique for using multiple wire overheats for separating velocity and temperature fluctuations was developed. Blair and Bennett [14] described the development of an instrumentation system designed for the simultaneous measurement of two components of instantaneous velocity (u and v) and temperature (T) in low-speed boundary layers. The three-sensor probe consisted of a vertical X wire, with a third wire mounted equidistant between the wires. The third wire was placed parallel to one of the wires in the X wire. Figure 11.50 shows the triple-wire probe with the sensor configuration used by Blair and Bennett [14]. Figure 11.51

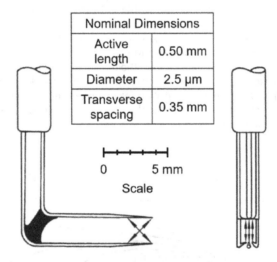

Nominal Dimensions	
Active length	0.50 mm
Diameter	2.5 μm
Transverse spacing	0.35 mm

FIGURE 11.50 Details of the triple wire probe.

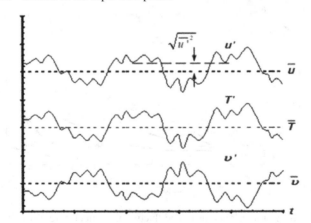

FIGURE 11.51 Typical simultaneous measurement of the instantaneous velocity (u and v) and temperature (T).

shows the typical simultaneous measurement of the instantaneous velocity (u and v) and temperature (T). These measured flow properties are useful to determine the turbulent stress and turbulent flux in a turbulent, boundary layer thermal flow field. Hot wires (single, X, and triple) have been used to measure fluctuating velocity and temperature components by several investigators. It is common to use hot wires for measuring turbulence intensity inside wind tunnels. Blair [15] in 1984, Kim and Simon [16] in 1987, and Zhou and Wang [17] in 1995 have also used hot wire probes similar to the one shown above for measuring free-stream turbulence and thermal structures in turbulent boundary layers. The temperature probe is also referred to as a cold wire, as it measures the temperature component.

PROBLEMS

1. You are asked to prepare experimental methods to evaluate flow and thermal fields for steady, incompressible, turbulent air flow through a 2D channel with protrusions on one wall, as sketched below (front view). The focus should be on the region of interest shown in the sketch. Use a miniature five-hole probe to map the flow field between two protrusions of the channel: Sketch the major experimental setup, describe the measurement principle, sketch and discuss the expected experimental results (mean velocity vector field), and perform an uncertainty analysis.

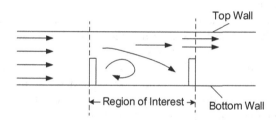

2. From Problem 1, use a hot wire anemometer (Constant Temperature Anemometry) to map the flow field between two protrusions of the channel: Sketch the major experimental setup, describe the measurement principle, sketch and discuss the expected experimental results (mean velocity vector field and turbulence intensity), and perform an uncertainty analysis.

3. From Problem 1, if the bottom wall with two protrusions is made of copper and heated to a certain temperature by heaters on the back side, use a cold wire anemometer (Constant Current Anemometry) to determine the flow temperature field between the two protrusions of the channel: Sketch the major experimental setup, describe the measurement principle, sketch and discuss the expected experimental results (mean and fluctuating temperature distributions), and perform an uncertainty analysis.

4. You are asked to prepare an experiment to evaluate flow and thermal fields for steady, incompressible, turbulent air flow through a two-passage, rectangular cooling channel with a 180-deg turn for heat exchanger applications, as sketched below. Focus on the region of interest shown in the sketch. Use a miniature five-hole probe to map the flow field before, during, and after the

180-deg turn of the channel: Sketch the major experimental setup, describe the measurement principle, sketch and discuss the expected experimental results (mean velocity vector field), and perform an uncertainty analysis.

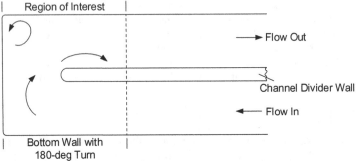

| Region of Interest |

Flow Out

Channel Divider Wall

Flow In

| Bottom Wall with 180-deg Turn |

TOP VIEW of Two-Passage Rectangular Channel Design

5. From Problem 4, use a hot wire anemometer (Constant Temperature Anemometry) to map the flow field before, during, and after the 180-deg turn of the channel: Sketch the major experimental setup, describe the measurement principle, sketch and discuss the expected experimental results (mean velocity vector field and turbulence intensity), and perform an uncertainty analysis.

6. From Problem 4, if the top and bottom walls of the channel are made of a copper plate and heated to a certain temperature by heaters on the back side, use a cold wire anemometer (Constant Current Anemometry) to determine the flow temperature field before, during, and after 180-deg turn of the channel: Sketch the major experimental setup, describe the measurement principle, sketch and discuss the expected experimental results (mean and fluctuating temperature distributions), and perform an uncertainty analysis.

7. You are asked to design an experiment to evaluate flow and thermal fields for steady, incompressible, turbulent air flow through a square channel with nine circular tubes between the top and bottom walls, as sketched below. Use a miniature five-hole probe to map the flow field on a plane downstream of the nine circular tubes: Sketch the major experimental setup, describe the measurement principle, sketch and discuss the expected experimental results (mean velocity vector field), and perform an uncertainty analysis.

TOP VIEW

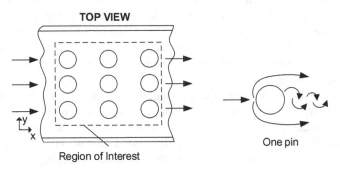

Region of Interest

One pin

8. From Problem 7, use a hot wire anemometer (Constant Temperature Anemometry) to map the flow field on a plane downstream of the nine circular tubes: Sketch the major experimental setup, describe the measurement principle, sketch and discuss the expected experimental results (mean velocity vector field and turbulence intensity), and perform an uncertainty analysis.

9. From Problem 7, if the top and bottom walls of the channel and the nine tubes are made of copper and heated to certain temperature by heaters on the back side, use a cold wire anemometer (Constant Current Anemometry) to determine the flow temperature field on a plane downstream of the nine circular tubes: Sketch the major experimental setup, describe the measurement principle, sketch and discuss the expected experimental results (mean and fluctuating temperature distributions), and perform an uncertainty analysis.

10. Prepare an experimental setup to evaluate flow and thermal fields for steady incompressible turbulent air flow through nine impinging jets for industrial applications. The required compressed air (25°C), heaters, piping, flow meters, materials, thermocouples, and equipment for calibrations and measurements are available for your use. Use a miniature five-hole probe to map the flow field on a downstream plane of the square channel with nine impinging jets: Sketch the major experimental setup, describe the measurement principle, sketch and discuss the expected experimental results (mean velocity vector field), and perform an uncertainty analysis.

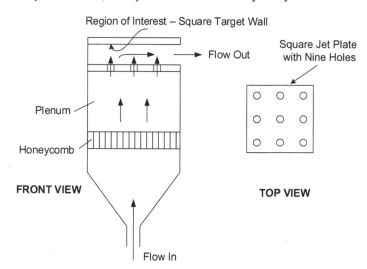

11. From Problem 10, use a hot wire anemometer (Constant Temperature Anemometry) to map the flow field on a downstream plane of the square channel with nine impinging jets: Sketch the major experimental setup, describe the measurement principle, sketch and discuss the expected experimental results (mean velocity vector field and turbulence intensity), and perform an uncertainty analysis.

12. From Problem 10, if the top wall of the channel is made of copper and heated to certain temperature by heaters on the back side, use a cold wire anemometer (Constant Current Anemometry) to determine the flow temperature field on a downstream plane of the square channel with nine impinging jet: Sketch the major experimental setup, describe the measurement principle, sketch and discuss the expected experimental results (mean and fluctuating temperature distributions), and perform an uncertainty analysis.

REFERENCES

1. Ligrani, P.M., Singer, B.A. and Baun, L.D., "Miniature Five-Hole Pressure Probe for Measurement of Three-Mean Velocity Components in Low-Speed Flows," *Journal of Physics Engineering: Scientific Instrumentation*, Vol. 22, 1989, pp. 868–876.
2. Morrison, G.L., Schobeiri, M.T. and Pappu, K.R., "Five-Hole Pressure Probe Analysis Technique," *Flow Measurement and Instrumentation*, Vol. 9, 1998, pp. 153–158.
3. Paul, A.R., Upadhyay, R.R. and Jain, A., "A Novel Calibration Algorithm for Five-Hole Pressure Probe," *International J. Engineering, Science and Technology*, Vol. 3, No. 2, 2011, pp. 89–95.
4. Takahashi, T.T., "Measurement of Air Flow Characteristics Using Seven-Hole Cone Probes," *AIAA Paper No. 97-0600*, 1997.
5. Gossweiier, C.R., Kupferschmier, P. and Gyarmathy, J.G., "On Fast-Response Probes: Part 1—Technology, Calibration, and Application to Turbomachinery," *ASME Journal of Turbomachinery*, Vol. 117, 1995, pp. 611–617.
6. Ainsworth, R.W., J. L Allen, J.L. and Batt, J.J.M., "The Development of Fast Response Aerodynamic Probes for Flow Measurements in Turbomachinery," *ASME Journal of Turbomachinery*, Vol. 117, 1995, pp. 625–634.
7. Goldstein, R.J., *Fluid Mechanics Measurements*, Hemisphere Publishing Corporation, New York, 1983.
8. Lee, T.-W., *Thermal and Flow Measurements*, CRC Press, Taylor & Francis Group, New York, 2008.
9. Young, C.D., Han, J.C., Huang, Y. and Rivir, R.B., "Influence of Jet Grid Turbulence on Flat Plate Turbulent Boundary Layer Flow and Heat Transfer," *ASME Journal of Heat Transfer*, Vol. 114, 1992, pp. 65–72.
10. Mehendale, A.B., Han, J.C. and Ou, S., "Influence of High Mainstream Turbulence on Leading Edge Heat Transfer," *ASME Journal of Heat Transfer*, Vol. 113, 1991, pp. 843–850.
11. Han, J.C., Zhang, L. and Ou, S., "Influence of Unsteady Wake on Heat Transfer Coefficient from a Gas Turbine Blade," *ASME Journal of Heat Transfer*, Vol. 115, 1993, pp. 904–911.
12. Teng, S., Sohn, D.K. and Han, J.C., "Unsteady Wake Effect on Film Temperature and Effectiveness Distributions for a Gas Turbine Blade," *ASME Journal of Turbomachinery*, Vol. 122, No. 2, 2000, pp. 340–347.
13. Kohli, A. and Bogard, D.G., "Fluctuating Thermal Field in the Near-Hole Region for Film Cooling Flows," *ASME Journal of Turbomachinery*, Vol. 120, 1998, pp. 86–91.
14. Blair, M.F. and Bennett, J.C., "Hot Wire Measurements of Velocity and Temperature Fluctuations in a Heated Turbulent Boundary Layer," *ASME Paper No. 84-GT-M 234*, 1984.

15. Blair, M.F., "Influence of Free-stream Turbulence on Turbulent Boundary Layer Heat Transfer and Mean Profile Development. Part I: Experimental Data. Part II: Analysis of Results," *ASME Journal of Heat Transfer*, Vol. 105, 1983, pp. 33–47.

16. Kim, J. and Simon, T.W., "Measurements of the Turbulent Transport of Heat and Momentum in Convexly Curved Boundary Layers: Effects of Curvature, Recovery, and Free-stream Turbulence," *ASME Paper No. 87-GT-199*, 1987.

17. Zhou, D. and Wang, T., "Effects of Elevated Free-stream Turbulence on Flow and Thermal Structures in Transitional Boundary Layers," *ASME Journal of Turbomachinery*, Vol. 117, 1995, pp. 407–412.

12 Flow Field Measurements by Particle Image Velocimetry (PIV) Techniques

12.1 INTRODUCTION TO PARTICLE IMAGE VELOCIMETRY (PIV)

Particle Image Velocity (PIV) emerged in the mid-1980s as a powerful method to obtain two-dimensional, velocity field measurements. The PIV technique expounded on traditional flow visualization techniques to provide quantitative velocity and turbulence measurements. While PIV is a particle-based technique similar to LDV, it has the advantage of providing simultaneous velocity distributions over an entire region of interest from a single test. Whereas LDV (and hotwire anemometers) are only capable of point measurements, and the hardware must be traversed across the fluid domain to provide two-dimensional or three-dimensional velocity distributions.

The PIV technique was implemented for the acquisition of instantaneous velocity distributions. With the acquisition and processing of images taking place on actual film, the technique was limited to single images or time-averaged data acquired over several minutes. However, the data processing and analysis often limited the technique to a single set of images. With the introduction of CCD cameras and increased computational power, researchers began considering both time averaged flow distributions (taken from hundreds of image pairs acquired over several seconds of recording) and instantaneous velocity fluctuations (u' and v'). While this digital PIV technique produced detailed velocity and turbulence data for a two-dimensional plane, it lacked the temporal resolution of the more traditional hotwire anemometers. Over the last decade, advances in CMOS cameras and laser repetition rates have opened the door for high-speed PIV. Combining high-speed cameras with high frequency lasers, researchers have developed high-speed PIV techniques capable of capturing thousands of image pairs per second. These hardware developments have allowed experimentalists to resolve highly turbulent flows both spatially (two-dimensionally) and temporally. With the digital CCD (charge-coupled device) camera replacing traditional film-based cameras in the early 2000s, PIV was commonly referred to as D-PIV (digital-particle image velocimetry). However, as digital CCD and CMOS (complementary metal-oxide semiconductor) cameras are commonplace in both research facilities and society, it is understood the vast majority of researchers are performing "D-PIV," although it is simply referred to as PIV.

Many optical techniques, rather on a surface or in a fluid, have their origin as a technique for visualization. Ludwig Prandtl's impact on the field of fluid mechanics and boundary layer theory is unquestionable. Dating back to the early 1900s, Prandtl devised a water tunnel capable of visualizing flow around objects such as cylinders and airfoils [1]. While his early tunnels were powered by manually turning a water wheel, the separation on the backside of an airfoil was easily observed within the tunnel. Prandtl and his colleagues attempted to produce subsequent images within the water tunnel. From a composite image of five successive frames, the separation and downstream vortex were captured for the world to see [2]. This composite set of images led to the development of particle tracking and velocity techniques, and more than a century later are mainstays in laboratories around the world.

12.2 FUNDAMENTAL PRINCIPLES OF PIV

With particle image velocimetry becoming widely available to many researchers, detailed references have been published to describe the nuisances of PIV instrumentation and measurements [3–5]. With these detailed sources available, only the basic, fundamental principles are discussed here to provide a broad overview of the requirements of the method.

When discussing PIV to obtain velocity distributions within a fluid, the use of seed particles cannot be overlooked. The fluid (rather gas or liquid) is seeded with particles capable of reflecting light to a camera and thus identifying the location of the seed particles at a specific instant in time. With the particles reflecting the light back to the camera, the particle velocity is actually being measured. A fundamental assumption associated with PIV is the velocity of the seed particles is equal to the velocity of the surrounding fluid. Therefore, the experimentalist must take care to ensure particles introduced into the fluid follow the motion of the fluid and do not alter the flow behavior.

Figure 12.1 shows a simple illustration of the kinematic motion of a single particle. The particle is initially located at (x_1, y_1), and after a given amount of time (Δt),

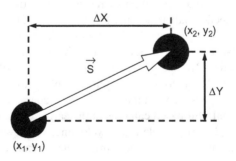

FIGURE 12.1 Single particle displacement.

the particle moves to (x_2, y_2). In the two-dimensional plane, the translation of the particle can be identified by the displacement vector, $\vec{S}$. Knowing the time required for the particle to move from location one, t_1, to location two, t_2, can lead to the calculation of the directional velocity of the particle.

For the planar velocity, the x- and y-velocity components can be calculated based on the directional displacements and the time difference for this displacement. Equations (12.1) and (12.2) represent the x- and y-velocity components, respectively. Combining the x- and y-components yields the velocity magnitude shown in equation (12.3).

$$u = \lim_{t_2 \to t_1} \left(\frac{x_2 - x_1}{t_2 - t_1} \right) \tag{12.1}$$

$$v = \lim_{t_2 \to t_1} \left(\frac{y_2 - y_1}{t_2 - t_1} \right) \tag{12.2}$$

$$V = \sqrt{u^2 + v^2} \tag{12.3}$$

Tracking the two-dimensional motion of a single particle is straightforward. With an accurate method for measuring the displacement of the particle and the time interval, a single velocity vector is easily calculated. Early attempts to generate velocity maps of a fluid were based simply on particle "tracking." Prandtl's early water tunnel experiments used superposition of images to visualize particle trajectories [2]. However, tracking individual particles is not an efficient way to characterize most flows. Tracking individual particles requires relatively sparse seeding of the flow, so individual particles can be identified. With increased turbulence, the particles are likely moving in or out of the planar light source, leading to additional error. Therefore, looking at collective groups of particles yields a more reliable representation of a two-dimensional flow field.

Whether considering a single particle or a grouping of like particles, the calculation of a velocity vector is the same. The displacement of the particle(s) is measured over an instant in time, and the velocity is calculated based on the displacement measured over a given time. With digital imaging techniques, relatively large fields of view (with ever increasing spatial resolution) can be recorded in subsequent images occurring over small times (micro- or nano-second separation times). With the increased spatial resolution of CCD and CMOS cameras available for today's researchers, millions of small particles can be captured on a single image. Therefore, it becomes a near impossible challenge to track individual particles from one image to the next. To ease the computational burden of linking particles between two images, it is more customary to track collective "groups" of particles and assign a displacement (velocity) for each group of particles, rather than each individual particle.

For a set of images (image one and image two), each image is divided into much smaller "interrogation regions." For the most basic calculations, each interrogation

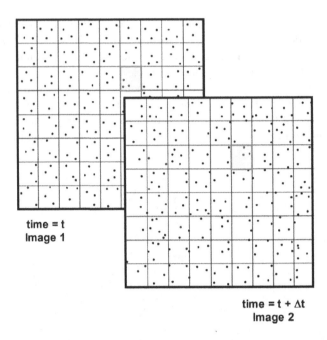

FIGURE 12.2 Sample image pair with individual interrogation regions.

region is isolated from the surrounding regions. Two general criteria are generally used when selecting the size of the interrogation regions: (a) The size is small enough such that all the particles contained within an isolated region are assumed to be moving in the same direction, and (b) the size is large enough to allow the particles to move a measurable distance (5–10 pixels) during the time (Δt) between the two images.

Figure 12.2 shows a sample image pair with each image divided into individual interrogation regions. While particles move over the entire image, the displacement of groups of particles contained within a single, small area is determined from image one to image two. The shaded interrogation region from Figure 12.2 is highlighted in Figure 12.3. As illustrated in Figure 12.3, it is necessary to determine the most probable displacement of all particles within a small region. In this simplified case, all particles within the interrogation region are clearly identified and have a match from image one to image two. In practical applications, particles may leave or enter the region between the two images, and for turbulent flows, particles are likely moving in an out of the light plane; therefore, not all particles have a direct match. Correlation statistics are used to determine the most probable path for the majority of particles in the interrogation region, and this displacement of the group is represented by a single vector for the region. Vectors cannot be assigned to all particles, as the paths for all individual particles cannot be identified. A velocity distribution for the entire flow field is determined by combining the vectors obtained for each individual interrogation region.

Single Interrogation Region

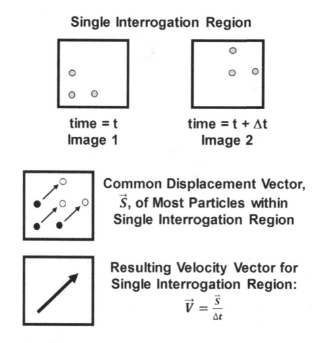

FIGURE 12.3   Single interrogation region showing particle displacement.

12.3   BASIC SYSTEM COMPONENTS FOR TRADITIONAL PIV EXPERIMENTS

In order to determine the displacement of seed particles, several basic hardware components are required to conduct a traditional PIV experiment. As shown in Figure 12.4, the fluid must be "seeded" with particles; therefore, a means to both generate particles and evenly disperse them in the flow is needed. Once the particles are added to the fluid, they must be visible to a camera. Therefore, the particles must reflect light. In the most basic PIV experiments, two-dimensional velocity fields are obtained on a single plane. A light source must be chosen to produce a planar light sheet with a minimal thickness. A suitable camera is required to capture the light reflected by the particles, and to ensure the camera captures images while the particles are illuminated, a synchronizer unit is required to properly time the light source with the camera. Finally, a computer is needed to save the images and process the images to yield velocity maps representative of the flow field. The basic requirements for each of these hardware components are discussed in the following sections.

12.3.1   SEED PARTICLES

As mentioned earlier in the chapter, the selection of proper seed particles is critical to ensure an accurate representation of the fluid velocity. The displacement of a group of particles is measured over a given time, and the measured velocity of these

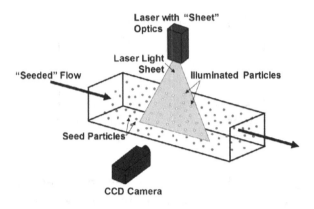

FIGURE 12.4   Basic setup for a traditional PIV experiment.

particles is assumed equal to the velocity of the surrounding fluid. Therefore, the seed particles must follow the fluid without disrupting its natural behavior.

When considering the selection of the ideal seed material, it is necessary to consider the Stokes number. The Stokes number represents how a foreign particle behaves when immersed in fluid. For example, when a large sphere is placed in moving fluid, the motion of the fluid changes to move around the sphere, while the "particle" (large sphere) remains relatively unchanged. Depending on the Reynolds number of the fluid (relative to the sphere size), the flow will exhibit the traditional separation and turbulent vortex street forming downstream of the sphere. The relatively large sphere clearly alters the fluid motion around it. For PIV applications, the foreign particle must not alter the flow field, and it must naturally follow the fluid.

The Stokes number represents the interaction between a foreign particle and its surrounding fluid. Non-dimensionally, the ratio of the particle's inertial force to the fluid's inertial force is represented with the Stokes number. More commonly, the Stokes number is presented as the ratio of the relaxation time of the particle compared to the characteristic time of the surrounding fluid. If the particle motion is independent of the fluid motion (the movement of the particle is not influenced by the fluid), both the relaxation time of the particle and the inertial force of the particle are large. In the extreme case of fixed foreign particles, the Stokes number approaches infinity. On the other hand, if the particle is identically following the motion of the fluid, it is not exerting unnatural force on the fluid, and its relaxation time approaches zero. Therefore, the Stokes number approaches zero. For PIV applications, it is necessary to select the seed particles so their Stokes number approaches zero, and they are neutrally buoyant in the surrounding fluid. Generally, a particle is acceptable if the Stokes number is less than 0.1; under such a condition, the error associated with deviation of the particle from the fluid is less than 1% [5].

Due to the relatively high density of liquids, the Stokes number criteria is easily achieved. A variety of solid particles can be dispersed in the fluid, and the density of the particles is comparable to that of the fluid. Table 12.1 provides several solid particles commonly seen in liquid PIV experiments. For liquid experiments, the size of the particle can be several orders of magnitude larger than typically used in gas flow

TABLE 12.1
Common Seed Particles for Liquid Flows

Particle Material	Approximate Density Range (g/cm³)	Approximate Particle Diameter Range
Polymer particles	1.03–1.05	10–100 μm
Coated, hollow glass beads	1.03–1.05	10–100 μm
Fluorescent particles	1.03–1.05	200–1000 nm

experiments. The size of the particle effects how much light is reflected by the particle; therefore, having the larger diameter particles is an advantage of liquid flows.

The density of gases is several orders of magnitude less than the density of liquids. Likewise, the density of solid- or oil-based particles is also much larger than the gas density. With the large variation of density between the potential seed particles and the surrounding fluid, the particle diameter must be sufficiently small to meet the Stokes number criteria. Comparing the seed particles listed in Tables 12.1 and 12.2 shows the particles dispersed in gas flows are much smaller than those used in liquid flows. With gas flows that might be open to the atmosphere, caution must be taken for the dispersion of particles into a contained laboratory space. DEHS oil is commonly used; it is a non-toxic, lightweight oil that will evaporate from surfaces. Most cooking oils are also suitable, as they are non-toxic, but a longer lasting residue can be left on exposed surfaces. The Silica and Titanium powders are desirable due to their white, reflective surface. However, these powders can cause irritation to the respiratory system, so proper precautions should be taken to minimize the exposure to the airborne particles.

When adding the seed particles to the fluid, it is important that particles remain distinctly separate from one another. Not only does this control the size of the particles (affecting their ability to follow the flow), but it also aids in the analysis of the PIV images. The seed density of the particles should be controlled in conjunction with sizing the interrogation regions for the image. The dispersion of too many or too few particles will lead to increased uncertainty in the velocity calculation. General guidelines point to approximately 10 distinct particles in each interrogation region. In both extreme cases (too many or too few particles), it is difficult to use the correlation functions to match the particles between the two images.

TABLE 12.2
Common Seed Particles for Gas Flows

Particle Material	Approximate Density Range (g/cm³)	Approximate Particle Diameter Range (μm)
Di-ethyl-hexyl-sebacat (DEHS) oil	0.9–1.0	1–3
Vegetable oil	0.91–0.93	1–3
Silica dioxide (SiO_2) powder	0.9	0.5–2
Titanium dioxide (TiO_2) powder	4.2	0.2–0.5

12.3.2 ILLUMINATION SYSTEM

To determine the motion of the particles moving with the fluid, the camera must be able to see the particles at specific instances in time. Therefore, the seed particles are illuminated. In gaseous flows, the seed particles must be sufficiently small to properly follow the fluid. However, as the diameter of the seed particles decreases, they become more difficult to see. With a Δt on the order of micro-seconds, and seed particles approximately 1 µm in diameter, a high-power light source is needed, so the small particles can quickly reflect sufficient light to be seen by the digital camera. Because of their high power, stable wavelengths, and fast repetition rates, lasers are the primary light sources for PIV experiments.

Varieties of laser types have made their way into the PIV systems. From dual-cavity, Nd:YLF high repetition lasers to traditional Nd:YAG with internal burst modes, lasers have evolved to meet the needs of modern fluid dynamists. Most lasers for PIV applications operate around the 532 nm wavelength. The bright intensity of the green light makes it very attractive for PIV applications, where limited light is reflected from the particles back to the camera. However, PIV is not limited to this specific wavelength. As digital cameras are capable of recording over a broader range of light than the human eye can see, wavelengths in both the UV and IR ranges have been used. In cases where fluorescing seed particles are used in the fluid, UV lasers transfer the necessary energy to cause the particles to fluoresce.

For PIV applications, most lasers either are equipped internally or attached externally at the output aperture, with a set of optics. As a minimum, a cylindrical lens is required to transform the laser beam into a planar light sheet. Having the ability to focus and adjust the light sheet thickness is desirable to minimize the thickness in application area. A "thick" laser sheet leads to increased error, as the particles could travel in the z-direction but not exit the light sheet from image one to image two. The planar PIV method cannot resolve this z-component of velocity; therefore, the accuracy of the results is reduced. The focusing of the light sheet into as narrow of a "line" as possible is achieved with a spherical lens. This region of minimum thickness is often referred to as the "waist" of the light sheet. Figure 12.5 shows a representation of the laser beam expansion with both spherical and cylindrical lenses. Based on the distance of the measurement area away from the laser output, these lenses can be customized to optimize the sheet expansion and thickness. Figure 12.5 shows the beam expansion occurring separately from the laser head. In most packaged systems, the light sheet optics are an integral component to attach directly to the laser head. These optic sets are easily focused and interchanged for a variety of applications.

12.3.3 CAMERAS FOR PIV

PIV has rapidly developed with the inclusion of digital cameras and digital photography. Film cameras were limited to large time changes between image one and image two, or they were used with long exposure times to create particle tracks. Highly sensitive CCD cameras advanced PIV development as they provided suitable images in low light environments. With the small seed particles reflecting minimal light to the sensor, the CCD sensor was capable of capturing the particles. The resolution of the

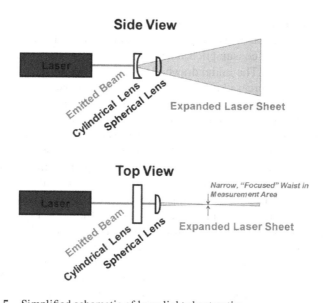

FIGURE 12.5   Simplified schematic of laser light sheet optics.

camera rapidly increased providing images with a resolution of several megapixels. An early drawback of CCD cameras was the limited frame rate of the sensor. Therefore, to be used for PIV applications, the camera required an "interframing" (double exposure) setting. With interframing, the overall speed of the camera is not increased, but at a given frequency, the camera records to consecutive images separated by the interframe time. The exposure time for each image is limited by the frame rate of the camera and the time required writing the images to memory. Figure 12.6 diagrams the basic timing structure for camera and PIV synchronization using interframing.

For years, cameras with CCD sensors were the primary camera for researchers. Although the sensors were relatively expensive to manufacture, the noise level of the digital images was relatively low. As technology developed, especially for physically small cameras for security and phone applications, CMOS cameras rapidly developed. As with a CCD sensor, the CMOS sensor also captures light (photons) and

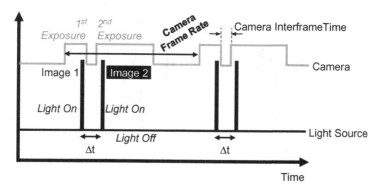

FIGURE 12.6   Sample timing structure for low-speed cameras.

converts the light to digital data. However, the CMOS camera requires significantly less power, and the digital conversion occurs much more quickly than with the CCD sensor. The low power consumption made this type of sensor very attractive for cell phone applications. The initial drawback of the CMOS sensor compared to the CCD sensor was the increased noise level in the digital images. Driven by multiple industries, the technology quickly matured, and the image quality has continuously improved. With the ability to convert and write digital data more quickly than with a CCD sensor, high-speed cameras with CMOS sensors became widely available.

Today, high-speed, time-resolved PIV systems are available using cameras equipped with CMOS sensors. When recording image pairs with high-speed cameras, interframing is not required. Pairs of images are created from subsequent images. The time separation between image one and image two is determined based on the frame rate of the camera and the pulse rate of the laser. Figure 12.7 shows a sample timing diagram for the synchronization of the laser with the high-speed camera. For these high-speed systems, the limiting factor is now the pulse rate of the laser. High-power, high-repetition rate lasers are costly to both build and operate. To provide the high pulse rates, dual cavity, pulsed lasers are typically utilized. With two laser heads incorporated into a single housing, the first laser pulse will be the light source for image one, and the second pulse provides light for image two. Therefore, each laser pulses at half the frequency of the camera.

12.3.4 HARDWARE SYNCHRONIZER

The time separation (Δt) between image one and image two of the PIV pair is set by the time between laser pulses. The seed particles are illuminated for an instant in time, and while the shutter of the camera may remain open, there is not light being reflected to the sensor if the light source is not active. Furthermore, the shutter of the camera must be open, when the light is switched on; otherwise, the camera does not record the position of the seed particles at a given time. The timing of the laser and camera is controlled by an external synchronizer. With modern systems, the actual timing is set using computer software, and the trigger signal is sent from

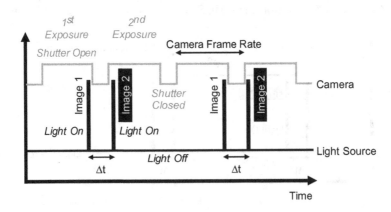

FIGURE 12.7 Sample timing structure for high-speed cameras.

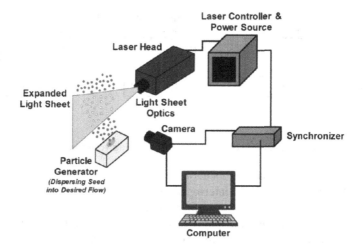

FIGURE 12.8   Assembled PIV system with hardware synchronizer.

the computer to the synchronizer. From there, separate signals are sent to the laser controller and camera to ensure the camera captures images while the particles are illuminated. In older systems, an external signal generator (separate from the computer) was used to trigger the camera and laser. The general setup of a PIV system, with all necessary hardware components, is shown in Figure 12.8.

12.4   GENERAL METHODOLOGY FOR IMAGE PROCESSING

The velocity distributions obtained using the PIV method are estimated based on the translation of a group of seed particles. Rather than identifying millions of individual particles, it is assumed all particles contained within a small, common area are moving in the same direction, and a single velocity vector represents the movement of this group. The particles are grouped together by the establishment of interrogation regions across the raw images. The assembly of vectors for each interrogation region leads to a vector field for a given instant in time. With hundreds or thousands of image pairs recorded for a flow field, the vectors for each interrogation region can be time averaged to yield a time averaged velocity distribution for the fluid.

Figure 12.9 shows sample particle distributions for the two images recorded at two instances in time. The seed particles are superimposed with one another in the middle sketch. Directly comparing the translation of the particles from the first image to the second image provides the displacement of each particle group. As shown in the right sketch, a separate vector is obtained for each interrogation region.

After the images are saved, each individual image is digitized. The image is divided into interrogation regions, usually beginning with each region 64 pixels by 64 pixels. Each interrogation region is isolated and initially analyzed separately from all regions. Within the first interrogation region, the location of discrete particles is identified. The relative intensity is measured for each pixel, and based on this intensity, the location of each particle identified (pixel location). For the given

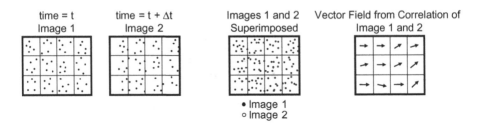

| time = t
Image 1 | time = t + Δt
Image 2 | Images 1 and 2
Superimposed | Vector Field from Correlation of
Image 1 and 2 |

• Image 1
○ Image 2

FIGURE 12.9   Idealized particle distributions and resulting vector field.

interrogation region, the particle locations from image one are compared with the locations of particles in the same interrogation region for image two. In an ideal scenario, every particle in image one can be matched with a particle in image two (all particles demonstrating the same displacement). However, due to particles entering or exiting the region and off-plane velocity components, there will not be a perfect match of particles between the two images. Therefore, the cross-correlation function is applied between the two images to identify the most probable displacement for the group of particles. This process is repeated for each interrogation region, and the complete procedure can be repeated for many image pairs.

Other sources provide detail of the statistical correlations relevant to PIV analysis [3,5], so only a top-level view is provided here. The use of correlations to relate two sets of images can be considered in the mindset of a jigsaw puzzle. Imagine a puzzle where most of the pieces have been assembled, so the overall picture on the puzzle is clear. However, there are a few remaining pieces that need to be added to the puzzle to make it complete. As one of the remaining pieces is picked up, it is moved around the puzzle to compare the details on the face of the piece with the surrounding details of the assembled puzzle. The piece can be placed in the correct location by fitting the details of the single piece with the details of the overall puzzle.

In a sense, the cross-correlation takes and isolated interrogation region from image one and moves it around image two to see where the particles align with one another (from image one to image two). There will be many places, where a few particles from the interrogation region directly overlap a few particles in image two. However, the goal is to find the location where most of the particles can be matched. The parameters of the experiment should be properly controlled, so the general search area can be identified. The time separation of the images should be set such that the seed particles move approximately ten pixels. Therefore, the search area for each interrogation region is limited, and this can allow for more efficient code development to analyze the PIV images.

The process that has been described is known as a "single-pass" data reduction scheme. This means, the interrogation region is sized, and the vectors are obtained for each region. Due to turbulence or other anomalies, it is possible that one or more of the regions yield a spurious vector. Figure 12.10 shows an embellished scenario where one of the calculated vectors clearly does not follow the trend inferred by the surrounding vectors. To rectify these vectors, most PIV algorithms run "multi-pass" schemes to compare individual vectors with their surrounding vectors. The cross-correlation could have returned a false peak for this interrogation region, so the

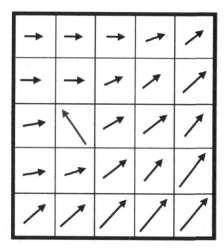

FIGURE 12.10   Representation of a spurious vector.

correlation can be checked for secondary peaks that potentially provide a better representation of the velocity in this region. In addition, the multi-pass scheme allows for improved spatial resolution by reducing the size of the interrogation region for subsequent passes. In many cases, an "overlap" of the interrogation regions is also assigned (50%–75%). This effectively reduces the size of the interrogation regions while still maintaining the displacement of ten pixels across the region.

The analysis of a single image pair provides a single, instantaneous velocity distribution on a single plane within the flow. Each interrogation region provides a measurement point. When the regions are combined, the planar flow field is realized (without traversing a probe across the flow). The analysis of a single image pair does not adequately describe the global behavior of the flow. Therefore, it is necessary to record hundreds or thousands of image pairs. The time average velocity for each interrogation region can be obtained by averaging the measured velocities from many image pairs. Also, with a sufficient number of image pairs, the turbulent statistics across the flow can also be considered. The later example problems further demonstrate the analysis of PIV data.

12.5   RECENT ADVANCEMENTS IN PIV

The majority of this chapter has been devoted to a high-level view of a traditional, low-speed, two-dimensional PIV experiment. With cameras becoming faster and computer capabilities rapidly improving (memory, storage, processing speed), PIV experiments have also developed in response to this technology. In many cases, the limiting factor for PIV applications is the laser pulse rate and power, but this industry has also seen rapid development to provide new opportunities for PIV experiments.

With dual cavity, high repetition rate lasers, time-resolved, two-dimensional flow fields can be obtained on a plane. Time-resolved data is typically needed in highly turbulent flows. Therefore, measuring these two-dimensional velocity components do not fully describe the flow. The off-component of velocity is not captured, and the accuracy of the two-dimensional measurements is reduced because particles are moving in and out of plane.

To better resolve these highly turbulent flows, it is necessary to capture the third velocity component. With a firm understanding of optics to create planar laser sheets, three-dimensional flow fields were first measured on a single plane. Often referred to as "Stereo-PIV," the basic hardware components are the same as for a traditional, two-dimensional setup. However, in order to gain an additional velocity component, a second camera is required. The two-cameras are offset from one another and not perpendicular to the light sheet (as with the two-dimensional setup). The viewing angle and image distortion for a single camera are described in terms of the Scheimpflug condition. Each camera as a distorted view of the plane. Much like two human eyes viewing a scene, due to the angle between the eyes, it is possible to capture depth. The brain is able to combine the views from each eye and create an undistorted scene. The images captured from each camera are mapped together relative to a known spatial calibration target. As a result, the two images are combined to create a depth to the image. Now the seed particles are tracked in both the x- and y-directions, along with through the thickness of the laser sheet.

For the Stereo-PIV technique, image one consists of two images (one from each camera), and images two also consists of two images. Before the PIV data processing begins, the image from camera A and camera B must be mapped together to create a single image for set one, and likewise the images are combined for image two at Δt. As the particles are identified in the image, it is possible to mark their locations in three-dimensions and track the particle group three-dimensionally.

Expanding the capabilities of multi-camera PIV methods led to the development of Volumetric or Tomographic PIV. With Tomo-PIV, the laser is further spread over a larger volume. Using four or more cameras, the volume of the flow field can be reconstructed and three-dimensional distributions are obtained through the entire volume. In the realm of Tomo-PIV, the volume is reconstructed in terms of voxels (planar images are composed of pixels). Interrogation regions are defined in terms of voxels across the illuminated volume. Again, the laser can be the limiting factor for volumetric measurements. Attaching a second cylindrical lens to the laser head, and misaligning it with the first cylindrical lens, the laser can be spread over a volume. With the beam now being spread over a volume, and lighting the discrete particles, a higher laser power is needed. Combining the tomographic concept with high-speed cameras and a high repetition rate laser provides a means to capture time-resolved, volumetric velocity distributions.

12.6   EXPERIMENTAL EXAMPLES OF PIV

Example 12.1: External, Film Cooling (Boundary Layer Mixing) with Traditional, 2D PIV [6]

PIV experiments have been conducted for decades to understand a wide variety of flows. Wright et al. [6] conducted a series of tests to understand how an inclined jet in cross-flow interacts with a mainstream flow over a flat surface. This type of flow in relevant to gas turbine cooling technology. High pressure turbine blades are cooled using film cooling jets which create an insulating film of air on the outer surface of the airfoil. A variety of flow conditions were investigated in this study, and this example demonstrates how the jet-to-mainstream interaction was captured using a traditional PIV experiment.

A low speed, open loop wind tunnel was used to investigate a fundamental film cooling geometry. The basic setup and the details of the film cooling geometry are shown in Figure 12.11. The floor of the wind tunnel was fitted with a film cooled flat plate. The film cooling holes were simple angle, cylindrical holes angled 35° to the mainstream flow. The jets were sufficiently spaced to eliminate hole-to-hole interactions. The mainstream velocity in the tunnel was 8.5 m/s, and a turbulence grid was placed upstream of the film cooling holes to increase the freestream turbulence from approximately 1.2%–12.5%.

For the basic setup of the PIV experiment, a Firefly FF0100 laser (from Oxford Lasers, Inc.) was mounted on a traverse above the wind tunnel. The laser was moved to create light sheets at various locations relative to the film cooling holes; for this example, the data shown is limited to the centerline of the of the hole $(Y = 0)$. The laser was a 200-Watt, pulsed diode laser operating with a wavelength of 808 nm. The laser separation time (time between image one and image two) varied in the experiment depending on the flow conditions. The data shown in this example cover a blowing ratio of 1.5, and for this velocity, the pulse separation time was 20 μs. This Δt was selected based on a five-pixel displacement of the particles between the images. The light reflected by the seed particles was captured using a PCO 1600, CCD camera from Cooke Corp. For the current setup, the interframing time was set to 0.180 μs. Images were captured at a resolution of 800 × 600 pixels. As shown in Figure 12.11, seed particles were only added to the cooling flow. An aerosol generated was added to the cooling line, and 1-micron DEHS oil droplets were added to the coolant. The camera and laser were triggered directly from the computer, so a separate, external synchronizer was not used.

For each flow condition, 500 image pairs were recorded to provide time-averaged velocity distributions and applicable turbulent statistics. Therefore, 500 separate velocity distributions were measured for each flow condition. Figure 12.12 shows sample raw images obtained for one flow condition. For a given instant,

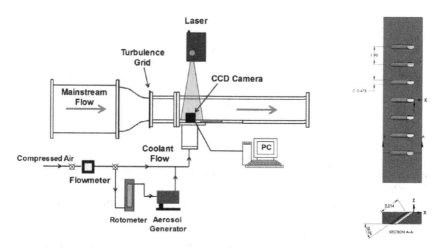

FIGURE 12.11   Experimental setup and film cooling geometry.

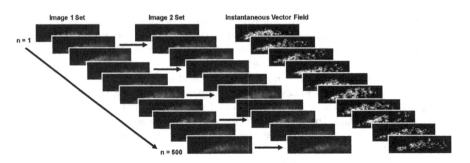

FIGURE 12.12   Sample images and instantaneous velocity distributions.

image one is very similar to image two, as the seed particles should have shifted approximately five pixels between the images. The images shown in the figure were preprocessed to remove surface reflections and background light from the images. Each of the 500 images sets were individually processed, and an instantaneous velocity distribution was obtained for each set. The data was processed using a multi-pass, cross-correlation. For the first-pass, the size of the interrogation region was 32 × 32 pixels. A second pass was completed at 16 × 16 pixels with a 50% overlap of the interrogation regions.

Sample velocity distributions are also shown in Figure 12.12. As shown in the figure, for a single image set, the vectors are scattered with it often appearing as if vectors are missing. This is the result of non-uniform seed dispersion, the lack of seed particles in the mainstream, the relatively high level of freestream turbulence, and post-processing filters (signal-to-noise and correlation peaks). Using a low frame rate camera, it is necessary to record a large number of image pairs to properly describe the flow field.

The instantaneous velocities for each interrogation region over the 500 images can be combined to yield the time average velocity across the measurement plane. The time averaged calculation of the x- and y-velocity components is shown in Equations (12.4) and (12.5).

$$\bar{u} = \frac{1}{N} \sum_{i=1}^{N} u_i \qquad (12.4)$$

$$\bar{v} = \frac{1}{N} \sum_{i=1}^{N} v_i \qquad (12.5)$$

While 500 image pairs are recorded, it is obvious from Figure 12.12, not every interrogation region returns a vector for every image. Therefore, care must be taken to "count" the number of valid vectors for each interrogation region, so the average is taken over the proper number of occurrences, N. In other words, an interrogation region near the core of the jet may return valid vectors for 475 of the image pairs, but near the edge of the jet, where mixing with the mainstream is prevalent, an interrogation region may only return 150 valid vectors. Therefore, the average of these two separate regions is 475 and 150, respectively.

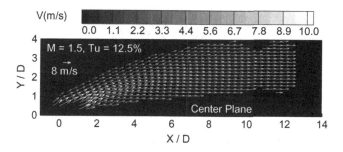

FIGURE 12.13 Sample time averaged velocity distribution.

Figure 12.13 presents a time averaged velocity distribution for the 500 image pairs partially represented in Figure 12.12. Both the vector length and the vector shading represent the two-dimensional velocity magnitude. Comparing this time-averaged distribution to the instantaneous distributions shown in Figure 12.12, the overall jet structure is more defined and populated across the domain.

By recording many image pairs, it is possible to glean additional flow field information from the PIV experiment. For this example, the freestream is intentionally elevated to model flow in the gas turbine engine. Using the recorded images, it is possible to consider both the velocity fluctuations. Rather considering the one-dimensional, u', or two-dimensional fluctuations, $u'v'$, understanding the how the velocity deviates from the mean velocity can offer additional insight into the flow. In addition, this information is also valuable to the validation of CFD codes and turbulence models.

The turbulent fluctuations in the x- and y-directions can be calculated as shown in Equations (12.6) and (12.7), respectively. u' and v' provide a measurement of the local turbulence intensity of the fluid. Figure 12.14 shows the turbulent fluctuation in the x-direction normalized by the freestream velocity. For this high velocity jet (1.5 times the speed of the mainstream), a significant velocity gradient exists along the edge of the jet. Throughout the entire cross-section of the jet, the turbulence is elevated above that of the mainstream. As the jet issues from the surface, the local turbulence is also increased due to secondary vortices forming near the wall as the mainstream moves around the discrete jet.

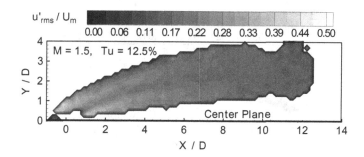

FIGURE 12.14 Sample distribution of the streamwise velocity fluctuations from PIV.

$$u'_{rms} = \sqrt{\frac{1}{N-1}\sum_{i=1}^{N}(u_i - \bar{u})^2} \qquad (12.6)$$

$$v'_{rms} = \sqrt{\frac{1}{N-1}\sum_{i=1}^{N}(v_i - \bar{v})^2} \qquad (12.7)$$

To aid in the development of turbulence models for CFD codes and to better understand how the jets interacts with both the wall and the mainstream fluid, the local Reynolds shear can be calculated using Equation (12.8). A sample distribution of the Reynolds shear is provided in Figure 12.15. For this application, it is interesting to note the sign change of the normalized Reynolds shear. Close to the surface, the Reynolds shear is "positive," but moving away from the wall, the quantity becomes "negative." With the current sign convention, the negative value indicates away from the wall, the jet is behaving as an unbound, free jet.

$$-\overline{u'v'} = -\frac{1}{N}\sum_{i=1}^{N}(u_i - \bar{u})(v_i - \bar{v}) \qquad (12.8)$$

In the work of Wright et al. [6], the effect of the film cooling jet was coupled with the surface film cooling effectiveness measurements. The jet-to-mainstream interaction impacts the ability of the coolant to remain attached to the surface and thus, the surface film cooling effectiveness. In Figure 12.16, the PIV measurements clearly show the high momentum jet lifting off the surface, and this lift-off yields a relatively low film cooling effectiveness on the surface. The surface film cooling effectiveness was measured using the PSP technique described in Chapter 9.

This simple example illustrates the power of PIV to obtain more than velocity distributions. By carefully considering the number of image pairs needed to fully describe a flow field, additional information can be gathered from the image sets. Preprocessing of the raw images can lead to improved quality and accuracy of the flow distributions. In addition, post-processing schemes can be implemented to further improve the quality of the results. As described throughout the chapter, both the hardware and software capabilities are critical to the success of a PIV experiment, and these must be selected to fit the specific needs of the experimentalist.

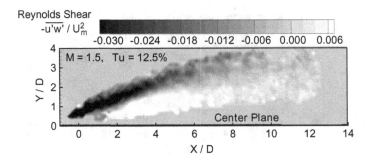

FIGURE 12.15 Normalized Reynolds shear distribution from PIV experiment.

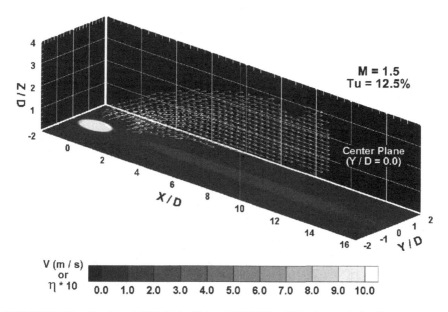

FIGURE 12.16 Combined PIV (Velocity) and PSP (Film Effectiveness) distributions.

Example 12.2: Internal Flow in a Channel Roughened with V-Shaped Dimples (Simplified Stereo-PIV) [7]

This example follows the work of Brown et al. [7] investigating heat transfer and flow in a rectangular channel roughened with V-shaped dimples. The published work included the use of a high-speed, stereoscopic PIV system to quantify the flow behavior on discrete planes relative to the dimples. The interrogation planes were perpendicular to the mainstream flow to visualize the secondary flows created by the dimples. This example will utilize the data from the investigation, but the details of the stereoscopic image transformation and superposition will not be presented (as those are beyond the scope of this book). Unlike the previous example that clearly captured the mainstream flow direction (dominant velocity component), this internal flow example focuses on the off components of velocity.

One wall of a rectangular channel was lined with V-shaped dimples. The performance of two different dimple arrays was considered: in-line and staggered arrangements. The V-shaped dimple was derived from a traditional hemispherical dimple. The traditional hemispherical dimples are most notably used to reduce drag on golf balls. The concept has been introduced as a mechanism for heat transfer enhancement. The periodic depressions effectively trip the boundary layer, resulting in increased heat transfer. With distinct areas of flow separation and reattachment, dimpled surfaces yield areas of both increased and decreased heat transfer (compared to a smooth surface). V-shaped concavities (dimples) were introduced en route to creating a more optimized dimple configuration. However, to fully realize the heat transfer potential, it is necessary to understand the flow development along a dimpled surface.

The test setup used for the PIV tests is shown in Figure 12.17. Upstream of the dimpled channel was a smooth entrance section, with the same rectangular cross

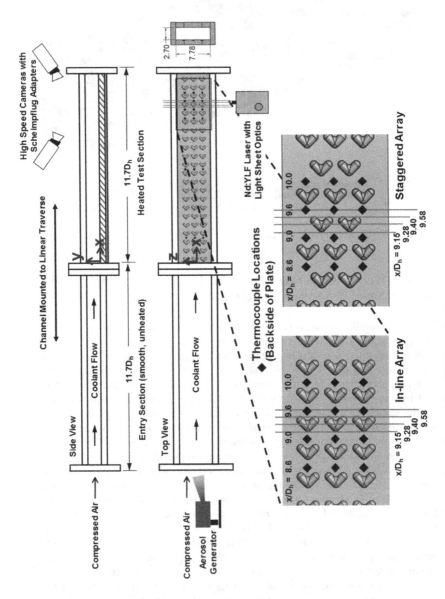

FIGURE 12.17 Experimental setup for the internal flow PIV investigation.

section as the dimpled channel. Air was supplied from a steady compressor. After flowing through an ASME orifice meter, the air flowed through a particular generator and into the test section. The particle generator created DEHS oil droplets with an average diameter of approximately 1 μm. Stereo-PIV requires the use of two cameras in order to obtain the third velocity component. As shown in the figure, the two cameras were mounted opposite of each other above the channel (the top surface was transparent Plexiglas to provide sufficient optical access into the channel). The rectangular channel was mounted to a linear traverse, so various planes relative to the dimples could be investigated. Moving the single test section was more efficient than moving the cameras and the laser simultaneously. Although this work also investigated the heat transfer performance of the dimple arrangements, the PIV tests were performed under adiabatic conditions (no heating of the dimpled surface).

For data acquisition, the following hardware components were utilized: a LaVision aerosol generator, two Phantom V211 12-bit cameras capable of recording 2190 frames per second at the full resolution of 1280 × 800 pixels with 100 mm lenses, a Photonics Nd:YLF dual cavity laser equipped with a cylindrical lens, an external hardware synchronizer, and a computer installed with DaVis from LaVision. Time averaged results were analyzed, and these flow results came from the recording of 1000 image pairs (cameras recording at 2000 frames per second). The lens on the laser allowed a laser plane of approximately 1 mm thick to be created on the dimpled surface. The laser separation time was adjusted based on the velocity of the coolant through the channel and must be adjusted for each Reynolds number under consideration. Equation (12.9) provides an equation for estimating the required separation time. With the thin laser sheet oriented perpendicular to the primary flow direction, Δt was selected; so approximately 75% of the particles remain in the laser plane for each image pair. With a laser plane thickness of 1 mm, Table 12.3 shows the required Δt for each Reynolds number of this investigation.

$$t_{ls} = \frac{w_l}{4V}$$

(12.9)

Before viewing the PIV results, the flow rates for each Reynolds number can be confirmed based on the data recorded for an ASME orifice meter (refer to Chapter 2). Using an orifice meter with a 1.5″ bore in a 2″ diameter pipe, the following pressure and temperature measurements were made at the meter for each flow rate.

TABLE 12.3
Nominal Reynolds Numbers, Mainstream Velocity, and Estimated Laser Separation Times for V-Shaped Dimple Example

Reynolds Number (D_h = 4.0 cm)	Coolant Bulk Velocity (m/s)	Laser Separation Time (μs)
10,000	3.78	66.1
20,000	7.56	33.1
30,000	11.3	22.0
37,000	14.0	17.9

Using Equation (2.1) from Chapter 2 with $y = 1.0$, $Y = 0.98$, and $K = 0.62$, the mass flow rate (and actual Reynolds number) for each condition can be calculated.

$$w = S \cdot D_2^2 \cdot K \cdot Y \sqrt{\frac{P_1}{T_1} \cdot G \cdot y \cdot \Delta P} \tag{12.10}$$

Combining the units as specified in Chapter 2, the mass flow rate for the first flow condition can be calculated along with the actual Reynolds numbers within the rectangular channel:

$$w = 0.1145 \cdot (1.5 \text{ inch})^2 \cdot 0.62 \cdot 0.98$$

$$\sqrt{\frac{111.4 \text{ in Hg (abs)}}{532°R}} \cdot 1.0 \cdot 1.0 \cdot (0.090 \text{ in } H_2O) = 0.0215 \frac{lb_m}{s} \tag{12.11}$$

$$\text{Re} = \frac{\rho V D_h}{\mu} = \frac{4\dot{m}}{P\mu}$$

$$= \frac{4 \left(0.0215 \dfrac{lb_m}{s} \right) \left(0.4536 \dfrac{kg}{lb_m} \right) \left(\dfrac{Ns^2}{kgm} \right)}{\left(2 \cdot (2.7 + 7.78) \right) cm \left(\dfrac{1m}{100 \text{ cm}} \right) \left(1.79 \times 10^{-5} \dfrac{Ns}{m^2} \right)} = 10,400 \tag{12.12}$$

Given the raw data in Table 12.4, the mass flow rate and actual Reynolds number for each nominal flow condition are given in Table 12.5.

With the bulk flow conditions set, a specified mass flow rate can be set, and raw PIV images acquired. After recording 1000 image pairs (2000 total images), the raw images were evaluated using the DaVis software, and similar to the previous example, only the time averaged results are presented with this example. In the DaVis software, a total of five passes of the cross correlation were preformed to

TABLE 12.4

Raw Data for Flow across and Orifice Meter

Reynolds Number ($D_h = 4.0$ cm)	Pressure Drop across the Orifice, ΔP	Upstream Static Pressure, P_1	Upstream Temperature, T_1
10,000	0.090 in H_2O	40 psig (111.4 in Hg abs.)	72°F (532°R)
20,000	0.36 in H_2O	40 psig (111.4 in Hg abs.)	72°F (532°R)
30,000	0.80 in H_2O	40 psig (111.4 in Hg abs.)	72°F (532°R)
37,000	1.14 in H_2O	40 psig (111.4 in Hg abs.)	72°F (532°R)

TABLE 12.5

Calculated Mass Flow Rate and Reynolds Number Based on Orifice Meter Measurements

Nominal Reynolds Number	Actual Coolant Mass Flow Rate (lb$_m$/s)	Actual Reynolds Number
10,000	0.0215	10,400
20,000	0.0430	20,800
30,000	0.0640	31,000
37,000	0.0765	37,000

generate the velocity fields. The first three passes were performed at interrogation regions of 32 × 32 pixels. For the final two passes, the regions were reduced to 16 × 16 pixels. For each pass, a 50% overlap was used, meaning each interrogation region overlapped the neighboring region by 50%. A post-processing filter was used, and vectors with a signal—to—noise ratio less than 1.1 were removed. Time averaged velocity components and the resultant velocity magnitude were calculated at the center of each interrogation region according to the following equations:

$$\bar{u} = \frac{1}{N} \sum_{i=1}^{N} u_i \tag{12.13}$$

$$\bar{v} = \frac{1}{N} \sum_{i=1}^{N} v_i \tag{12.14}$$

$$\bar{w} = \frac{1}{N} \sum_{i=1}^{N} w_i \tag{12.15}$$

$$V = \sqrt{\bar{u}^2 + \bar{v}^2 + \bar{w}^2} \tag{12.16}$$

For the high mass flow rate case of Re = 37,000, the laser separation time of approximately 18 μs was required. Reducing this time would limit the distinct movement of the DEHS droplets through the plane, and increasing the time would allow too many particles to leave the laser plane between the laser pulses. Considering a single, two-dimensional interrogation region for this high Reynolds number, the velocity vector for this region can be determined. Along the edge of the V-shaped dimple at its downstream edge (x/D_h = 9.40, see Figure 12.16), the particle displacement is estimated based on the cross-correlation of successive images. Table 12.6 provides this two-dimensional displacement for several image pairs from the 1000 pair population for the sample region shown in Figure 12.18.

Based on the displacements shown in Table 12.6 and the laser separation time of 17.9 μs, the respective velocity components can be calculated for each image set (Table 12.7).

TABLE 12.6

Sample Particle Displacements for a Single Interrogation Region

Image Pair Number	Δy (μm)	Δz (μm)
100	23.8	3.94
200	22.4	4.48
300	36.0	9.49
400	27.9	6.27
500	26.9	10.1
600	37.6	6.62
700	30.4	6.80
800	34.5	7.34
900	31.5	7.80
1000	20.8	7.16

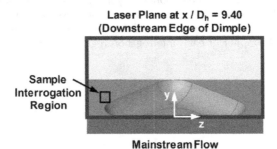

FIGURE 12.18 Location of sample interrogation region relative to a single inline dimple.

TABLE 12.7

Sample Particle Displacements for a Single Interrogation Region

Image Pair Number	v (m/s)	w (m/s)
100	1.33	0.22
200	1.25	0.25
300	2.01	0.53
400	1.56	0.35
500	1.50	0.56
600	2.10	0.37
700	1.70	0.38
800	1.93	0.41
900	1.77	0.44
1000	1.16	0.40
Average	**1.63**	**0.39**

$$v = \frac{\Delta y}{\Delta t} = \frac{23.8 \, \mu m}{17.9 \, \mu s} = 1.33 \frac{m}{s}$$

(12.17)

$$w = \frac{\Delta z}{\Delta t} = \frac{3.94 \, \mu m}{17.9 \, \mu s} = 0.22 \frac{m}{s}$$

This particular interrogation region has a strong vertical, v, component of velocity due to the upwash coming out of the leg of the dimple. The magnitudes of these velocity components are much less than the bulk velocity of approximately 14 m/s. The components presented are the off-components of velocity and represent the secondary flow induced by the dimples. The primary velocity component (the streamwise, x-direction) would capture the bulk motion.

From all 1000 image pairs, the time averaged velocity components can be calculated. This procedure can be repeated for every interrogation region across the channel. Figure 12.19 shows the time averaged velocity distribution for this flow condition. The contour shown in the background shows the local velocity magnitude compared to the bulk velocity. The velocity decreases near the surface and in the recirculation region at the outer edge of the dimple. The white box represents the interrogation region used in the previous calculations.

With many image pairs used to arrive at the time averaged profile shown in Figure 12.19, other turbulent statistics can be computed as with the previous example. Turbulence and vorticity are commonly used to further quantify the flow behavior. Sample turbulence and vorticity distributions from Brown et al. [7] are shown in Figure 12.20. The high turbulence intensity and circulation at the edge of the dimple are very clear in these figures.

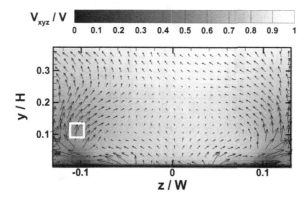

FIGURE 12.19 Time averaged velocity distribution at the downstream edge of a V-shaped dimple in an inline array (Re = 37,000).

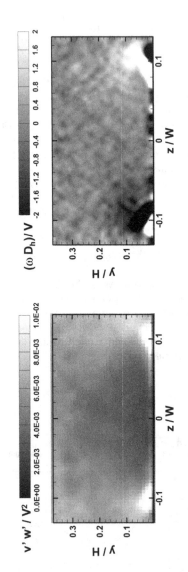

FIGURE 12.20 Turbulent velocity fluctuations and vorticity at the downstream edge of a V-shaped dimple in an inline array (Re = 37,000).

Example 12.3: Flow Behavior in the 180° Turn of a Two-Pass Channel with Smooth Walls and Ribbed Walls Using Traditional 2D, PIV [8]

Many heat exchanger and gas turbine cooling channels utilize serpentine channels to maximize the cooling/heating area. The fluid flow through the 180° turns connecting the passages of these channels is very complex. In the turn regions, areas of both low and high heat transfer exist, and these regions must be considered by heat transfer designers. In addition, the flow in these regions becomes very three-dimensional with additional turbulence generated as the fluid is forced to change directions. Son et al. [8] used planar (2D) PIV to quantify the flow behavior in the turning region of multi-pass channels.

A two-pass, square channel (D_h = 50.8 mm) was used in this study. Figure 12.21 shows an overview of the test section dimensions and coordinates. Compressed air was metered through an ASME orifice meter located upstream of the section. Tests were completed at Reynolds numbers (based on the channel hydraulic diameter) of 12,000, 30,000, and 55,000. Both passages of the test section were fabricated of 12.7 mm thick Plexiglas. To reduce laser light reflections, the inner surfaces of the walls were painted with non-gloss, black paint (except for the surfaces where optical access was required).

For the PIV measurements, an Nd:YAG laser from Spectra Physics was used (Model PIV-400). A frequency-doubled (532 nm) pulsed emission of 400 mJ/pulse was provided by the laser. The separation time between two successive laser pulses varied from 5 to 40 µs, depending on the Reynolds number of the flow. Two cylindrical lenses and a spherical lens were used with the laser to create a laser light sheet approximately 1 mm thick at the measurement regions. A 12-bit CCD camera with a maximum resolution of 1280 × 1024 pixels was used to record the PIV images. A Nikon 55 mm manual lens was used with the camera. A Laskin-type nozzle was used to generate seeding particles from corn oil. The average diameter of the oil particles was approximately 1 µm. The commercial LaVision, FlowMaster-3 software was used for image recording, synchronization of the hardware, and processing of the images.

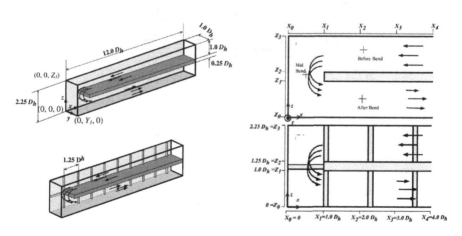

FIGURE 12.21   Square, two-passage test section used for PIV measurements and locations of the PIV measurement planes.

A multi-pass cross-correlation algorithm was used to process the raw PIV images. The first pass used interrogation regions of 64 × 64 pixels, and the interrogation regions for the second pass were reduced to 32 × 32 pixels. With both primary and secondary flows being considered in the study, the laser separation time was carefully considered and adjusted based on the laser sheet orientation and the probable out-of-plane particle displacement. For all cases, 1000 PIV image pairs were recorded and analyzed to meet the statistical requirements for the mean and fluctuating velocity components.

Figure 12.22 shows a sample mean velocity distribution and turbulent kinetic energy (TKE) distribution in the turn region of the two-pass channel with smooth walls. From the PIV measurements, the flow through the first pass clearly impinges on the tip wall before being forced to redirect in the 180° turn. As the coolant begins the turn, there is another impingement zone on the outer wall leading into the second pass. These impingement zones lead to increased heat transfer. There is also a distinct region of separation captured on the inner wall of the second pass. The separation bubble yields flow acceleration near the outer wall, and reduced heat transfer on the inner surface. The TKE in the second pass is significantly higher than that in the first pass. The turn generates increased turbulent mixing, and the result is increased heat transfer in the second pass, compared to the first pass.

The effect of rib turbulators was also considered in this study. One wall of the square channel was lined with orthogonal (90°) ribs. Figure 12.23 shows a sample of the velocity and TKE distributions over the surface of these ribs. The upward motion of the fluid over the ribs is clearly captured by the PIV. In addition, the

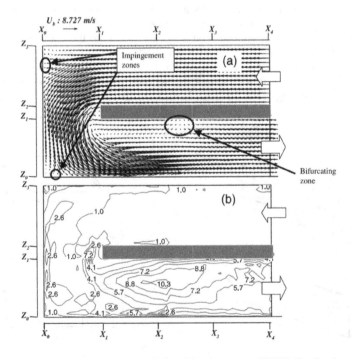

FIGURE 12.22   (a) Velocity and (b) turbulent kinetic energy (TKE) distributions along the mid-plane of a two-passage channel with smooth walls.

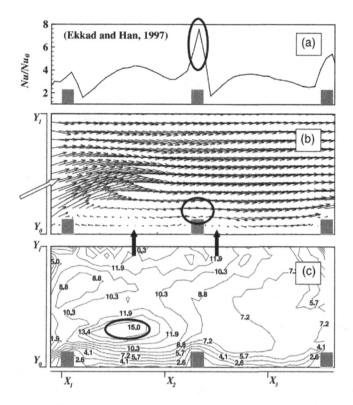

FIGURE 12.23   (a) Heat transfer, (b) velocity, and (c) TKE distributions in a channel with orthogonal ribs.

reattachment of the flow on and downstream of the ribs is also shown. The TKE is elevated midway between two turbulators, and this leads to high heat transfer in this region (as also represented in the figure).

The authors of this work present many other planes and flow distributions through the two-pass channels, and only a small sample has been shown here. The PIV experiments completed in this study were able to provide the detailed and localized flow behavior in two-pass, square channels with smooth and ribbed walls. Areas where high and low heat transfer are expected were clearly seen based on the measured flow patterns within the channel. The researchers demonstrated the ability to look at different regions within the channel to fully characterize the three-dimensional, turbulent flow patterns.

12.7   OTHER RESOURCES

With an established history of convective heat transfer measurements in turbulent flows, for years researchers were left to imply how the fluid was moving near the surface. PIV techniques have allowed researchers to first visualize the turbulent flow behavior and now quantify the fluid motion. With improved imaging capabilities, faster and more powerful lasers, and greater computing power (memory and processing speed),

PIV techniques have developed rapidly over the past decades. Internal and external flows, high-speed and low-speed flows, stationary and rotating flows have all been investigated using different adaptations of particle image velocimetry. Different challenges arise with each specific flow domain; to understand how to overcome these specific challenges, additional details are available in open literature.

Stationary, Internal Flows: Schabacker et al. [9], Suden [10], and Sharma et al. [11]

Rotating, Internal Flows: Bons and Kerrebrock [12], Elfert et al. [13], and Coletti et al. [14]

External Flows: Gogineni et al. [15], Bernsdorf et al. [16], Aga et al. [17], and Wright et al. [18]

PROBLEMS

1. A two-dimensional velocity field is to be captured along the centerline of a square duct ($2'' \times 2''$) in the streamwise direction. If the desired Reynolds number range is $Re = 5000$ to $Re = 45,000$, what is the maximum Δt between two images so the particles move approximately five pixels between the two images? The working fluid is air. The lens and position of the camera allow for the scaling of $0.25'' = 100$ pixels.

2. The figure provides the x- and y-pixel displacements for select interrogation regions for the channel used in Problem 1. At $Re = 45,000$, calculate the velocity components and plot the velocity vectors (use the corresponding Δt calculated in Problem 1).

Interrogation Region	x-Displacement (Pixels)	y-Displacement (Pixels)
1	3.9	−0.1
2	3.7	−1.3
3	3	−2.5
4	1	−3.8

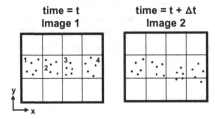

3. For a single interrogation region, the following instantaneous velocity components are calculated from a 2D PIV experiment. Based on the velocity components, calculate the time averaged velocity and the turbulence intensity of the flow at this specific location.

u (m/s)	v (m/s)	u (m/s)	v (m/s)	u (m/s)	v (m/s)
10.10	0.200	10.4	0.194	11.0	0.182
9.25	0.217	9.65	0.209	10.67	0.189
8.90	0.224	9.75	0.207	10.49	0.192
10.0	0.202	9.42	0.213	10.3	0.196
10.53	0.191	10.17	0.199	10.0	0.202
11.10	0.180	10.14	0.199	11.0	0.182
10.1	0.200	10.33	0.195	9.99	0.202
9.37	0.214	10.46	0.193	9.66	0.209
9.45	0.213	10.9	0.184	9.76	0.207
10.0	0.202	11.1	0.180	8.9	0.224

4. A wind tunnel is used investigate the flow around an airfoil. The desired Reynolds number, based on the chord length of the airfoil (10.5″), is 1.0×10^6. What Δt between images should be used so the mainstream particle motion is limited to seven pixels? The physical scaling of the images is $1″ = 100$ pixels.

5. Given the following x- and y-pixel displacements in the wake of the airfoil from Problem 4, calculate the velocity distribution and plot the velocity vectors. Use the calculated Δt from Problem 4.

Interrogation Region	x-Displacement (Pixels)	y-Displacement (Pixels)
1	5	−4
2	6	−1
3	5.75	2.5
4	5	4

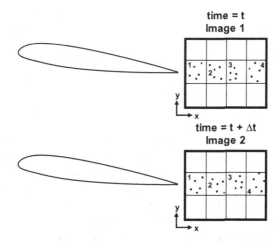

6. Computational fluid dynamics (CFD) analysis shows the potential performance of a newly proposed airfoil. You have been asked to verify the CFD

Experimental Methods in Heat Transfer and Fluid Mechanics

results by measuring the aerodynamic performance of the airfoil. While searching through the storage room of a laboratory within your facility, you find the following items (in addition to basic flow/pressure meters, thermocouples, and traverse systems):

1. Pitot-Static Probe
2. Cobra Probe
3. Five-Hole Probe
4. Differential Pressure Transducers
5. Small-Diameter Tubing
6. Single-Wire, Hot-Wire Probe
7. X-Wire, Hot-Wire Probe
8. Connection Leads for both Probes
9. Data Acquisition Hardware to Power and Monitor Response of Hot Wire Probes
10. A Low-Speed CCD Camera
11. Liquid Oil
12. Oil Atomizer
13. Laser with Light Sheet Optics
14. Hardware Synchronizer

You have the necessary hardware to use either multi-hole probes, a hot wire, or a PIV system to quantify the flow field around the new airfoil design within an available wind tunnel.

An open-loop wind tunnel with the airfoil model is available to meet your needs. Furthermore, regardless of which technique you are considering, you must return an accurate representation of the flow field to your manager and your group.

Your group is in a time-critical situation, and you need to obtain your measurements and process the results as quickly (and cheaply—time is money) as possible.

a. What method would you propose for acquisition of time-averaged flow data? Sketch the overall setup for proposed experiment. What hardware and instrumentation is needed for the experiment?
b. What flow field quantities could be obtained with the method? Sketch expected results you expect to obtain from the method you have chosen in this time-sensitive situation.
c. What factors have a significant effect on the uncertainty of your results for the method you have chosen?

7. As your company sees the benefit of designing based on CFD simulations, it is vital to develop turbulence models that accurately represent turbulent flows. Referring to Problem 6, it is desired to not only obtain mean data, but to obtain enough raw data to make a meaningful comparison of the predicted and measured Reynolds stresses over a large area/volume.

a. In this case, what experimental method would you propose? Sketch the overall setup for proposed experiment. What hardware and instrumentation is needed for the experiment?

   b. What are the disadvantages of this method compared to other techniques available for measuring flow field quantities?

   c. What can be done to reduce the experimental uncertainty of this method?

8. For turbine blade, film cooling applications, researchers are interested to understand how the film cooling flow interacts with the mainstream flow passing over a surface. To model leading edge film cooling, a cylinder with multiple rows of film cooling holes near the leading edge is placed in a low-speed wind tunnel (as shown in the figure). The required compressed air (25°C), piping, flow meters, materials, and a full 2D PIV system are available for your use.

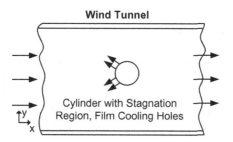

Wind Tunnel

Cylinder with Stagnation
Region, Film Cooling Holes

   a. Show an overall sketch of your experimental test loop and the associated equipment and instrumentation required to conduct the proposed PIV test, i.e., sketch your overall experimental setup along with a suitable flow measuring device, detail your test section design with proper materials, describe the 2D, PIV measurement principle, and the detailed measurement procedures at the steady-state condition.

   b. Based on the wind tunnel flow and cooling flow capability, one steady-state blowing ratio (cooling to mainstream mass flux ratio) will be tested: $M = 1.0$. What measurement planes should be selected to gain the best understanding of the coolant-to-mainstream interaction? Sketch and discuss the expected results from the PIV method.

   c. Discuss the factors influencing the experimental uncertainty for the experiment.

9. Design an experimental setup to evaluate the flow development for steady, incompressible, turbulent air flow through a two-dimensional channel with protrusions on one wall, as sketched below. Focus on the region interest shown in the sketch. Use the 2D, PIV method to map the velocity and turbulence distributions within the flow.

   a. Show an overall sketch of your experimental test loop and the associated equipment and instrumentation required to conduct the proposed PIV test, i.e., sketch your overall experimental setup along with a suitable flow measuring device, detail your test section design with proper materials, describe the 2D, PIV measurement principle, and the detailed measurement procedures at the steady-state condition.

   b. Sketch and discuss the expected experimental results (velocity, turbulent kinetic energy).
   c. Discuss the factors influencing the experimental uncertainty for the experiment.

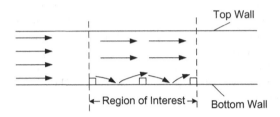

10. You are asked to prepare an experiment to evaluate flow fields for steady incompressible turbulent air flow through a square channel with nine circular tubes between the top and bottom walls, as sketched below. Focus on the region of interest shown in the sketch. Use a PIV method to map the flow field at a downstream plane of the square channel with nine circular tubes: Sketch the major experimental setup, describe the measurement principle, sketch and discuss the expected experimental results (mean velocity vector field, turbulence intensity).

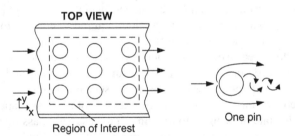

REFERENCES

1. Bodenschatz, E. and Eckert, M., "Prandtl and the Göttingen School," *A Voyage Through Turbulence*, pp. 40–100, Cambridge University Press, Cambridge, UK, 2011.
2. Willert, C. and Kompenhan, J., "PIV Analysis of Ludwig Prandtl's Historic Flow Visualization Films," *63rd Annual Meeting, American Physical Society, Division of Fluid Dynamics (DFD)*, Long Beach, CA, 2010.
3. Raffel, M., Willert, C., Wereley, S. and Kompenhans, J., *Particle Image Velocimetry, A Practical Guide*, 2nd ed., Springer-Verlag, Berlin, Germany, 2007.
4. Westerweel, J., 1993, *Digital Particle Image Velocimetry—Theory and Application*, Delft University Press, Delft, the Netherlands.
5. Tropea, C., Yarin, A. L. and Foss, J. F. (eds.), *Springer Handbook of Experimental Fluid Mechanics*, Springer-Verlag, Berlin, Germany, 2007.
6. Wright, L.M., McClain, S.T. and Clemenson, M.D., "Effect of Freestream Turbulence on Film Cooling Jet Structure and Surface Effectiveness Using PIV and PSP," *ASME Journal of Turbomachinery*, Vol. 133, No. 4, 2011, Article No. 041023, 12 p.

7. Brown, C.P., Wright, L.M. and McClain, S.T., "Comparison of Staggered and In-Line V-Shaped Dimple Arrays using S-PIV," *ASME Paper No. GT2015-43499*, 2015.

8. Son, S.Y., Kihm, K.D. and Han, J.C., "PIV Flow Measurements for Heat Transfer Characterization in Two-Pass Square Channels with Smooth and 90° Ribbed Walls," *International Journal of Heat and Mass Transfer*, Vol. 45, 2002, pp. 4809–4822.

9. Schabacker, J., Boelcs, A. and Johnson, V., "PIV Investigation of the Flow Characteristics in an Internal Coolant Passage with Two Ducts Connected by a Sharp 180 deg. Bend," *ASME Paper No. 98-GT-544*, 1998.

10. Suden, B., "Convective Heat Transfer and Fluid Flow Physics in Some Ribbed Ducts using Liquid Crystal Thermography and PIV Measuring Techniques," *International Journal Heat and Mass Transfer*, Vol. 47, 2011, pp. 899–910.

11. Sharma, N., Tariq, A. and Mishra, M., "Experimental Study of Detailed Heat Transfer and Fluid Flow Characteristics in a Rectangular Duct with Solid and Slitted Pentagonal Ribs," *ASME Paper No. GTINDIA2017-4651*, 2017.

12. Bons, J.P. and Kerrebrock J.L., "Complementary Velocity and Heat Transfer Measurements in a Rotating Turbine Cooling Passage with Smooth Walls", *ASME Journal of Turbomachinery*, Vol. 121, 1999, pp. 651–662.

13. Elfert, M., Voges, M. and Klinner, J., "Detailed Flow Investigation Using PIV in a Rotating Square-Sectioned Two-Pass Cooling System with Ribbed Walls", *ASME Paper No. GT2008-51183*, 2008.

14. Coletti, F., Cresci, I. and Arts, T., "Time-Resolved PIV Measurements of Turbulent Flow in Rotating Rib-Roughened Channel with Coriolis and Buoyancy Forces," *ASME Paper No. GT2012-69406*, 2012.

15. Gogineni, L., Goss, D., Pestian, D. and Rivir, R., "Two-Color Digital PIV Employing a Single CCD Camera," *Experiments in Fluids*, Vol. 25, 1998, pp. 320–328.

16. Bernsdorf, S., Rose, M.G. and Abhari, R.S., "Modeling of Film Cooling – Part I: Experimental Study of Flow Structure," *ASME Journal of Turbomachinery*, Vol. 128, 2006, pp. 141–149.

17. Aga, V., Rose, M. and Abhari, R.S., "Experimental Flow Structure Investigation of Compound Angled Film Cooling," *ASME Journal of Turbomachinery*, Vol. 130, 2008, Article 031005, 8 p.

18. Wright, L.M., McClain, S.T., Brown, C.P. and Harmon, W.V., "Assessment of a Double Hole Film Cooling Geometry Using S-PIV and PSP," *ASME Paper No. GT2013-94614*, 2013.

Index

Printed in the United States
by Baker & Taylor Publisher Services